Goat Production and Supply Chain Management in the Tropics

Goat Production and Supply Chain Management in the Tropics

Pramod Kumar Rout[1]
Ashok Kumar[1]

and

Basanta Kumara Behera[2]

[1]*ICAR-Central Institute for Research on Goats, Makhdoom, Farah, Mathura-281122, India*

[2]*Formerly of M D University, 1031/23 DLF Colony, Rohtak-124001, Haryana, India*

CABI

CABI is a trading name of CAB International

CABI
Nosworthy Way
Wallingford
Oxfordshire OX10 8DE
UK

Tel: +44 (0)1491 832111
Fax: +44 (0)1491 833508
E-mail: info@cabi.org
Website: www.cabi.org

CABI
745 Atlantic Avenue
8th Floor
Boston, MA 02111
USA

Tel: +1 (617)682-9015
E-mail: cabi-nao@cabi.org

A catalogue record for this book is available from the British Library, London, UK.

Library of Congress Cataloging-in-Publication Data

Names: Rout, Pramod Kumar, author.
Title: Goat production and marketing in the tropics / Pramod Kumar Rout, ICAR-Central Institute for Research on Goats, Ashok Kumar, CAR-Central Institute for Research on Goats, and Basanta Kumara Behera.
Description: Wallingford, Oxfordshire, UK ; Boston, MA : CABI, [2019] | Includes bibliographical references and index.
Identifiers: LCCN 2019014193 (print) | LCCN 2019013228 (ebook) | ISBN 9781789240139 (hbk : alk. paper) | ISBN 9781789240146 (ePDF) | ISBN 9781789240153 (ePub)
Subjects: LCSH: Goats--Tropics. | Goat farming--Tropics. | Goats--Breeding--Tropics. | Goats--Feeding and feeds--Tropics. | Goat industry--Tropics.
Classification: LCC SF383 .R68 2019 (ebook) | LCC SF383 (print) | DDC 338.1/7639--dc23
LC record available at https://lccn.loc.gov/2019014193

ISBN-13: 9781789240139 (hardback)
9781789240146 (ePDF)
9781789240153 (ePub)

Commissioning Editor: Alex Lainsbury
Editorial Assistant: Emma McCann
Production Editor: Tim Kapp

Typeset by SPi, Pondicherry, India
Printed and bound in the UK by Severn, Gloucester

Contents

List of Tables

List of Figures

About the Book

Smallholder and family farmers play leading roles in promoting more resilient and sustainable global agricultural food systems. Goats as a species are highly suitable for smallholder systems, and they contribute significantly to family nutrition and income in many Asian and African countries. The present estimates show that 815 million people are suffering from chronic malnourishment. Goats as a species have the potential to fulfil the nutritional requirements in many disadvantaged regions and can increase the income of poor people in most agro-ecological conditions. Asia, Africa and small island developing states (SIDS) are highly impacted by climate change and hunger. The region has a high density of goat population, and goat farming is gaining momentum all over the world; therefore, we have written a household manual for goat production for better productivity and income. The book describes the tools for selecting a genotype with respect to particular agro-climatic conditions and feed formulations based on available crop and fodder resources. The book also provides targeted preventive measures based on disease occurrence to improve product quality and reduce environmental pollution. Chapters deal with the optimized package of practices to be adopted by goat farmers with respect to climatic conditions and available natural resources, and gives guidance for setting up goat entrepreneurship groups, running a developmental project and strengthening livelihoods by disseminating information for income generation. The book provides an insight to devise strategies to overcome challenges and tap opportunities by way of technological know-how in product diversification and marketing. It also gives ideas for the sustainable use of natural resources and addresses the socio-economic development of goat farmers of different regions.

Preface

Goats are adapted to all types of ecological conditions across the globe and are among the 'big five' livestock species recognized by the Food and Agriculture Organization of the United Nations. The global preference for animal protein consumption has increased over the last decade. The global milk market is estimated to reach revenues of around US$15 billion by 2024, growing at a compound annual growth rate of more than 7% during 2018–2024. Similarly, total goat meat imports amounted to US$404 million in 2017. During the last decade the total import value increased at an average annual rate of 3.8%. The 2017 data on the global market show that goat meat importing reached its maximum level in 2017, and is supposed to retain its growth in the near future. Currently, the life science industries are very much inclined to use goats as a bioreactor for the production of therapeutic proteins due to its low nurturing cost and low-budget housing requirements. This book is designed to acquaint readers with updated information on goats and goat farming technologies.

Chapter 1 begins with basic information on tropical livestock unit calculation at different geographic locations in order to easily access and compare the goat population and adaptability under a variety of climatic zones. It is also explains how the livestock unit acts as a reference unit to clearly understand the aggregation of livestock from various species, age and genetic diversity perspectives.

Chapter 2 explains how to improve the production performance of goats using different breeding methods. Breeding strategies can produce highly adaptable goats suitable to a variety of local agro-climatic conditions, socio-economic statuses and market demands. The availability of good quality bucks and conservation of recognized indigenous breeds by preventing them from rampant genetic dilution is also discussed.

Chapter 3 narrates the significance of reproduction management practices for enhancing reproductive efficiency by using new biotechnological approaches, ultimately resulting in faster turnover of the flock in a production environment. Starting from the fundamentals of reproduction criteria, the chapter then presents the latest information on optimization of male fertility, breeding of young does, oestrus detection, state of breeding and mating, overcoming seasonality of breeding, pregnancy diagnosis, and recording of reproduction.

Chapter 4 is framed on how to manage the nutritional budget for better growth of goats, on the basis of market demand. The provision of the required level of energy in goats is an unavoidable part of maintaining a healthy herd. Protein is one of the basic units for the building blocks of cells. If the level of protein in a goat's diet is inadequate, then foetal development can be affected. In addition, it could lead to reduced growth and development and milk production. It also explains how the nutritional requirement is fixed on the basis of age, sex, breed, production system (dairy or meat), body size, climate and physiological stage. The formulation feed in different agro-climatic regions is also explained.

The success of small ruminant health management is an essential practice for commercialization of goats. Good feeding and proper breeding needs to be complemented with good health management to ensure that goats remain healthy and produce quality milk and meat. Chapter 5 presents methods to develop an effective healthcare plan and the implementation of biosecurity measures, vaccination, deworming, body condition and disease diagnosis. The methods to maintain perfect records for each animal regarding medication, production, breeding and culling are also discussed.

Although the goat is a highly adaptive small ruminant, it still needs simple shelter facilities to survive in adverse climatic conditions. Chapter 6 covers the design aspects of goat shelters that are easily affordable to rural farmers with low budgets. In addition, the plans for modernized farm facilities for larger sized goat flocks (e.g. milking parlours, kidding pens, quarantine pens, storage facilities, etc.) are presented.

The main objective of Chapter 7 is to explain the different goat production systems, and how to identify major constraints and opportunities for achieving better productivity. The options and strategies for planning and implementing goat improvement projects and assessing goat production systems are some of the important factors highlighted in this chapter.

Supply chain management as applied to the goat industry is a new concept so very little detail related to it is available. The last chapter of this book deals with the supply chain management of diversified products derived from goats (e.g. goat milk, goat meat and transgenic goat products). In addition, the chapter explains how supply chain management serves as a baseline for information for the government, private sector and industry on the development of an integrated dairy industry linked with meat processing units and biopharmaceutical units for the production of diversified products. This book provides insights on all related information on goat production to all stakeholders to ensure economic and safe goat farming.

Acknowledgements

We are highly grateful to all those individuals who have encouraged us to take up the assignment. We are really indebted to our family members for their constructive remarks and encouragement during the preparation of the manuscript. We are grateful to Professor G.F.W Haenlein for his support during various stages of the book preparation. We are also thankful to Dr M.S. Chauhan, Director, CIRG for his good words and encouragement and grateful to Dr A.K. Goel, Principal Scientist, ICAR-CIRG, and Dr U.B. Chaudhary, Principal Scientist, ICAR-CIRG for their kind help and encouragement. We are also really grateful to Mr Manvendra, Dr Mahima and Mr Rakesh who extended their helping hands to make this book a reality. Finally, we would like to thank Ms Caroline Makepeace, Executive Editor at CABI for her constructive criticism and for believing in us to carry out the work. We are also thankful to Tabitha Jay and her team for enthusiastic support for the time bound effort.

1 Goat Production Challenges to Food Security

Livestock production makes a significant economic contribution to developing countries by supporting food security, making use of animal by-products (e.g. skin, fibre, manure) and contributing to capital accumulation. In addition, livestock improvement programmes are immensely helpful in improving the social and cultural lives of several million resource-poor farmers, enabling sustainability in farming management practice and economic stability. Both human and livestock populations are increasing in magnitude, but at different rates. Surprisingly, the current growth rate of ruminant populations is half that of the human population growth rate. At present, the total goat population is estimated to be about 1 billion. Goats inhabit a wide range of ecological systems (Skapetas and Bampidis, 2016). The goat population has undergone the largest increase (+34%) since 2000, compared to pigs (+15%), cattle (+14%) and sheep (+14%) (Bertolini *et al.*, 2018; for current information, see www.fao.org/faostat/en). The distribution pattern shows that 90% of goats are distributed in Asia and Africa, followed by the Americas, Europe and Oceania (Skapetas and Bampidis, 2016).

1.1 Tropical Livestock Units

Comparing Livestock Species

The major food-producing animals are cattle, buffaloes, sheep, goats, pigs, chickens, ducks, turkeys and other poultry. When comparing different livestock species, each species is converted to livestock units (LSU), which is obtained by converting body weight into metabolic weight ($Bw^{0.75}$). The grazing equivalent of one adult dairy cow producing 3000 kg of milk annually, without additional feedstuffs, is used as the reference unit (1 LSU).

A tropical livestock unit (TLU) is a convenient method for quantifying different livestock types and sizes from the tropics. Using the exchange ratio concept, different species with a range of average sizes can be compared to a common unit, the tropical livestock unit (TLU); 1 TLU is one cow with a body weight of 250 kg. TLU values applied to different species vary by country as a function of prevailing production systems, type of breeds and feed requirements.

1.1.1 Livestock unit calculations

Livestock unit coefficients have been compiled by the Food and Agriculture Organization of the United Nations (FAO, 2005) and can be used for international comparison (Table 1.1). When estimating livestock units it is assumed that a cow in the US has the highest value and is assigned the value of 1 LSU. The term tropical livestock unit (TLU) is used for livestock raised in the tropics and is used for livestock production system analysis in the tropics. The common TLUs for the tropical regions are cattle = 0.7, sheep = 0.1, goats = 0.1, pigs = 0.2 and chickens = 0.01. The livestock unit coefficients with respect to different age groups in different species is presented in Table 1.2.

For example, in North America the TLU of 1 cow is 1 TLU and 1 goat is 0.1 TLU. If you have 3 cows and 20 goats, then the total TLU is 5 (cow: 3 × 1 = 3; goat: 0.1 × 20 = 2; so total is 3 + 2 = 5 TLU). Similarly, the livestock coefficient for different regions will be used for calculation of TLU based on the population numbers of different species.

1.1.2 Animal unit coefficient

The animal unit (AU) is a standard unit used in calculating the relative grazing impact of different classes of livestock (Table 1.3). One AU is defined as a 450 kg beef cow (approximately 1000 lb) with or without a nursing calf, with a daily dry matter forage requirement of 11.8 kg.

Table 1.1. Livestock units coefficients that can be used for international comparisons.

Region	Cattle	Buffalo	Sheep	Goats	Pigs	Asses	Horses	Mules	Camels	Chickens
Near East North Africa	0.70	0.70	0.10	0.10	0.20	0.50	0.40	0.60	0.75	0.01
North America	1.00		0.15	0.10	0.25	0.50	0.80	0.60		0.01
Africa, south of the Sahara	0.50		0.10	0.10	0.20	0.30	0.50	0.60	0.70	0.01
Central America	0.70		0.10	0.10	0.25	0.50	0.50	0.60		0.01
South America	0.70		0.10	0.10	0.25	0.50	0.65	0.60		0.01
South Africa	0.70		0.10	0.10	0.20	0.50	0.65	0.60		0.01
OECD	0.90	0.70	0.10	0.10	0.25	0.50	0.65	0.60	0.90	0.01
East and South-East Asia	0.65	0.70	0.10	0.10	0.25	0.50	0.65	0.60	0.80	0.01
South Asia	0.50	0.50	0.10	0.10	0.20	0.50	0.65	0.60		0.01
Transition Markets	0.60	0.70	0.10	0.10	0.25	0.50	0.65	0.60		0.01
Caribbean	0.60	0.60	0.10	0.10	0.20	0.50	0.65	0.60		0.01
Near East	0.55	0.60	0.10	0.10	0.25	0.50	0.56	0.60	0.70	0.01
Other	0.60	0.60	0.10	0.10	0.20	0.50	0.65	0.60		0.01

Source: FAO (2005).

Table 1.2. Livestock unit coefficients with respect to age group and body size in different species.

Type of animal	Age group	Coefficient
Bovine	<1 year of age	0.4
	>1 year of age and <2 years of age	0.7
	Male >2 years of age	1.0
	Heifers >2 years of age	0.8
	Dairy cow	1.0
	Other cows (>2 years of age)	0.8
Sheep & Goat		0.10
Equine		0.80
Pig	Piglet body weight <20 kg	0.027
	Breeding sows >50 kg	0.50
	Other Pigs	0.30
Poultry	Broilers	0.007
	Laying Hens	0.014
	Ostriches	0.350
	Other poultry	0.0350
Rabbit		0.020

Source: Chilonda and Otte (2005).

Livestock units is the reference unit that uses specific coefficients on the basis of nutritional or feed requirement of each type of animal. The reference unit used for the calculation of livestock units (=1 LSU) is the grazing equivalent of one adult dairy cow producing 3000 kg of milk annually without additional concentrated stuffs. The animal unit coefficient is used as a standard measure for animals based on size and manure production.

1.1.3 Stocking density in the tropics

Stocking rate is defined as the number of animals that can be supported by a particular pasture or rangeland. Therefore, it is necessary to know the amount of forage consumed by the livestock and the forage production capacity of the pasture and rangeland. Generally, the tropical animal unit month is used to determine the carrying capacity of grazing animals on pastureland or rangeland during a particular period. The basis of comparison is

Table 1.3. Animal unit coefficients for different types of animals.

Type of animal	Animal unit coefficient
A Dairy Cattle	
Mature cow over 1000 lb (450 kg)	1.4
Mature cow under 1000 lb (450 kg)	1.0
Heifer	0.7
Calf	0.2
B Beef Cattle	
Slaughter steer or stock cow	1.0
Feeder cattle or heifer	0.7
Cow and calf pair	1.2
Calf	0.2
C Swine	
Over 300 lb (136 kg)	0.4
Between 55 and 300 lb (25–136 kg)	0.3
Under 55 lb (25 kg)	0.05
D Horse	1.0
E Sheep and lambs/goats	0.1
F Chickens	
Laying hen or broiler (liquid manure system)	0.033
Chicken over 5 lb (dry manure system)	0.005
Chicken under 5 lb (dry manure system)	0.003
G Turkeys	
Over 5 lb	0.018
Under 5 lb	0.005
H Ducks	0.01
Animals not listed in items A to H	Average weight of the animal in pounds divided by 1000 lb

the TLU, which is the dry matter feed and forage requirement of a 250 kg cow.

The stocking density is calculated as total livestock units divided by the area (in hectares). For example:

10 cows maintained in5 ha for 3 months,
the annual stocking density10/5 = 2 LU/ha.

10 cows or 5 ha for 3 months =
2 LU/4 = 0.5 LU/ha

Stocking rate determination by average animal weight method

Using the average animal weight for stocking rate determination seems to be a more accurate method. A 250 kg cow with a calf needs to consume 2.667% of its body weight each day to meet its metabolic requirements. This percentage is used regardless of breed or species, and determines the daily forage requirement for their body size by using a conversion factor of 2.667% (multiplying 0.02667).

1.1.4 Goat population trends in the tropics

South-East Asian countries

Myanmar, Thailand and Vietnam recorded an overall growth rate of more than 200% in the goat population between 1997 and 2014, whereas Brunei, Malaysia and Singapore recorded an overall population growth rate between 21.82% and 111.99% over the same period. The Philippines and Cambodia had a negative growth trend in the goat population during that period (Fig. 1.1). In Myanmar, the population growth rate was 5.88% per year from 1997– 2006 and 20.63% per year from 2007–2014. Brunei reported a negative population growth trend during the period 1997–2006, however it was –18.28% per year during the 2007–2014. Cambodia had a 1.02% population growth rate from 1997–2006; however, it was –1.96% per year from 2007–2014. The Philippines had a negative population growth trend (–41.42%) during 1997–2006 and the population growth rate was –3.35% per year from 1997–2006 and again it was –1.35% per year during 2007–2014. Although Singapore had a 27% growth rate from 1997–2006, the growth rate was negative (–4.0%) from 2007–2014 (FAO, 2018).

South American countries

In this region, French Guyana, Bolivia and Paraguay recorded overall goat population growth rates between 17.18% and 118.75% during the period 1997 to 2014. Suriname and Ecuador recorded negative population growth rates of –46.53% and –92.75%, respectively (Fig. 1.2). Bolivia showed an annual population growth rate of 2.25% between 1997 and 2006, and an annual growth rate of 2.56% for the 2007–2014 period. Ecuador had a negative population growth rate, reporting annual rates of –4.94% and –9.52% during 1997–2006 and 2007–2014, respectively. Similarly, Suriname also had a constant negative population growth between 1997 and 2014, which was reported to be about –2.51% per year (Fig. 1.2).

Central American countries

Costa Rica observed a 222.50% population growth rate in the goat population between 1997

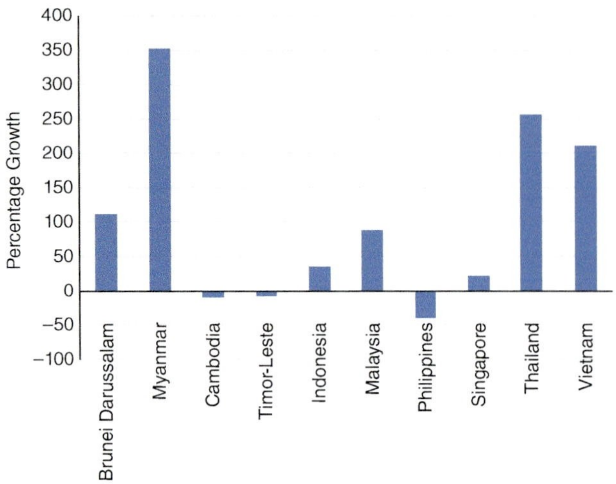

Fig. 1.1. The overall population growth trend between 1997 and 2014 in different countries of South-East Asia.

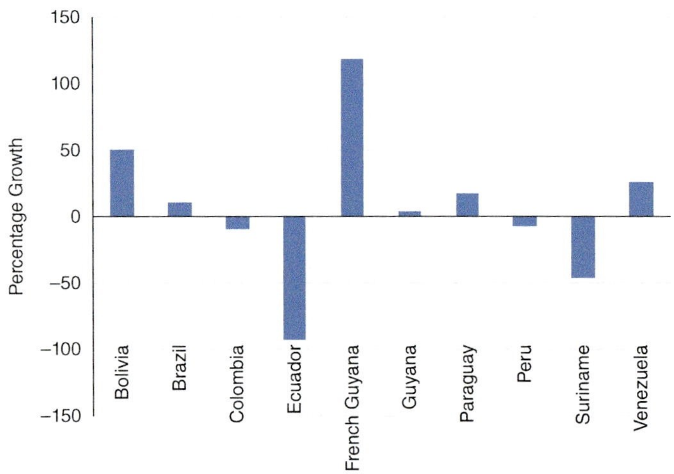

Fig. 1.2. The overall population growth trend between 1997 and 2014 in different countries of South America.

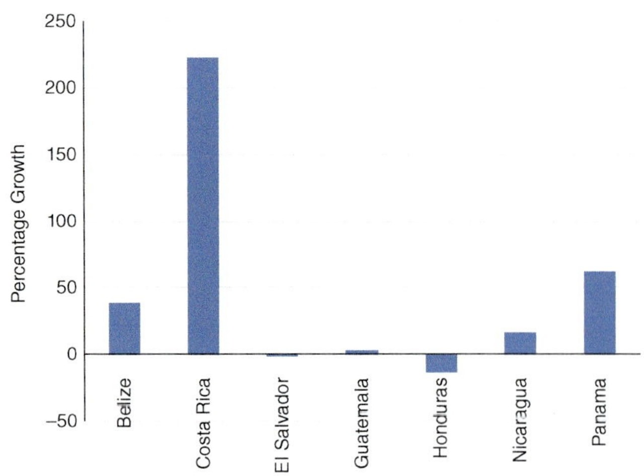

Fig. 1.3. The overall population growth trend between 1997 and 2014 in different countries of Central America.

and 2014, whereas El Salvador reported a marginal reduction in goat population during the same period (Fig. 1.3). Most countries in this region recorded overall positive population growth rates, except Honduras, which had a negative (–13.79%) population growth rate during the period, and El Salvador. Costa Rica, Panama and Belize reported population growth rates between 38.46% and 222.5% during the period 1997–2014. Belize and Panama recorded a constant annual growth rate of 1.85% and 2.85%, respectively, from 1997 to 2014. Costa Rica exhibited the highest population growth rate in the region during the 1997–2006 period, recording an annual rate of 23.44%; however, the population growth was almost static, tending towards deceleration between 2007 and 2014.

Caribbean countries

Cuba and the Cayman Islands recorded a significant increase in goat population (threefold to sixfold increase, respectively) between 1997 and 2014. Nine countries had a negative population growth, including Montserrat (–100%) and Guadeloupe (100%) (Fig. 1.4).

The Cayman Islands had a 45.32% population growth rate per year during 1997–2006 and 4.20% growth during 2007–2014. Cuba recorded a population growth rate of 74.52% per year during 1997–2006, but there was a decelerating growth rate of –5.33% per year from 2007 to 2014 (Fig. 1.4).

Central African countries

This is one of the best regions for goat rearing and all the countries showed positive overall population growth rates between 1997 and 2014. Angola, Zambia and the Central African Republic had an approximate twofold increase in goat population during that period (Fig. 1.5). Sudan and the Democratic Republic of Congo reported a negative population growth of –13.4% and –10.68%, respectively, during 1997–2014. Sudan had a negative annual population growth rate of –3.05% between 2007 and 2014, whereas Zambia had the highest annual population growth of 17.86% between 1997 and 2006 and 3.70% between 2007 and 2014.

East African countries

All but three of the countries (Uganda, Somalia and Mauritius) in this region reported a positive overall growth rate between 1997 and 2014. Burundi, Malawi, Ethiopia and Rwanda showed twofold to fourfold increases in goat population during 1997–2014 (Fig. 1.6). The region recorded the highest

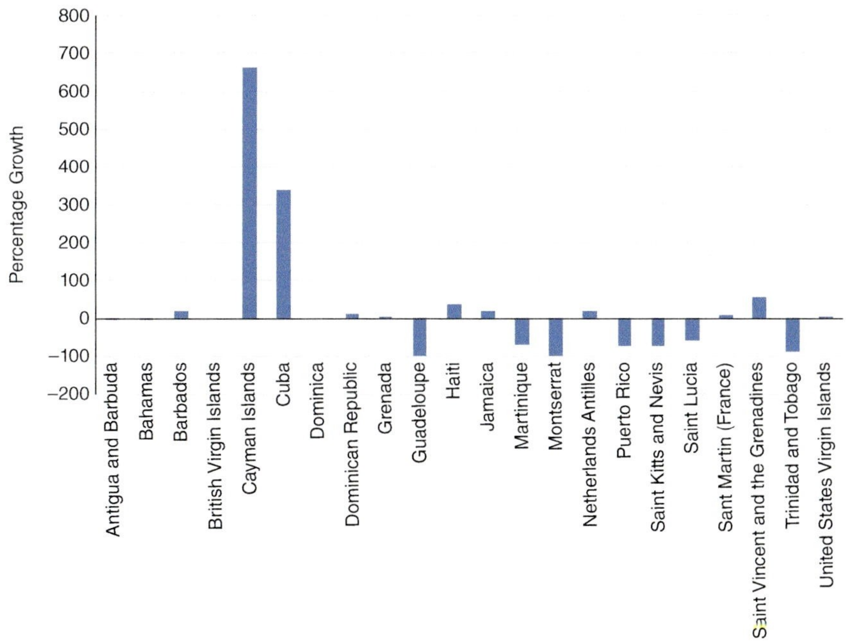

Fig. 1.4. The overall population growth trend between 1997 and 2014 in different Caribbean countries.

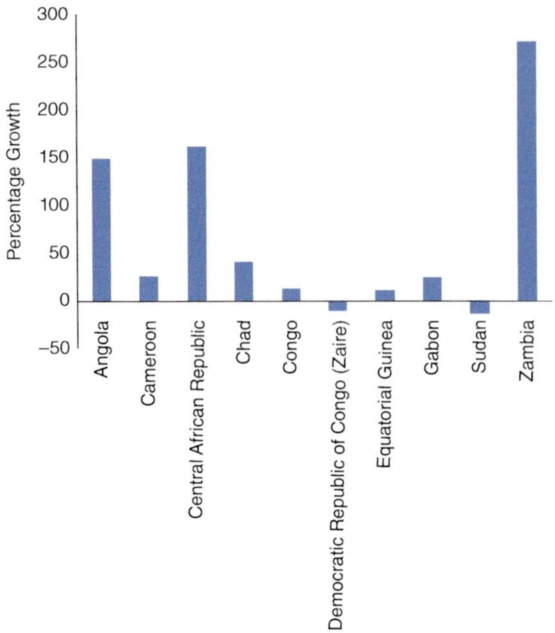

Fig. 1.5. The overall population growth trend between 1997 and 2014 in different countries of Central Africa.

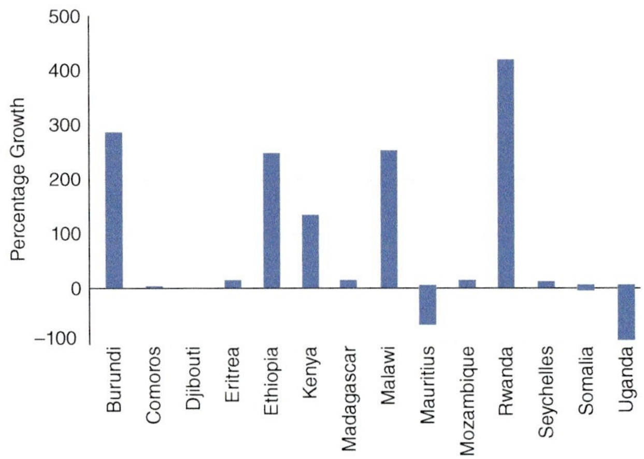

Fig. 1.6. The overall population growth trend between 1997 and 2014 in different countries of South-East Africa.

average growth rate between 1997 and 2014. Mauritius had a negative annual population growth rate of −7.39% during 1997–2006, but reported a positive annual population growth rate of 1.32% during 2007–2014.

West African countries

This region is one of the best areas for goat rearing. All of the countries reported a positive overall population growth rate between 1996 and 2014 (Fig. 1.7). The population growth varied from 3.68% (Saint Helena) to 373.68% (Sierra Leone). There were accelerating annual population growth rates in all the countries from 1997, which varied from 1.07% to 10.11% per annum during 1997–2014.

South Asian countries

A significant increase in goat population was reported in Nepal, Bangladesh and Bhutan (Fig. 1.8). Sri Lanka had an overall negative growth rate in goat population growth between 1997 and 2014,

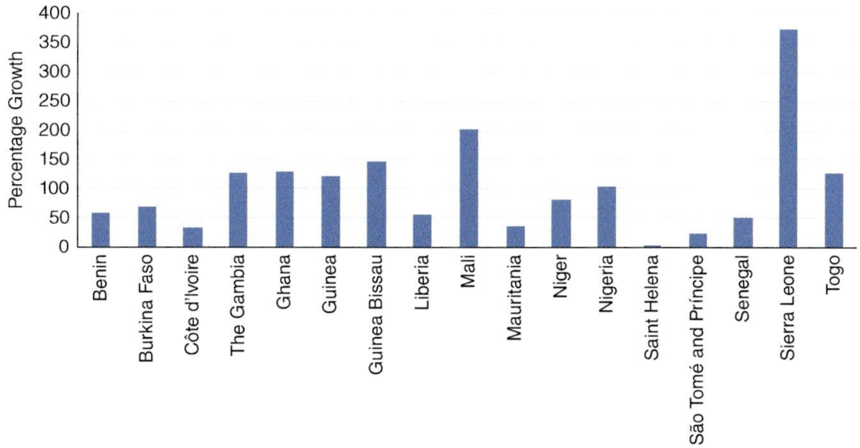

Fig. 1.7. The overall population growth trend between 1997 and 2014 in different countries of West Africa.

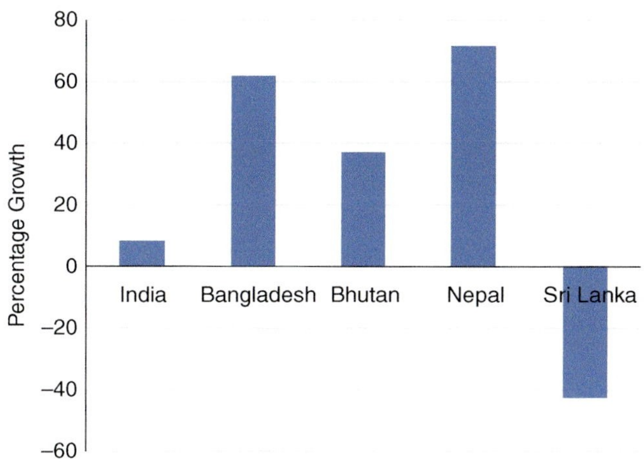

Fig. 1.8. The overall population growth trend between 1997 and 2014 in different countries of South Asia.

with an annual population growth of −2.67% and −2.40% per year during 1997–2006 and 2007–2014, respectively. India recorded an annual growth rate of −2.0% during 1997–2006 and 11.1% during 2007–2014.

South American and North American countries

In this region, Ecuador and Suriname reported significant decreases in overall population growth between 1997 and 2014. Mexico also reported a negative overall population growth trend (−5.65%) during the same period.

1.1.5 Impact of goat population on tropical livestock units

The overall goat population in South America reported a negative growth rate between 1997 and 2014, which corresponds to the negative overall TLU (−5.64%) of the region for the same period. The overall TLU percentage was 99.94 %, 105.5%, 21.84%, 37.51% and 29.0% in West Africa, East Africa, South Asia, South-East Asia and the Caribbean, respectively. Central America and South America showed 4.46% and 10.34% overall TLU growth between 1997 and 2014, respectively.

During 1997, the percentage of goats in the total livestock population was 18.27, 16.15, 15.54 and

14.68% in Central Africa, South Asia, West Africa and East Africa, respectively. During 2014, the percentage of goats in the total livestock population was highest in East Africa followed by West Africa, Central Africa and South Asia. The percentage of goats in the total livestock population was 1.32%, 0.88%, 0.92%, 0.17% and 2.11% in South-East Asia, the Caribbean, South America, Central America and North America respectively. During 2014 the percentage of goats in the total livestock population was highest in East Africa (18.6%) followed by West Africa (16.77%), Central Africa (15.26%) and South Asia (12.05%). The goat population also increased in Central America.

1.1.6 Factors affecting livestock units of goat populations

The livestock units represent the state of health of goats reared in smallholdings, commercial farms and at co-operative farming level. The main goal of goat farmers is to increase the population with animals of better body weight and health. Thus the socio-economic condition of goat owners is of primary importance. The availability of proper nutrients and grazing land, breeding programmes and the climatic conditions of specific geographical locations are some of the important factors affecting the livestock units suitable for profitable goat farming. For example, selection of Damascus goat, known for both the quality and quantity of milk production, is a factor in developing sustainable production for better business profit. Improved nutrition and managerial methods are also essential factors for developing a well-grown goat model suitable for milk and meat production for commercial purposes. Natural vegetation available for browsing and crop residue stubble provides grazing for free-range goats. In commercial farming, government subsidies support the importation of feed grains, which helps to develop sustainable goat farming practice. As skilled labour costs are high, it is better to train the local unskilled people for effective utilization of the labour pool. In goat farming, the level of adoption of breeding methods, disease and parasite control, and access to markets influence the productivity of goats in smallholder production systems.

1.2 Livestock Units in Different Tropics

1.2.1 Change in livestock population growth in tropics

Comparing the total livestock population in the different regions of the tropics, the population growth percentage was significantly higher in West Africa, Central America, East Africa and South Asia (Fig. 1.9). South-East Asia and South America

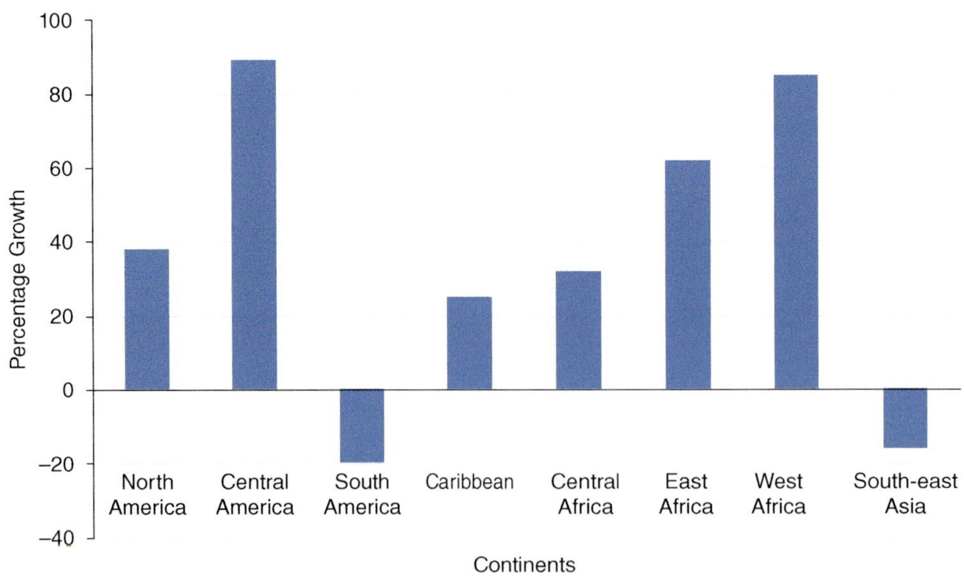

Fig. 1.9. Percentage livestock population growth trend between 1997 and 2014.

had negative livestock population growth during the period 1997–2014.

The livestock population growth based on tropical livestock unit (TLU) during the period 1997–2014 is presented in Fig. 1.10. There was no negative TLU growth rate in any of the regions. The percentage change in TLU in the tropics during 1997–2014 was positive and presented in Table 1.4.

The livestock population growth for South America had a negative overall growth rate, indicating that cattle and other large animals are preferred in the region. The TLU growth rate for the region was 20.4% between 1997 and 2014, Similarly, South-East Asia had a negative population growth (-15.95%), but the TLU increased by 11.01% between 1997 and 2014.

Out of total livestock, the percentage of cattle population increased significantly in West Africa (5.43%), East Africa (12.194%), South Africa (3.84%) and Central Africa (2.73%) (Fig. 1.11). However, cattle TLU showed a decreasing trend between 1997 and 2014 in Central Africa, East Africa and West Africa. Sheep population growth

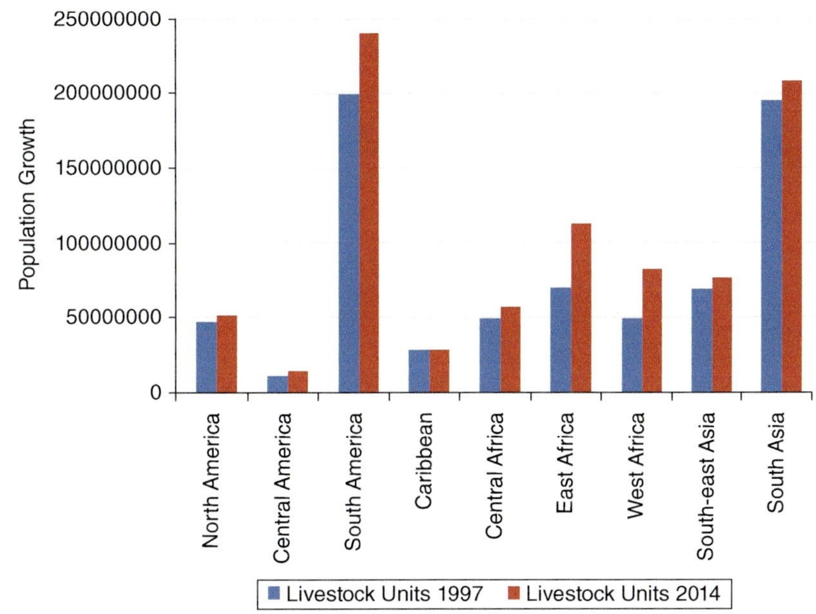

Fig. 1.10. The livestock population growth based on tropical livestock unit (TLU) between 1997 and 2014.

Table 1.4. The percentage change in livestock population and TLU in the tropics between 1997 and 2014.

Regions	Livestock population (1997)	Livestock population (2014)	Percent population Growth	Livestock Units (1997)	Livestock Units (2014)	Percent TLU growth
North America	437201500	602780931	37.873	46809250	51112297	9.193
Central America	99063243	187397265	89.169	10042991	13331201	32.741
South America	1558034645	1253683898	−19.534	196901409	237204112	20.468
Caribbean	418233807	523800121	25.241	28149279	28181899	0.116
Central Africa	296633748	391953634	32.134	47926735	55925040	16.689
East Africa	330824112	535740876	61.941	69210718	111594780	61.239
West Africa	473605907	877290992	85.236	48128753	81485077	69.306
Southeast Asia	1725628131	1450359020	−15.952	67941264	75424759	11.015
South Asia	1013691610	1654351923	63.201	193046947	205600604	6.503

rates reported similar trends to cattle in African countries (Fig. 1.12).

Based on TLU, cattle comprised the highest livestock percentage in North America (64.45%), Central America (73.82%) and South America (84.71%). Similarly, cattle comprised the dominant TLU in the Caribbean, Central Africa, East Africa and West Africa between 1997 and 2014. Cattle, pigs and buffaloes were also the dominant species with respect to TLU in South-East Asia, whereas cattle, buffaloes and goats were dominant species in South Asia. Goats comprised 10.69%, 8.93% and 18.05% of TLU in Central Africa, East Africa and West Africa, respectively. Goats had the highest TLU percentage with respect to total livestock species in the tropic regions of West Africa.

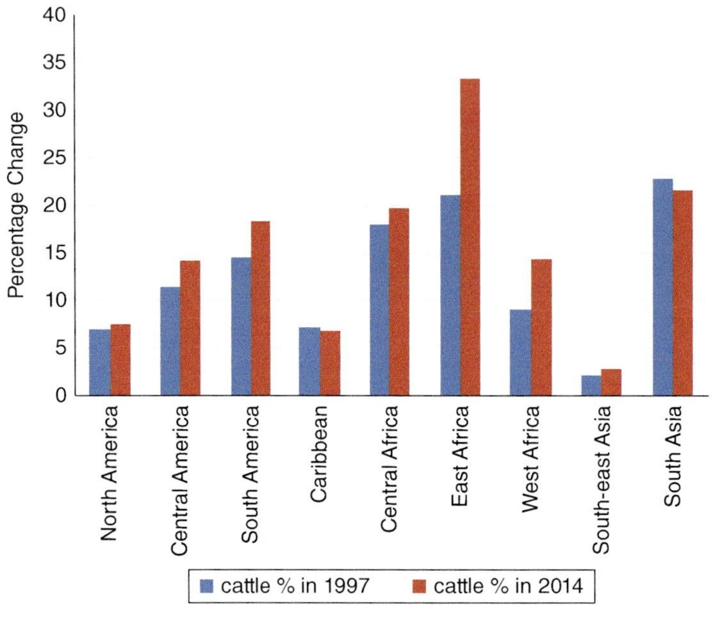

Fig. 1.11. Percentage change in the cattle population in comparison to total livestock population in different regions of the tropics between 1997 and 2014.

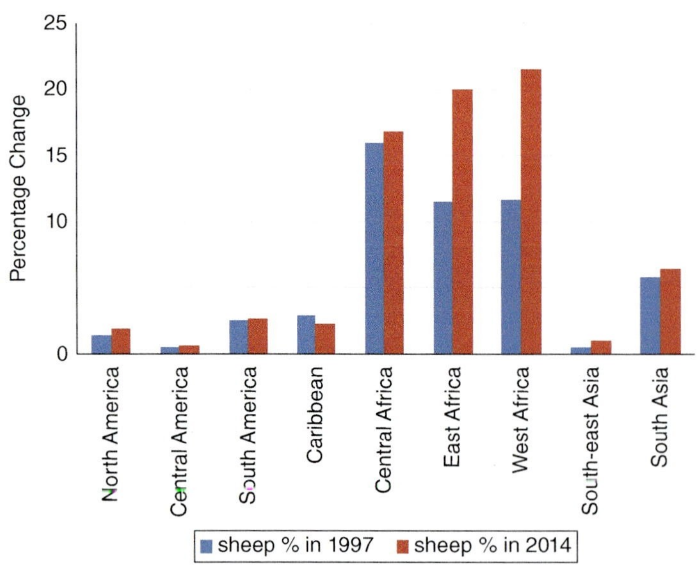

Fig. 1.12. Percentage change in the sheep population growth in comparison to total livestock population in the tropics between 1997 and 2014.

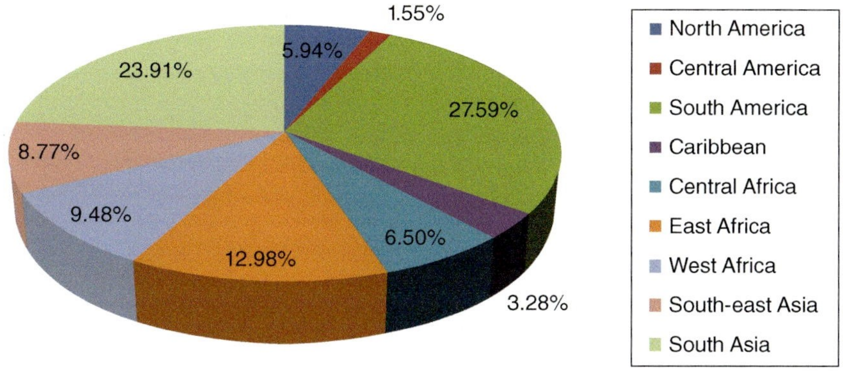

Fig. 1.13. Percentage of livestock units in different continents during 2014.

Cattle, buffalo, sheep, goats, camels, horses, pigs, asses and chickens are livestock types that are available in all the regions of the world. Apart from South America and South-East Asia, all the regions of the tropics showed an increase in livestock population during 1997–2014. However, the population growth rates based on TLUs showed increasing population growth in all regions. Population growth based on TLUs showed a significant increase in West Africa and East Africa during 1997–2014. South America and South-East Asia reported negative livestock population growth rates, but they recorded an increase in TLUs of 20.4% and 11.01%, respectively. In the tropics, South America, South Asia and East Africa are the major livestock harbouring regions (Fig. 1.13).

1.3 Goat Population with Respect to Herd Health Monitoring

The main target of herd health monitoring is to improve the herd's productivity through general husbandry, nutrition management, parasite control, vaccination and environmental management. This approach emphasizes the targeted management practices and preventive measures to manage each herd for optimum productivity, better utilization of resources and fulfilling environmental norms for sustainable growth.

Each herd is different; therefore, a herd health plan should be developed in consultation with production and health experts. All the records such as breeding, kidding, weaning, lactating, vaccination, and so on, should be maintained and utilized to develop the herd health plan. The details of recording observations and specific formats are discussed in Chapters 2 and 5. The herd health calendar initiation point is 6 weeks before breeding, and should include specific measures for adult does and bucks. The first steps should be checking the body condition score, determining the parasitic status by FAMACHA, verifying the vaccination status, hoof trimming, and so on. The mandatory *peste des petits ruminants* and enterotoxaemia vaccines are administered, depending on the agroclimatic region. The bucks and does should also be screened for brucellosis, caprine arthritis encephalitis, Johne's disease, and so on, and they should be given the desired nutritional care. Similarly, evaluation of the animal's anatomical features should be performed. The animal evaluation includes udder and teat checking for does and testicle checking for males. The mature body weight of does should be checked before breeding and it should be at least 65–70% of their adult body weight.

The breeding season should be kept to a minimum for better management; it should not exceed 42 days with the bucks. Ideally, a buck-to-doe ratio of 1:25 should be maintained. Pregnancy detection should be carried out within 60 days and open does should be culled from the flock. The nutritional management, including trace mineral and vitamin supplementation, should be in place for two months before kidding. Kidding kit and bedding should be prepared in advance prior to parturition. The herd health procedure for new-born kids should be followed, specifically the examination for any congenital defects after birth. Moreover, proper housing to manage hyperthermia in kids and environmental stress for lactating does should be available. Prevention measures for diseases are usually less expensive than treatment and achieve better economic returns with minimum health problems. The emphasis should be on nutritional quality and quantity at different stages of development. Fat goats are more prone to go off feed and have problems at kidding as well as pregnancy toxaemia.

The major aspect of herd health monitoring is to provide appropriate shelter to protect the goats from adverse weather and variable climatic conditions. Shelter development and management are discussed in Chapter 6. Goats should be kept away from wet and marshy land in order to protect their hooves from damage. Goat hooves should be trimmed regularly to prevent the overgrowth that leads to foot problems. Herd health programmes for lactating goats should be optimized to maintain population growth at herd level. Various biological pathogenic agents may cause unexpected loss in a dairy goat herd. Thus, the dairy goat herd should be checked for zoonotic diseases such as brucellosis. There are several zoonotic diseases transferred from goats to humans, the most common of which is *Brucella melitensis*. The effective management of goat herds (e.g. movement control and sanitation care) should be adapted to reduce the possibility of exposure to pathogenic organisms and parasites. The pathogenic organisms infecting goats are discussed in Chapter 5. Nutritional supplement is a critical factor for goat health management, as is access to water.

The general herd health procedures include record keeping, biosecurity and sanitation, colostrum feeding, nutritional supplements, foot care, parasitic control, disease resistance and culling. Therefore a database of goat production management information should be developed and used to enable effective decision making in order to optimize management for better profitability, risk management and sustainable utilization of resources.

1.4 Goat Production Contributions and Challenges to Food Security

Goats produce meat, milk, fibre, skin and manure. Goat meat has both a large export market and local market. Fibre from goats (mohair and cashmere) is considered to be one of the most luxurious fibres in the world. Goat skin also has high export value due to its role in value-added products. Low-income, food-deficit countries have about 85% of the goat population. The largest goat population is in Asia, followed by Africa. Oceania is the highest exporter of goat meat. North America has the smallest goat population, but it has impressive population growth rates.

The biophysical environment of a particular area determines the types of crops that can be grown and feed resources available for maintaining livestock (Devendra, 2007a). The European goat production system is quite different from developing countries. The livestock production system in developing countries is mainly a pastoralist, range-based, mixed crop-based and landless system, and is therefore a sustainable process. These types of production systems are found in Asia, North Africa, Central Asia, West Asia, Latin America and the Caribbean regions. Agro-pastoralist systems are very specific to Africa and Asia (see Chapter 7). This production system is characterized as extensive grazing where households are nomadic. A similar system is also observed in India in different agro-climatic regions. A range-based system is essentially an extensive system characterized by rainfall of less than 150 mm/year, chronic drought and minimal feed resources. The range-based system is mainly observed in nomads and migratory farmers.

Four categories of goat production system have been identified in the existing livestock systems (Devendra, 2007b): rural landless systems; extensive systems; systems combining cropping and grazing; and integrated systems with tree cropping.

The rural landless production system is an extensive management system where livestock migrate from one place to another for better forage to manage drought effects and availability of water. This system is observed in resource-poor nomads, migratory farmers, poor agricultural labourers and migrant workers. This type of production system is observed in arid and semi-arid areas (Devendra, 1999), for example, in India, Pakistan and the Hindu Kush Himalayan region in South Asia. The integrated system involving grazing and tree lopping is observed in areas where grazing land is scarce.

It is essential to know the production environment of goat production as it will identify the constraints limiting goat production and enable farmers to address issues and formulate strategies to resolve problems. The diverse constraints in different production systems need to be analysed and a system approach should be adopted to provide an important pathway for increasing ruminant productivity. Constraints such as traditional

Table 1.5. Summary of livestock systems, priority production systems and major issues across regions.

Type of livestock systems	Priority production system	Region: Asia	SSA	CA	WANA	LAC	Major issues
Landless	Peri-urban/urban dairy	/	/	/	/	/	Surface water
	production	/	/	/	/	/	contamination
	Peri-urban/urban Poultry	/	/	/	/	/	Zoonosis
	and pig production	/	/		/	/	
	Feedlot (cattle or small						Waste Disposal
	ruminants)						Nutrient flows
	Goat and Sheep production						Overgrazing
Crop-based mixed	Integrated systems with	/	/	/	/	/	Food –Feed systems
	Annual crops (ruminants)	/	/	/	/	/	All –year –round feeding
	and non –ruminants	/	/	/	/	/	systems
	plus fish) integrated	/	/			/	Nutrient flows/soil fertility
	systems with perennial						Productivity
	crops (ruminants)						enhancement
	Beef and dairy production						Intensification and
	Goat and Sheep						specialization
	production						Overgrazing
Agro-pastoralist	Cattle		/	/	/		Feed supplies/drought
	Goat and Sheep		/	/	/		strategies
	production						Property regimes
							Overgrazing
							Trypanosomiasis
Range-based	Sheep and goat	/	/	/	/	/	Drought strategies
	production						Overgrazing
							Property regimes
							Marketing

Abbreviations: SSA, sub-Saharan Africa; CA, Central Asia; WANA, West Asia and North Africa; LAC, Latin America and the Caribbean; / indicates that both the production systems and animal species are the most important within the region.
Source: Devendra *et al.* (2005).

methods of farming, crop–animal interactions and the role played by the animals in a specific environment should be resolved by a multidisciplinary approach.

It is important to accelerate the productivity potential of different genotypes in specific agro-ecological conditions such as a dry cold region, a drought-prone region and the high humid and rainfall areas. The technologies specific to different ecozones need to be developed and demonstrated to farmers to increase productivity. It is also required to address large-scale innovation to optimize the management and decrease the loss in specific environments due to various factors. The major issues in the different livestock production systems in a range of geographical regions are presented in Table 1.5. The details of challenges and opportunities in goat production systems are discussed in Chapter 7.

1.5 Protocols

Protocol 1: Stocking rate determination by average animal weight method (Dida, 2017)

The average animal weight method for stocking rate determination seems to be the more accurate method. The metabolic requirement of a cow of 250 kg with a calf requires consuming 2.667% of its body weight each day. This number was derived by using the metabolic requirement of a cow with a calf on a daily basis. This method is used regardless of breeds or species and determines the daily forage requirement for their body size by using a conversion factor of 2.667% (multiplying 0.02667).

For the average animal weight method, use the following steps:

1. Determine total production of area.
2. Calculate the total 'available' forage. First you need to determine the percentage of use you would like in the

area. This number varies based on your management objectives. A conservative figure often used is the 'take half, leave half' (or 50%) rule of thumb. Calculate your available forage by multiplying total forage by your percentage of use (0.5 in in the case of 50% of use).

3. The average animal weight method allows you to calculate the required forage for the animal, regardless of the breed or species, and also determine the daily and monthly forage requirement for their size by using the conversion factor of 2.66%.

a. Estimate your average size of animal (in kilograms).

b. Multiply this number by the average animal weight method conversion factor (0.02667).

c. Multiply this figure by 30 days/month to get your herd animal unit month (AUM) consumption: Monthly forage requirement = average animal size × 0.02667 × 30 days/month

4. Calculate proper stocking rate for class of livestock: Stock Rate =

5. Determine the number of animals you can graze over the allotted time:

Number of animals =

Example:

Information: If natural pasture DM yields from 0.5 m^2 plot of Abarnosa rangeland was 0.6 kg and the total area of Abarnosa rangeland is 4000 m^2, what is the total available forage of the range land?

Step 1: Determine total production of the area.

Step 2: Calculate the total available forage.

$$\text{Total available forage} = \frac{0.6\text{ kg}}{0.5\text{ m}^2}\,4000\text{ m}^2 = 4800\text{ kg}.$$

Information: If the average animal weight is 300 kg, how much forage is consumed per day and per month?

Step 3a: Determine average animal size in kilograms: 300 kg

Step 3b: Multiply this number by the conversion factor to determine amount of forage consumed per day:

Forage consumed per day= 300 kg × 0.02667 = 8.001 kg

Step 3c: Multiply this figure by 30 days/month to determine the amount of forage consumed per month

Monthly intake = 8.001 kg × 30 days = 240.03 kg

Step 4: Calculate the proper stocking rate:

$$\text{Stocking rate} = \frac{4800\text{ kg}}{240.03\text{ kg}} = 19.998 \text{ animals per month}$$

Step 5: Determine the number of animals that can be grazed over an allotted time. If the allotted time is 4 months of grazing, then

$$\text{number of animals} = \frac{19.998}{4} = 4.999\text{ animals}$$

Protocol 2: Indicators to monitor livestock production at national, regional and international levels

Indicators for comparison between livestock resources in different regions can be compared based on the following aspects:

- livestock production index (livestock output and productivity);
- livestock resources in relation to human population;
- annual growth rate;
- densities on total or agricultural land;
- number of livestock per person;
- number of livestock units per person;
- total number of agricultural households;
- livestock dependent households;
- human development index (UNDP);
- consumption of livestock products;
- calories consumed;
- proteins and fat/day;
- annual growth rate; and
- per capita consumption of meat, milk and eggs.

Protocol 3: Selected indicators of livestock resources derived from livestock units (Chilonda and Otte, 2005)

Indicator	Definition
Density total land (LU/sq. km)	'Livestock units' divided by the total land area, expressed as livestock units per square kilometre
Density agriculture land (LU/sq. km)	'Livestock units' divided by the agricultural area, expressed as livestock units per square kilometre
Livestock units per 100 people	Total livestock units divided by total population, expressed as livestock units per 100 people
Livestock units per 100 people in agriculture	Total livestock units divided by agricultural population, expressed as livestock units per 100 people in agriculture
Proportion of monogastrics	The proportion of monogastric animals in total livestock resources expressed as a percentage
Proportion of ruminants	The proportion of ruminants in total livestock resources expressed as a percentage

Protocol 4: Selected indicators to monitor production of livestock products (Chilonda and Otte, 2005)

Indicator	Description
Total production	Quantity of meat/milk/eggs produced by species slaughtered within national boundaries expressed in metric tonnes, irrespective of their origin, whether indigenous or foreign
Production density	Quantity of meat/milk/eggs produced in metric tonnes ('Total prod. (mt)') divided by agriculture land and/or total land expressed in kilograms of meat per square kilometre of agricultural land per year
Per capita production	Per capita meat/milk/eggs production is the quantity of meat produced ('Total prod. (mt)') divided by agricultural population and/or total human population expressed as kilograms of meat per person per year
Production per animal	Quantity of meat/milk/eggs produced ('Total prod. (mt)') divided by total number of animals of the species in the country expressed as kilograms of meat per animal per year
Production per animal	Quantity of meat/milk/eggs produced ('Total prod. (mt)') divided by total number of animals slaughtered of the species in the country expressed as kilograms of meat per animal slaughtered per year. This gives an indication of the carcass weight,
Number of animals slaughter/milked/laying	Total number of animals slaughtered/milked/laying of the species within national boundaries, irrespective of their origin
Proportion of animals slaughtered/milked/laying	Number of animals slaughtered/milked/laying divided by the total population of species, expressed as a percentage

Protocol 5: Selected indicators to monitor trade trends in livestock and livestock products (Chilonda and Otte, 2005)

Indicator	Definition
Exports/imports – quantity (mt)	Both report foreign trade (imports and exports) in quantitative forms. The unit of measure is weight (metric tonnes) for all commodities except for live animals which are reported in units (heads). As a general rule, trade data refer to net weight, excluding any sort of container
Exports/imports – value	Both express foreign trade in value terms
Value of exports/imports per person	Value of exports/ imports divided by total human population
Quantity of exports/imports per person	Quantity of Export/imports divided by total human population
Net trade (quantity)	Quantity of exports less quantity of imports
Net trade (quantity)	Value of exports less value of imports
Exports, % total production/Imports, % total production	Quantity exported/quantity imported divided by total production of the livestock product expressed as a percentage
Trade orientation	Net exports i.e. Exports - quantity (mt) minus Imports - quantity (mt) divided by total production expressed as a percentage
Exports/Imports, % total population	Number of livestock of given species exported or imported divided by total population of the respective species expressed as a percentage

Protocol 6: Indicators to monitor livestock health during a five year period (Chilonda and Otte, 2005)

Indicator	Description
Cases per 10,000	Number of animals affected by the disease (sick animals + animals that died from the disease) during the period divided by the species population times 10,000
Cases, total	Number of animals affected by the disease (sick animals + animals that died from the disease) during the period
Deaths, total	Number of animals that died from the disease during the period
Deaths per 10,000	Number of animals that died from the disease during the period divided by the species population times 10,000
Destroyed, total	Number of animals killed for disease control purposes during the period whose carcasses were destroyed
Number of years with disease	The number of years in which the disease in question occurred within a country during the period
Outbreaks, total	Number of occurrences of the disease in question within a country during the period
Vaccination, total	The number of animals vaccinated against the disease in question during the period
Vaccination coverage	Number of animals vaccinated against the disease in question during the period divided by the species population expressed as a percentage

1.6 References

Bertolini, F., Servin, B., Talenti, A., Rochat, E., Kim, E.S. *et al.* (2018) Signatures of selection and environmental adaptation across the goat genome post-domestication. *Genetic Selection Evolution* 50, 57.

Chilonda, P and Otte, J. (2005) Indicators to monitor trends in livestock production at national, regional and international levels, Livestock information sector Analysis and policy branch. Food and Agriculture Organization of the United Nations, Rome, Italy.

Devendra, C. (1999) Goats: challenges for increased productivity and improved livelihoods. *Outlook on Agriculture* 28, 215–226.

Devendra, C. (2007a) *Goats: Biology, Production and Development in Asia*. Academy of Sciences Malaysia, Kuala Lumpur, Malaysia, p. 246.

Devendra, C. (2007b) Perspectives on animal production systems in Asia. *Livestock Sciences* 106, 1–18.

Devendra, C., Morton, J.F. and Rischkovsky, B. (2005) Livestock systems. In: Owen, E., Kitalyi, A., Jayasuria, N. and Smith, T. (eds) *Livestock and Wealth Creation. Improving the Husbandry of Animals Kept by Resource Poor People in Developing Countries.* Nottingham University Press, Nottingham, UK, pp. 29–52.

Dida, M.F. (2017) Review paper on determining stocking rate in tropical countries by the use of tropical animal unit month (Taum). *International Journal of Microbiology and Biotechnology* 2(1), 48–51.

Food and Agriculture Organization (FAO) (2005) FAO stat data. Available from http://www.fao.org/faostat/en/#home (accessed 10 February 2018).

Food and Agriculture Organization (FAO) (2018) Goat population of the world, Available at: http://www.fao.org/faostat/en/#home (accessed 10 February 2018).

Skapetas, B. and Bampidis, V. (2016) Goat production in the world: present situation and trends. *Livestock Research for Rural Development* 28, 200.

2 Prospects and Strategies of Genetic Resources

Goats are multifunctional animals, providing nutritional supplementation, creating prosperity during adversity and serving as companion animals. Goats are one of the best adaptogenic animals (i.e. adapting to difficult climatic condition) and have successfully maintained themselves in many adverse locations. Goats adapt to different climatic conditions and are also able to acclimatize to different agro-ecological conditions. They have particular significance in resource-poor areas because of their biological attributes.

Goats were the first animal to be domesticated by humans, and have accompanied human beings during migration. Goats have played a role in religious ceremonies, folklore, tradition, nutrition, livelihood and economics (Boyazoglu *et al*., 2005). Goats were domesticated first in Jericho (Jordan) around 8000-9000 BC and in the Zagros Mountains in Gangi Darch (Iran) around 10,000 BC (Zeuner, 1963). Domestication of goats was associated with the Nile civilization in North-East Africa, the Tigris–Euphrates in West Asia and the Indus in the Indian subcontinent.

Goats are distributed in varied agro-ecological conditions. They were carried out in ships to supply fresh milk in the West Indies, which is how goats from India were introduced to Trinidad, Guyana and Jamaica. Goats also accompanied immigrants to Australia during the eighteenth century. Asia, West Africa and the Indian subcontinent were important as goats migrated along two routes from these places (Devendra and Nozawa, 1976). The major route of goat migration was from Iran, Afghanistan and Turkistan to Mongolia and northern China through the Silk Road. The second route was from the Indian subcontinent to Asia and Europe through the Khyber Pass. This is the oldest route.

2.1 Phenotypic and Genotypic Adaptation

2.1.1 Phenotypic evaluation of the goat

Anatomy

The body system consists of a number of organs that work together to make the body functional. It is therefore essential to know the different body parts and functions in order to understand the animal's behaviour, production efficiency and welfare. Goat farmers need a working knowledge of the external caprine anatomical parts and their functional characteristics to evaluate the phenotype in order to select superior animals. The first step is choosing a morphologically sound individual that is able to lead an adaptable life within the population in a particular geographical region. It is important to select superior adaptable individuals as replacements to existing stock or for establishing a new stock of population. Anatomical knowledge helps the farmers to identify the best animals for better productivity; it also guides the farmer to evaluate the individuals for animal show purposes. The external body parts for male and female goats are shown in Figs 2.1 and 2.2 and listed in Tables 2.1 and 2.2.

2.1.2 Factors affecting phenotypic evaluation

Body size

Body size is a key breeding characteristic and is dependent on the type of environment and geographical region of the breed. Skeletal frame size or weight and bone and muscle development are indicators of body size. Frame is commonly referred to as the height of the animal at the hips. Volume or capacity of the animal is an important trait. Volume

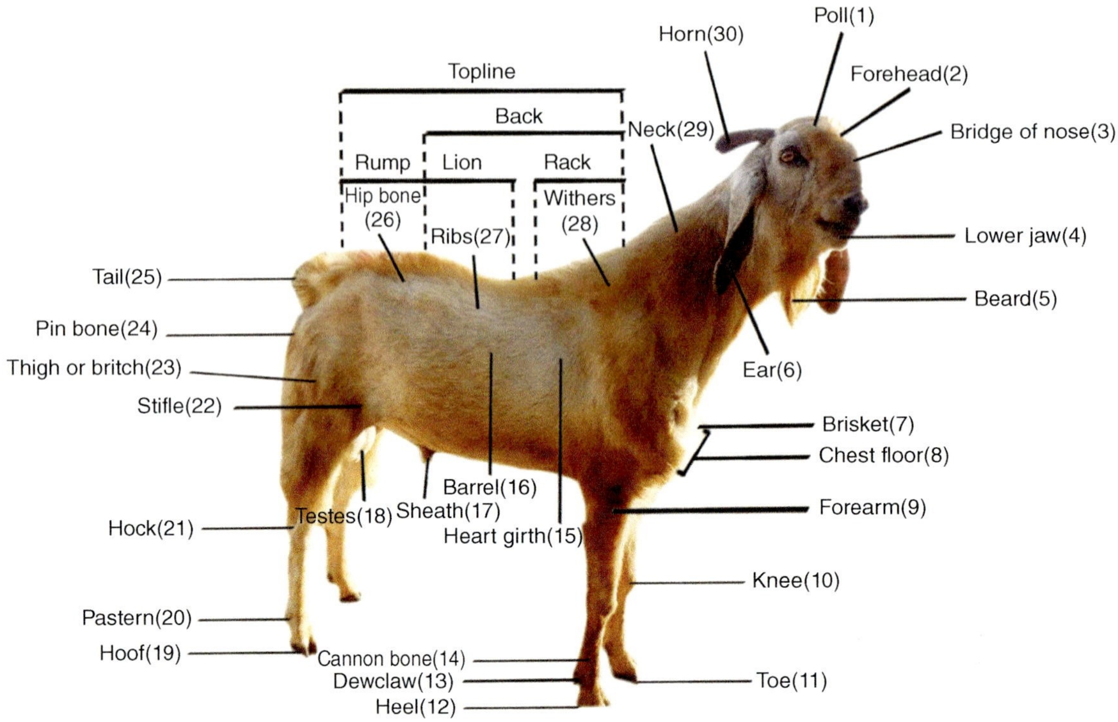

Fig. 2.1. External body parts of a male goat.

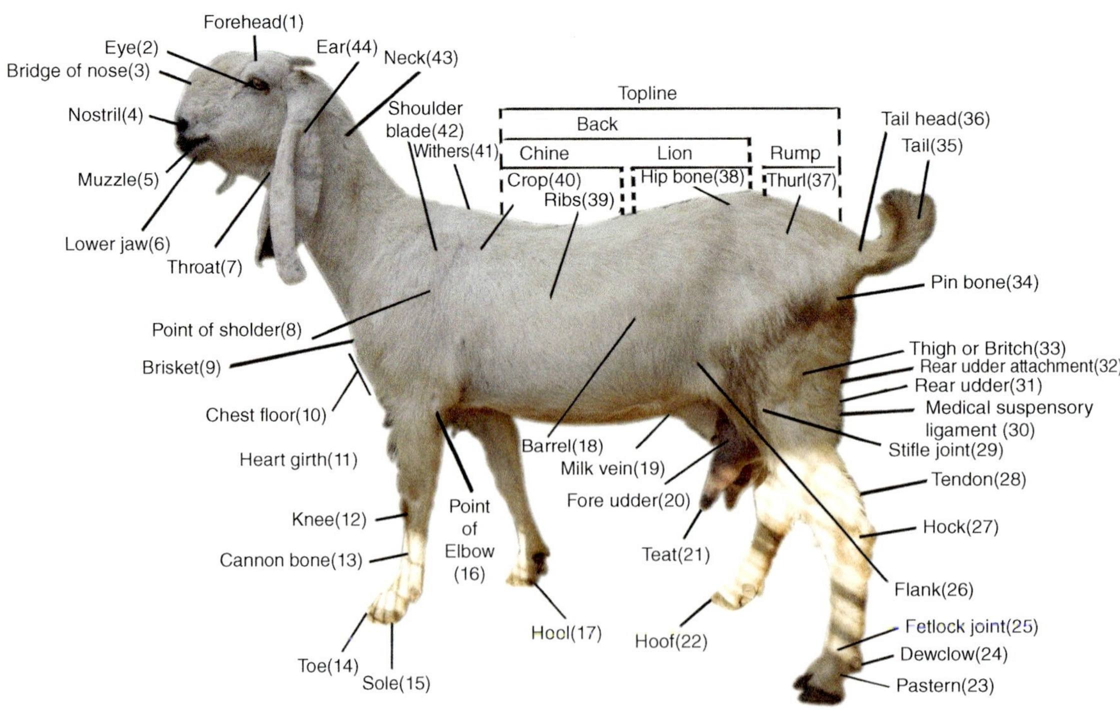

Fig. 2.2. External body parts of a female goat.

shows the internal capacity of the heart, lungs, rumen and uterus size. Animals with a broad chest and a long-sided, barrel-shaped body are preferred. The body size can be evaluated based on body weight, body length and body girth (see Protocol 1 at the end of the chapter for details).

Sexual organs and reproductive parts

MALE Bucks should have two large testicles that are well shaped, of equal size with an evenly single scrotum. Masculinity is an important trait for male evaluation and show-ring judging. Bucks with one testicle (cryptorchidism) or testicles that hang unevenly, vary in size, are hard or soft to touch may be infertile or sub-fertile.

FEMALE Does should have attractive features defined by a long head and neck. The udder should be soft, smooth, well shaped and balanced. The female external genitalia should develop properly.

Table 2.1. Male body parts (see Fig. 2.1).

1. Poll	11. Toe	21. Hock
2. Forehead	12. Heel	22. Stifle
3. Bridge of nose	13. Dew claw	23. Thigh or britch
4. Lower jaw	14. Cannon bone	24. Pin bone
5. Beard	15. Heart girth	25. Tail
6. Ear	16. Barrel	26. Hipbone
7. Brisket	17. Sheath	27. Ribs
8. Chest floor	18. Testes	28. Withers
9. Forearm	19. Hoof	29. Neck
10. Knee	20. Pastern	30. Horn

Table 2.2. Female body parts (see Fig. 2.2).

1. Forehead	16. Point of elbow	31. Rear udder
2. Eye	17. Heel	32. Rear udder attachment
3. Bridge of nose	18. Barrel	33. Thigh or britch
4. Nostril	19. Milk vein	34. Pin bone
5. Muzzle	20. Fore udder	35. Tail
6. Lower jaw	21. Teat	36. Tail head
7. Throat	22. Hoof	37. Thurl
8. Point of shoulder	23. Pastern	38. Hip bone
9. Brisket	24. Dewclaw	39. Ribs
10. Chest floor	25. Fetlock joint	40. Crop
11. Heart girth	26. Flank	41. Withers
12. Knee	27. Hock	42. Shoulder blade
13. Cannon bone	28. Tendon	43. Neck
14. Toe	29. Stifle joint	44. Ear
15. Sole	30. Medial suspensory ligament	

Specific morphological characteristics/distinct anatomical structures are specific for breed.

Selecting the best animal based on anatomical characteristics

Breed characteristics are defined as the specific anatomical features in different agro-climatic regions. Therefore, breed characteristics are important parameters and farmers should be aware of typical characteristics of the particular breed in a specific area. The animals should be evaluated for breed characteristics as a first step towards genetic improvement. If the animal is fulfilling the breed characteristics, then look for structural correctness and the ability to withstand environmental pressure. Perfect standing position, hock joint, pastern joint and mammary gland (in the case of does) should be assessed. There should be no congenital deformities or atypical anatomical structures.

MALE Muscling should be apparent throughout the body of the animal. Muscling indicative points include the thigh, stifle and back region of the animal. The width from stifle to stifle should be broader, and the circumferences of the cannon bones and the foot or hoof should have increased muscle mass.

FEMALE A good wedge-shaped body with a wide pelvic cavity and developed mammary glands are desirable features. There should be width between the pin bones. It is essential to have structural correctness of the feet and legs. The udder should be symmetrical in shape and soft to touch. There should not be any fibrous mass at the teat canal.

A prominent milk vein in front of the udder is a desirable characteristic.

Structural soundness/correctness

Animals with better morphological and functional characteristics are essential for producing seed stock in a herd. Moreover, structural correctness is the major criteria used to assess the general appearance of show-ring livestock. Production traits are directly dependent on the structural soundness of an animal. The animal should be examined from the side view for structural correctness. It is necessary to evaluate the length and depth of the body and the volume of the animal. Again, the correct hind leg position is an angulated hock and strong pastern. The correct position is straight down from the pins to hock and dewclaw. These animals have strong pasterns by which they can browse or graze and walk for longer periods (Fig. 2.3). The different defective hind leg positions include a straight leg with no pastern, sickle-hocked animals and angulation to the curvature of the hind legs (Fig. 2.4).

Similarly, the correct position of the front leg will be approximately straight down from the shoulder to the knee and the dewclaw. Deformities in the front leg include 'calf-kneed', a condition where the forelimbs have reduced shock-absorbing ability and a reduction in grazing and tree-climbing efficiency, and 'buck-kneed' (Fig. 2.5) (Ebert and Solaiman, 2010; Constantinescu and Constantinescu, 2010).

REAR-VIEW EVALUATION The most common structural problem of the rear leg is a narrow distance between hocks, called 'cow-hocked'. A 'bow-legged' stance is less commonly observed in the population (Fig. 2.6).

FRONT-VIEW EVALUATION The correct form of leg position in the front view, known as the ideal structure (Fig. 2.7), has the knee structure straight up and down, in-line with the forearm. This provides a straight line to the pastern, enabling the animal to climb trees and walk long distances. Abnormalities include 'splay-footed', which is a common defect observed in the population, and 'pigeon-toed' (Fig. 2.7) (Ebert and Solaiman, 2010).

Other evaluation parameters

Other anatomical features that should be examined for structural soundness include the eyes, ears, mouth, teat and hooves. A blind animal should not be included in the selection programmes and should be culled. Similarly, cracked or split hooves affect the productive life span of the animal.

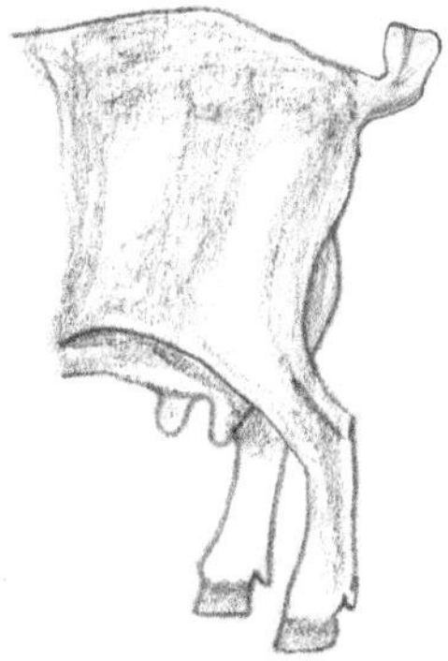

Ideal

Fig. 2.3. Correct position of the hind leg.

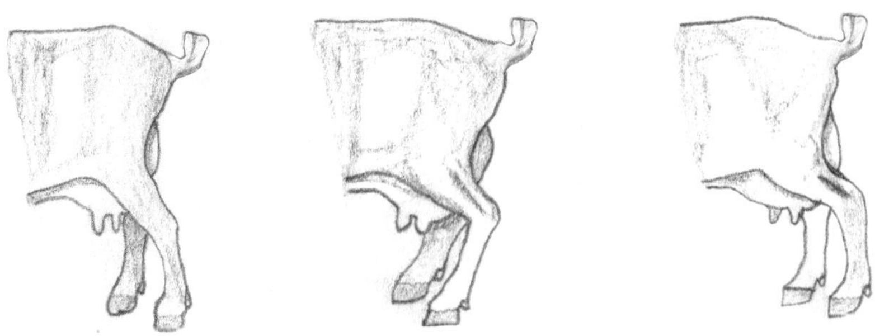

Weak rear-pastern | Sickle-hocked | Post-legged

Fig. 2.4. Defective positions of the hind leg.

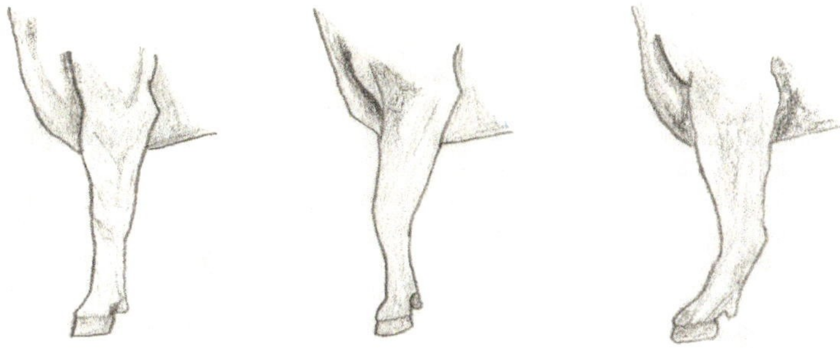

Ideal Buck-kneed Calf-kneed

Fig. 2.5. Defective positions of the foreleg.

Ideal Cow-hocked Bowlegged

Fig. 2.6. Defective positions in rear-view

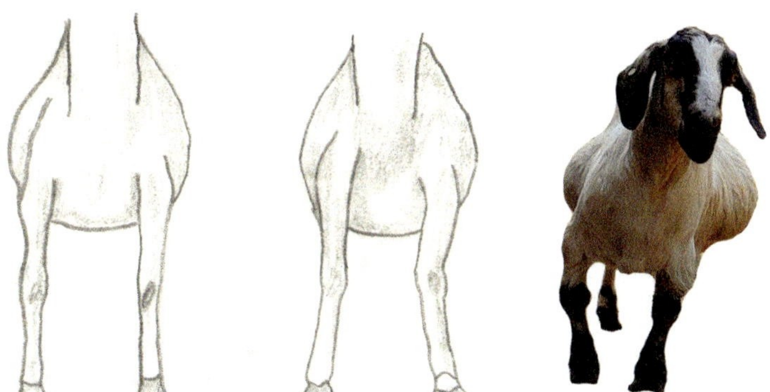

Ideal Splay-footed Pigeon-toed

Fig. 2.7. Defective front leg structure

MOUTH There are some common defects observed in the jaw. Examples include 'parrot mouth' (the bottom jaw is shorter than upper jaw) and 'monkey mouth' (the bottom jaw is longer than the upper jaw, i.e. an overshot bottom jaw). Both of these conditions affect the alignment of the teeth and the dental pad, thus hindering mastication (Fig. 2.8) (Ebert and Solaiman, 2010).

TEAT The teat structure is of critical importance. Does should have two functional teats and the udder should be pendulous. Goats with a bulbous or oversized teat should be culled. One functional teat is acceptable; however cluster or split teats are not ideal in a productive herd. The does with teat defects should not be retained in the breeding flock.

Determination of age

The approximate age of the goat can be determined by observing the dentition pattern (teeth development). Goats should attain specific weight by a certain age. Age and size are important in evaluating an animal, although the environment can play a significant role in the developmental process. The plane of nutrition and health can also influence the growth of teeth in goats. The age of a goat can be estimated based on their eight lower incisor teeth. The loss of the first two 'baby teeth' occurs at approximately 1 year old. Annually, two baby teeth are replaced with permanent teeth, until all the baby teeth are replaced, at approximately 4 years of age (Fig 2.9 and Table 2.3).

2.1.3 Genetic adaptability

A breed should be selected based on the adaptability to particular environmental conditions, purpose (e.g. milk, meat, fibre) and production performance (Gipson, 1996). Goats are comparatively adaptable to all types of agro-ecological conditions and found in most regions in tropical countries.

2.1.4 Selection practices

The phenotype is the expression of the genes possessed by the individuals under the influence of the environment (non-genetic factors). Phenotype is observed in an animal in different forms, such as morphological variation and adaptive characteristics of

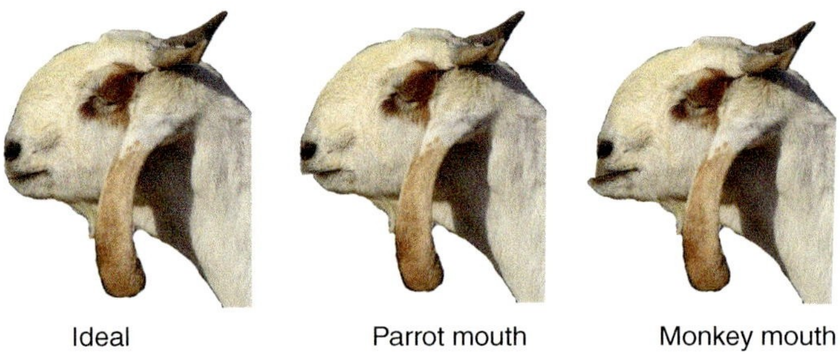

Fig. 2.8. Defective jaw conditions.

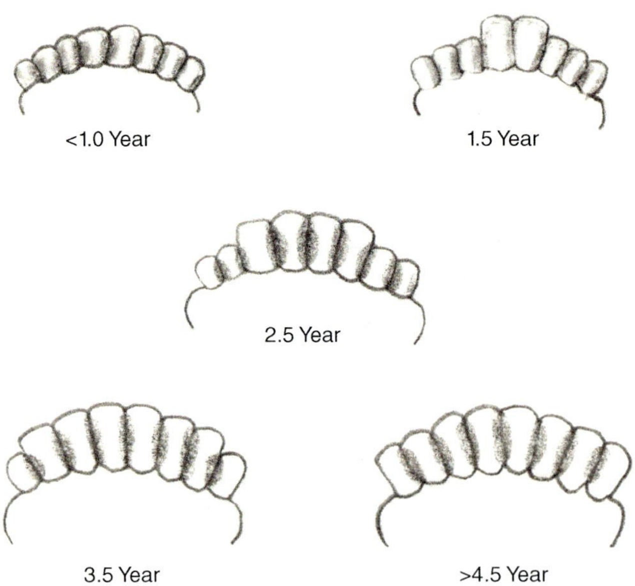

Fig. 2.9. The dentition pattern of the goat at different ages.

Table 2.3. Age of goats by dentition pattern.

Type of teeth	Age (months)
8 baby teeth (milk teeth)	0–14
2 large teeth in the middle (one pair of permanent incisors)	14–19
4 large teeth and 4 milk teeth (two pairs of permanent incisors)	19–24
6 large teeth and 2 milk teeth (three pairs of permanent incisors)	24–30
8 large teeth (four pairs of permanent teeth)	>30

an individual. Phenotype can be better described as the specific appearance and adaptive behaviour of the individual in a particular geographical condition. Genotype is affected by the environment and the expression of genotype varies in different environmental conditions. Environmental factors influence the phenotype as soon as the individual is conceived, and continues after birth with respect to climate and management. Intrauterine environment plays a significant role on development of the embryo's physiological and psychological condition, fitness and immune status. Genotype refers to the genetic potential of individuals. As the individuals need specific environmental conditions for the expression of genes to attain the desired trait, it is necessary to understand the genetic potential of the individual or breed first. Individuals cannot exceed the genetic limit even in the best environmental conditions. Similarly, if a high-yielding animal is exposed to poor environmental conditions then it is not able to attain the desired growth during a particular stage. It is therefore important to understand the genetic potential and production environment to produce the optimum expression of genotype. Another aspect is genotype by environmental interaction, and epigenetic control for the expression of different traits:

$$P = G + E$$

where P is the phenotypic value, G is the genotypic value and E is the environmental deviation.

Phenotypic variance is the expression of genotype by environmental interaction. It is necessary to determine the proportion of total variation in the population due to genetic component. Generally, the effect of G and E are expressed as deviation from the population mean (flock average):

$$P = \text{population mean}(\mu) + \text{genotype deviation} + \text{environmental deviation.}$$

The major characteristics of goat production are manifested by growth at different ages, milk production and fibre yield. The trait expression will be dependent on genetic and non-genetic factors (environment). For example, a 12-month-old goat's body weight is influenced by a combination of factors such as birth weight, mother's milk yield, birth status (single or twin), growth rate, feeding status, disease incidence and shelter conditions. Similarly, milk production depends on feeding status, parity, season and birth status.

Breed characteristics, genetic make-up and production performance are determined by genes transmitted by parents to the offspring. Genetic change in a population can be in a positive or negative direction. Variation of a trait in a population is the basis of selection. The main objective of the selection is to improve the performance of desired traits and to reduce the degree of variation in the population over the years. The genetic variation of body weight in 6-month-old Jamunapari goats is presented in Fig. 2.10.

Concept of selection

Selection is the process of choosing the best animals from a population to become parents for the next generation, and how many progenies will be kept for producing future generations. Selection is used as a management tool in any livestock production system to introduce the desired genetic change. Genetic variation in a population is the basis of selecting the best individuals to carry out a genetic improvement programme (Fig. 2.10). A breed or local genotype performance can be improved by selection and it provides the opportunity to improve the performance for optimal productivity. Identification of selection traits for a population or breed should be accurate and based upon characteristics for which the animal is valued. Selection and mating are the major tools to carry out breed-improvement programmes.

Selection is the tool which can improve the production performance as well as maintain productivity over the period. It brings a permanent change in a population and maintains productivity over the period. Selection is a complex decision-making process that selects the best males for breeding

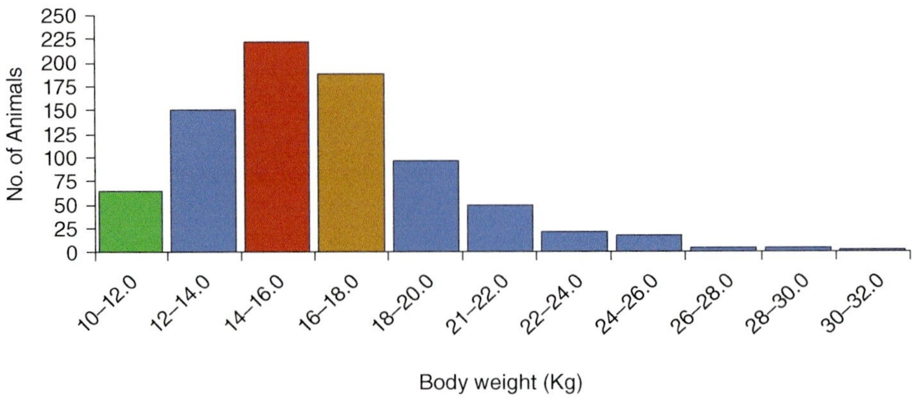

Fig. 2.10. Distribution pattern of body weight in 6-month-old Jamunapari goats in a semi-intensive production system.

purposes and culls the undesired females (does) from the flock.

Before starting the selection programme in a flock it is important to know the essential requirements. The primary information required to start the selection programme are:

- trait recording;
- population structure; and
- data record formats (data register) for growth, milk or fibre yield, mortality, feeding, etc.

2.1.5 Trait recording and traits for selection

The genetic variability of traits under selection is the first criterion to consider. If the desired trait has a high degree of genetic variability, then the improvement in that trait will be faster. It is important to define certain criterion before the selection process begins.

Production traits for goats are:

- Milk yield traits
  - yield per day;
  - yield in a defined period of lactation (e.g. 60 days, 90 days and 140 days);
  - yield per lactation;
  - fat and protein content;
  - total fat and protein production per lactation;
  - lactational milk yield (LMY); and
  - fat protein ratio (FP ratio).
- Growth traits:
  - birth weight;
  - body weight during a particular age (e.g. 3, 6, 9 or 12 months);
  - body size (body height, body length, heart growth) over a defined period (e.g. 3–6 or 3–9 months' of age);
  - average daily growth (g/day);
  - pre-weaning growth rate;
  - post-weaning growth rate;
  - number of kids born and survived per adult female per year; and
  - carcass weight compared to body weight before slaughter.
- Fibre/hair:
  - amount per clipping;
  - fineness and length;
  - fleece weight;
  - average fibre diameter; and
  - mature body weight of female.

2.1.6 Criteria for selection

Basis of selection

Selection of individual goats for mating can be made on the following basis:

- *Individual selection/mass selection:* Individuals are selected according to their own phenotypic value, that is, individuals are selected on the basis of their own performance. This is the most simple method and is better for traits with a high degree of heritability. It seems to be the most practical approach for selection at village level in tropical countries.
- *Family selection:* The selection is based on the mean phenotypic value of the family and whole families are selected or rejected as units. In this method, the within-family deviations are given

zero weight. The family selection is preferred when the trait for selection has a low degree of heritability.

- *Pedigree selection:* The individuals are selected on the basis of the performance of their parents and grandparents. This method is useful when there is no information on the performance of individuals. This method is useful if the traits are expressed in a later period of life such as carcass characteristics or milk yield.
- *Progeny testing:* The selection of the individual is based on the mean phenotypic value of the individual's progeny. This is also another form of family selection. This is useful when the degree of heritability of the trait is low. This method has limited value for the improvement of goat production in the tropics.

Multi-trait selection

Economically important traits need to be improved either as a single trait at a time or including more than one trait at a time. This depends on the correlation between traits and other parameters.

- *Tandem selection:* One trait is selected over generations until the desired improvement has been achieved. After this, a second trait is improved. This selection may be useful for two or three traits that are not correlated.
- *Independent culling level:* In this method, a threshold level needs to be defined for each trait in the selection programme. The individuals that satisfy the threshold value of each trait is selected as parents for the next generation. Again this method of selection is useful for three traits only. The method of selection can be implemented without the knowledge of heritability and economic worth.
- *Selection index:* This is the most effective method when the genetic parameter and economic worth of the trait are known. The method utilizes the information from individuals, parents, half siblings or from relatives. The aim of the method is to combine all the information in order to select individuals. This method has some drawbacks in that it is necessary to maintain all the individuals until the evaluation is complete.

Genetic progress

Genetic progress occurs in the form of change in performance of individual goats from one generation to the next.

Genetic progress (ΔG) and expressed as:

$$\Delta G = SD \times h^2 \text{ or } \Delta G = h^2\left(\overline{X}s_f + \overline{X}s_m\right)$$

Selection response

Selection response (R) is the difference of mean phenotypic value between the offspring of selected parents and the whole of the parental generation before selection.

$$R = h^2 S$$

The rate of improvement for the traits under selection over the generation is dependent on the following aspects:

- *Accuracy of selection:* It is the correlation between the criteria on which selection is based and the objective of selection. The accuracy of selection depends on:
  - using additional measurements for traits from the same animal;
  - using measurements of correlated traits from the same animal; and
  - using measurements of traits from relatives (family selection).
- *Intensity of selection:* The standardized selection differential is called selection intensity and is denoted by '*i*'; higher selection intensity means a higher proportion of individual average breeding value available for transmission to the next generation.
- *Generation interval* is the average age of the parents when their offspring are born. For goats kept under village conditions this is probably 3–3.5 years. This is dependent on breed and age of sexual maturity.

Genetic improvement can be expressed as the expected genetic progress per year:

$$\text{Annual genetic progress} = \frac{\text{accuracy} \times \text{intensity} \times \text{genetic variation}}{SD} / GI$$

In an effort to make the most rapid progress possible over a given time period, accuracy, intensity and genetic variation should be made as large as possible and generation interval reduced as much as feasible.

2.1.7 Genetic parameters and correlated traits

Selection differential

The selection differential (SD) is the mean phenotypic value of the individuals selected as parents expressed as a deviation from the population mean. It is the degree of superiority of the selected individuals over the mean phenotypic value of all the individuals in the parental generation.

This is expressed as:

$$SD = \overline{X}_s - \overline{X}$$

where $\overline{X}_s$ is the mean of the selected individuals and $\overline{X}$ is the population mean. This is expressed in the units the traits are measured. The selection differential should be calculated separately for bucks and does, and then the average selection differential can be calculated.

Heritability

Heritability in the broad sense is expressed as:

$$h^2 = \frac{\sigma^2 G}{\sigma^2 P}$$

that is, it is the ratio of the genetic variance ($\sigma^2 G$) to total phenotypic variance ($\sigma^2 P$). This measure is not used in practice, because the total genetic effect or variance is not transmitted from parents to their offspring.

Heritability in narrow sense is expressed as:

$$h^2 = \frac{\sigma^2 A}{\sigma^2 P}$$

that is, it is the ratio of additive genetic variance ($\sigma^2 A$) to total phenotypic variance ($\sigma^2 P$). Heritability can be measured using several methods, for example:

- offspring and parent regression;
- half-sib correlation;
- hull-sib correlation;
- intra-sire regression of offspring on dam;
- least squares method; or
- mixed model method.

Repeatability

Some traits are expressed repeatedly throughout an animal's lifetime, for example, milk production in does is repeated from first lactation to sixth lactation during a lifetime. Repeatability (R) is defined as the ratio of genetic ($\sigma^2 G$) and permanent environment ($\sigma^2 PE$) variance for a trait to total phenotypic ($\sigma^2 P$) variance:

$$R = \frac{\sigma^2 G + \sigma^2 PE}{\sigma^2 P}$$

Milk yield and litter size are also examples of characteristics repeated over time. The within-individual variation is entirely environmental in origin and caused by temporary differences in environmental conditions at a particular time. Between-individual variation is partly environmental and partly genetic.

The heritability and repeatability estimates of different traits in different breeds are presented in Tables 2.4 and 2.5, which gives an idea for selecting traits in different environments. The heritability also varies from flock to flock in different production environments. The estimates of heritability for body weight, average daily gain and reproductive traits and the genetic correlation between the traits are presented in Tables 2.4 and 2.5.

Similarly, a study has been carried out to determine the unweighted and weighted average of direct heritability for growth, milk production and reproductive traits based on the genetic parameter estimation reports from different countries in different breeds. The details of heritability or repeatability of different parameters have been presented in Tables 2.6, 2.7 and 2.8 (Jembere *et al.*, 2017).

Table 2.4. Estimates of heritability and repeatability for reproductive traits, body weight and daily gain, and direct and maternal heritability and their correlation.

Trait/ Breed	Heritability	Repeatability
Age at first kidding		
Alpine × Beetal, Beetal	0.56–0.48	
Litter size		
Beetal × Black Bengal	0.15–0.09	
Alpine × African common	0.02	
Multiple births		
Egyptian Baladi	0.25	0.29
Beetal	0.15	0.22
Black Bengal	0.09–0.17	0.15
Birth weight		
Black Bengal	0.07–0.40	
Boer	0.18–0.36	
Jamunapari	0.46–0.55	
Body weight at 12 months		
Black Bengal	0.32	
Jamunapari	0.13–0.31	
Boer	0.31	
Daily gain (birth to 150 days)		
West African Dwarf	0.38–00.63	0.21–0.38

Source: Shrestha and Fahmy (2007)

Phenotypic and genetic correlation

Calculating the genetic, phenotypic and environmental correlations is necessary as they are useful for practical goat breeding. A breeder wants to improve many traits simultaneously, for example, fertility, growth, milk yield and muscle growth. Therefore it is necessary to know which traits have positive or negative genetic correlations. If there is a significant genetic correlation, then improving one trait will also improve the associated traits in future generations.

Table 2.5. Estimates of heritability for production traits and direct and maternal correlation.

Breed/Trait	Heritability: Direct	Heritability: Maternal	Correlation: Direct × Maternal
Birth weight			
Boer	0.16–0.33	0.14–0.36	–0.31
Turkish Angora	0.02	0.10	0.18
Weaning weight			
Boer	0.18–0.27	0.05–0.06	–0.15
Turkish Angora	0.03	0.10	0.91
Yearling fleece weight			
Turkish Angora	0.06	0.04	–0.91
Litter Size			
Turkish Angora	0.06	0.04	–0.07

Source: Shrestha and Fahmy (2007).

Phenotypic correlation is denoted by r_P and is estimated by:

$$r_P = \frac{\sigma_{XY}}{\sqrt{\sigma_X^2 \sigma_Y^2}}$$

It is the association between two traits measured in an individual and is the result of genetic effects that are common for the two traits; it is also influenced by environmental effects.

Similarly, genetic correlation is denoted by r_A or r_g and is usually calculated by using the additive genetic effects:

$$r_A = \frac{\sigma_{A_{XY}}}{\sqrt{\sigma_X^2 \sigma_Y^2}}$$

The genetic correlation between two traits is a result of the genes that have effects on both the traits due to a pleiotropy effect. Genetic correlation also occurs when genes affecting two traits

Table 2.6. Unweighted average direct heritability (ha²) and common environmental effect (c²) for milk and reproductive traits in goats

Traits	*n*	ha² ± SD	PSE	*n*	c² ± SD
Litter size at birth	16	0.13 ± 0.10	0.10	3	0.07 ± 0.02
Litter size at weaning	6	0.06 ± 0.05	0.07	1	0.04
Litter weight at birth	8	0.08 ± 0.06	0.03	2	0.05 ± 0.02
Gestation length	2	0.14 ± 0.06	0.02		
Age at first kidding	9	0.29 ± 0.14	0.02		
Kidding interval	11	0.06 ± 0.04	0.15		
1st Kidding interval	4	0.08 ± 0.05	0.04		
Average daily milk yield	8	0.26 ± 0.09	0.14	1	0.31
90 days milk yield	5	0.31 ± 0.15	0.22		
150 days milk yield	2	0.33 ± 0.21	0.19		
270 days milk yield	1	0.19			
Lactation milk yield	13	0.32 ± 0.17	0.14		
1st lactation milk yield	11	0.41 ± 0.08	0.17		
Protein yield/lactation	12	0.43 ± 0.19	0.16		
Fat yield/lactation	12	0.40 ± 0.15	0.15		
COMB P & F	6	0.45 ± 0.09	0.17		
Fat percentage	6	0.51 ± 0.06	0.18		
Protein percentage	6	0.53 ± 0.09	0.19		
Ratio of protein to fat	6	0.39 ± 0.10	0.19		
Lactation length	7	0.24 ± 0.27	0.14		
Survival rate to weaning	2	0.10 ± 0.06	0.03		

n, number of studies; SD, observed standard deviation; PSE, mean of predicted standard error of heritability estimates; COMB P&F is fat and protein components.

Table 2.7. Weighted average of direct heritability (ha^2), maternal genetic effect (hm^2) and common environmental effect (c^2) for growth traits in goats.

Traits	*n*	ha^2 ± SE	*n*	hm^2 ± SE	*n*	c^2 ± SE
Birth weight	52	0.16 ± 0.014	15	0.12 ± 0.013	9	0.09 ± 0.009
3 months weight	40	0.22 ± 0.011	12	0.08 ± 0.011	12	0.09 ± 0.011
Six months weight	22	0.28 ± 0.015	3	0.13 ± 0.010	3	0.08 ± 0.016
9 months weight	15	0.38 ± 0.018	3	0.09 ± 0.009	3	0.10 ± 0.030
12 months weight	21	0.31± 0.018	3	0.05 ± 0.024	3	0.05 ± 0.024
18 months weight	3	0.17 ± 0.319				
Pre-weaning daily gain	16	0.17 ± 0.012				
Daily gain from 3–6 months	7	0.28 ± 0.014	3			0.11 ± 0.138
Daily gain from 6–12 months	2	0.03 ± 0.289	1			0.02 ± 0.020

SE, pooled standard errors over studies; *n*, number of studies from which averages were calculated.

Table 2.8. Weighted average of direct heritability (ha^2) and common environmental effect (c^2) for reproductive and milk production traits in goats.

Traits	*n*	ha^2 ± SE	*n*	c^2 ± SE
Kidding interval	11	0.09 ± 0.010		
First kidding interval	4	0.002 ± 0.018		
Litter size at birth	16	0.05 ± 0.004	3	0.06 ± 0.015
Litter size at weaning	3	0.06 ± 0.017	1	0.04 ± 0.020
Litter weight at weaning	8	0.04 ± 0.003	2	0.06 ± 0.018
Gestation length	2	0.10 ± 0.010		
Age at first kidding	9	0.17 ± 0.012		
Average daily milk yield	6	0.31 ± 0.015		
90 days milk yield	4	0.23 ± 0.076		
150 days milk yield	2	0.36 ± 0.129		
Lactation milk yield	12	0.33 ± 0.011		
First lactation milk yield	11	0.43 ± 0.048		
Protein yield/lactation	11	0.40 ± 0.018		
Fat yield/lactation	11	0.36 ± 0.018		
Fat percentage	6	0.52 ± 0.066		
Protein percentage	6	0.54 ± 0.068		
Comp F&P	6	0.41 ± 0.069		
Lactation length	7	0.15 ± 0.035		
Survival rate to weaning	2	0.09 ± 0.020		

n, number of studies; SE, pooled standard error; COMP F&P, fat and protein components.

are linked. The knowledge of genetic correlations can be applied in selection programmes. For example, the body weight at different ages (e.g. birth, weaning and mature weights) have positive and medium-to-high genetic correlation. This indicates that selecting for early growth in a breeding population will also increase the mature body weight in future generations. Similarly, growth rates of young goats has a negative genetic correlation with feed conversion efficiency (feed intake per unit body weight gain).This again signifies that kids selected for rapid growth require less feed per unit gain.

A number of traits expressed during the early life period are controlled by both the individual's own genetic capability and the maternal parent effect in the form of milking ability or nurturing ability. These two effects are termed the direct effect and maternal effect, respectively. Both these effects are heritable, and there is also a genetic correlation between direct and maternal effects.

The environmental correlation between two traits can be estimated as:

$$r_E = \frac{\sigma_{E_{XY}}}{\sqrt{\sigma^2_{E_X}\sigma^2_{E_Y}}}$$

2.1.8 Selection for growth, milk yield and fibre production

Genetic improvement programmes have focused on increasing various economically important traits for higher productivity and better income. The knowledge of breed characteristics and genetic parameters for economically important traits are necessary to increase the productivity and to bring the desired improvement in selection programmes. The genetic parameter estimates for milk, growth and fibre production traits form essential inputs for future realistic breeding programmes. The knowledge of the (co)variance components and heritability estimates for production traits will support breeding strategies by selecting animals with superior genetic merit to optimize the response to selection and to improve desired traits (Sullivan *et al.*, 1986; Barillet, 2007).

Growth and feed conversion efficiency

The global preference for animal protein consumption has increased over the last decade. Goats can utilize fibrous plant materials to produce meat. Goat is the preferred source of meat in developing countries and presently goat meat consumption is increasing in developed countries (Webb *et al.*, 2005; Devendra, 2010). Goats have great potential to fulfil the protein requirement of the growing population. Moreover, the carcass fat content of goat meat is low and virtually all the parts of the carcass can be consumed (Webb, 2014). Body weight and average daily gains are important selection traits for improving production performance.

Growth is a function of the life cycle of each animal that begins with embryo fertilization and ends with death. Cells are the basic unit of growth and development. Growth and developmental rates are governed by both genetic potential and environmental factors. The foetal phase of growth is from differentiation to parturition. The factors that affect post-natal growth and development are genetic potential and the influence of environment and nutrition to attain the genetic make-up. The rate of growth is expressed in terms of weight or weight gain per day or weight gain during a particular time period or average daily gain. The average daily gain is the change in weight between two time periods. Efficiency of growth and development is expressed as the output divided by the input; this is expressed as the weight gain per unit feed consumption. There are three main phases of growth: pre-weaning, post-weaning and adult growth periods. Weaning age should be determined for each breed in their native home tract because weaning shock affects production performance and early weaning causes heavy mortality. The digestive system should adapt to convert forage or grains into nutrient required for growth. Weaning age is the age generally after 8–9 weeks after birth or when kids attain 2.5 times their birth weight.

Generally, kids grow at least 80–125 g/day during the first 3 months of age in the tropics, dependent on body size. Pre-weaning growth depends on sex, parity and season of birth. Kids from early-maturing breeds have faster growth compared to late-maturing breeds. Male kids have higher growth rates than female kids. Pre-weaning growth rate will be affected by factors such as parasitic infection (e.g. coccidiosis) or insufficient milk from mothers. Comparison of pre-weaning growth rates is based on the weight at 90 days. The growth rate can also be compared between kids weaned at 6 weeks or 10 weeks of age. Correction factors need to be used for comparing the growth rate of kids and correction factors are based on litter size, sex and parity of the doe (see Protocol 3 at the end of this chapter). If birth weight is available, then the average daily gain can be expressed and weaning weight at 90 days of age can be determined. If birth weight is not available, then the weaning weight at 90 days can be calculated as (weaning weight/weaning age) × 90. The average daily gain of Jamunapari goats maintained in a semi-intensive production system in a semi-arid climate is presented in Fig. 2.11.

Post-weaning growth rate is less than the pre-weaning growth. Post-weaning average daily growth is a measure of an individual animal's performance and rearing environment. Growth after weaning depends on energy intake by an individual. Both the rate of tissue accumulation and growth rate decline as maturity is approached. Average daily gain and feed conversion ratio (FCR) in different breeds at different ages are presented in Table 2.9.

The heritability estimates for the body weight at 9 months of age ranged from 0.11 ±0.04 to 0.30 ±0.11 in different breeds and locations (Snyman and Olivier, 1996; Roy *et al.*, 2008, 2011; Hermiz *et al.*, 2009; Thiruvenkadan *et al.*, 2009; Alade *et al.*, 2010; Ekambaram *et al.*, 2010). On the other hand, the higher estimates of heritability (>0.33) at 9 months of age were observed in the Markhoz goat

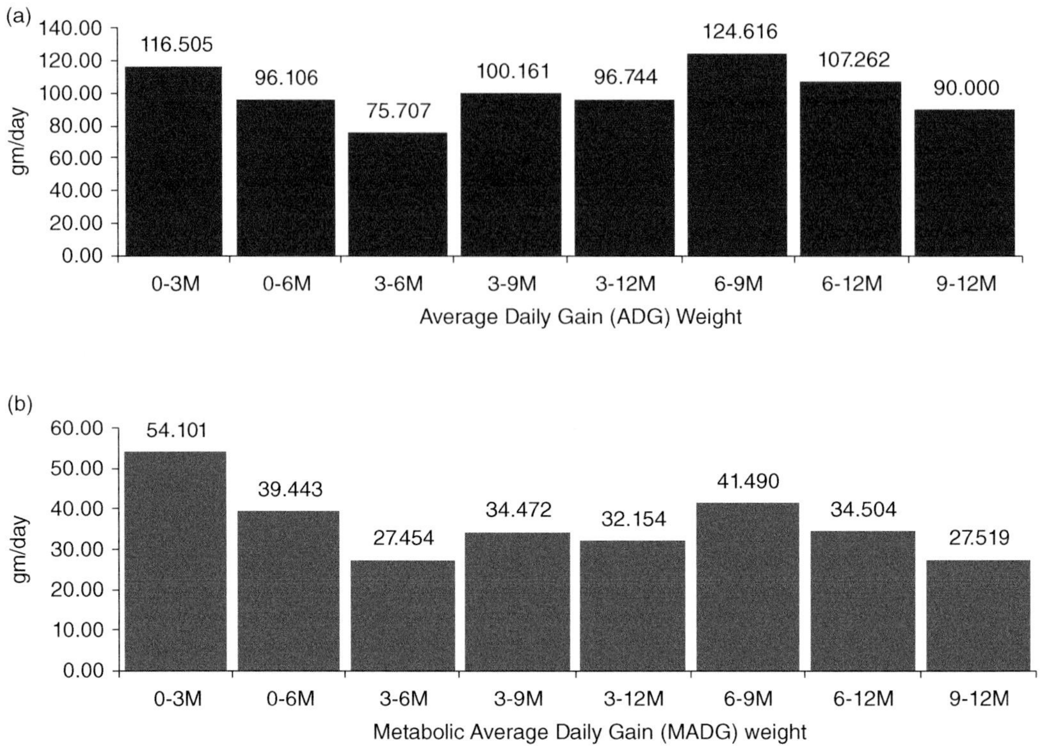

Fig. 2.11. The average daily gain (a) and metabolic average daily gain (b) of Jamunapari goats in semi-arid regions.

breed in Iran (Rashidi *et al.*, 2008), Boer goat (Schoeman *et al.*, 1997) and Black Bengal goats (Faruque *et al.*, 2010). The heritability estimates for the weight at 12 months of age has been reported in different goat breeds from different countries ranging from 0.07 to 0.11 ±0.03 (Zhou *et al.*, 2002; Sawalha and Tabbaa, 2004; Gowane *et al.*, 2011). Higher estimates of heritability at 12 months of age ranged from 0.18 to 0.88 ±0.06 in different breeds (Mukundan and Bhat, 1978; Madeli and Patro, 1984; Jin and Zhang, 1995; Bishop and Russel, 1996; Snyman and Olivier, 1996; Schoeman *et al.*, 1997; Rashidi *et al.*, 2000, 2008; Roy *et al.*, 2008, 2011; Hermiz *et al.*, 2009; Thiruvenkadan *et al.*, 2009; Ekambaram *et al.*, 2010; Faruque *et al.*, 2010).

In general, low estimates of heritability for average daily growth have been reported in the literature with estimates of 0.01 for post-weaning daily gain in Angora goats (Gerstmayr, 1988), 0.04 for daily weight gain at 6 months in the Sirohi goat breed (Gowane *et al.*, 2011) and 0.08 for daily weight gain from 3 to 6 months of age in Raeini cashmere goat (Mohammadi *et al.*, 2012). The estimates of heritability for 0–3 months, 3–6 months, 3–12 months, 6–12 months and 9–12 months of age in Jamunapari goats in semi-arid tropical regions ranged from 0.04 to 0.47 (Rout *et al.*, 2018). On the other hand, a higher estimate of heritability (0.86) of post-weaning body weight was reported in Beetal goats (Shafiq and Sharif, 1996).

The genetic trends obtained for weight at all ages were positive in Jamunapari goat (0.144 kg/year at 9 months of age and 0.199 kg/year at 12 months of age) (Rout *et al.*, 2018). The genetic trend for birth weight was positive but almost constant in nature. Ali and Khan (2008) reported that overall genetic trend for birth weight was static and estimated breeding values ranged from −0.61 to 0.60 kg for bucks. The direct genetic trend at 16-month body weight was 0.11 kg/year, weaning weight increased by 0.06 kg/year while 8-month body weight increased by 0.04 kg/year (Snyman, 2012)

Milk production

Goat milk has potential human health benefits and plays a significant role in nutraceutical, ayurvedic medicine formulation and the fashion industry.

Table 2.9. Average daily growth and feed conversion ratio (FCR) in different breeds at different ages.

Breed	Country	Age	FCR	Average daily growth (g)	Reference
Local goats	Iran	9 months	9.8	77–89	Zamiri *et al.* (2012)
Local goats	Sweden	5–6 months	11.2–16.2	29–55	Phengvichith and Ledin (2007)
Omani and Dhofai goats	Oman	3–5 months	6.3–7.5	80–96	Mahgoub *et al.* (2005)
Shami goats	Jordan	2–5 months	4.8–7.6	141–221	Titi (2003)
Cross breeds of Jamunapari goat (Bachthao × Barbari)	Vietnam	4–6 months	10.3–13.2	59–67	
Cross breeds (Alpine × local goats)	Turkey	3–6 months	6.1–10.9	90–140	Darcan and Cankaya (2008)
Local goats	Italy	Birth to 56 days	8.0	148–170	Agazzi *et al.* (2007)
Caninde goats	Brazil	5 months	9.7	74	
Black Bengal goats	India	6–8 months	8.3–9.4	30–36	
Black Bengal goats	India	9–12 months	10.4	28–32	
Cross breed (BachthaoxBarbari or Jamunapari)	Vietnam	3–6 months	7.6–13.6	40–70	
Black goats	Pakistan	2–5 months	4.76	210	
Barbari goats	India	3 months	9.0	58	Chaudhary *et al.* (2013)
Barbari goats	India	4–5 months		51–88.7	Chaudhary *et al.* (2015)
Barbari goats	India	3–4 months	8–10	56.6–78	Kumar *et al.* (2015)

Goat milk is attracting more attention due to its health-promoting properties and can be used as a safe infant food (Haenlein, 2004). Goat milk is an important source of protein, phosphate and calcium for people of resource-poor countries (Park and Haenlein, 2007). Goat milk is mainly consumed in households and supports protein supplement in the most disadvantaged places for women and children (Haenlein, 2004; Devendra, 2012). Dairy goats are of economic and social importance in the tropics due to their ability to use poor vegetation in varied agro-climatic regions (Devendra, 2012). Goats are favoured as diary animals in low-input systems and marginal environments due to special characteristics such as low capital investment, low production costs and optimum use of meagre resources (Devendra and Liang, 2012).

Saanen is best known for milk production with low levels of protein and fat (Haenlein and Caeccese, 1984). The Saanen, Nubian, Alpine, Toggenburg, La Moncha, Canaria, Norwegian Nordic and Murciana-Granadina are higher milk-producing breeds. There are at least 22 recognized breeds with medium milk yield, including Jamunapari, Damascus and Italian Maltese (Devendra and Haenlein, 2003). The fat and total solid content of milk from Alpine, Nubian and Saanen breeds are lower than native breeds in tropical environments. Similarly, Jamunapari, West African Dwarf and Red Sokoto breeds have richer milk composition but lower yield compared to Saanen and Alpine breeds. After parturition, hormones cause an increase in milk secretion. Colostrum is secreted for about 5 days and contains a higher concentration of solids. The milk yield increases up to 8 weeks after parturition and decreases after peak yield is attained (Fig 2.12a). Similarly, milk yield in does increased up to fourth parity and declined thereafter (Fig. 2.12b). The decline in milk yield is calculated each month as a percentage of the previous months. This measure is called lactation persistency, which is dependent on the breed and individual. The milk yield and quality traits are significantly affected by genetic and environmental components. Parity, birth type and season also have significant effects on milk yield and composition traits. Young goats produce lower milk yield, whereas more mature goats attain higher milk yield during third and fourth parity.

Heritability estimates of 0.29 to 0.39 for milk yield in French Alpine and Saanen primiparous goats, respectively, have been reported by Boichard *et al.* (1989). Heritability estimates for milk yield ranging from 0.34 to 0.37 in Alpine goats and from 0.32 to 0.40 in Saanen were reported in France (Belichon *et al.*, 1999). The heritability estimate

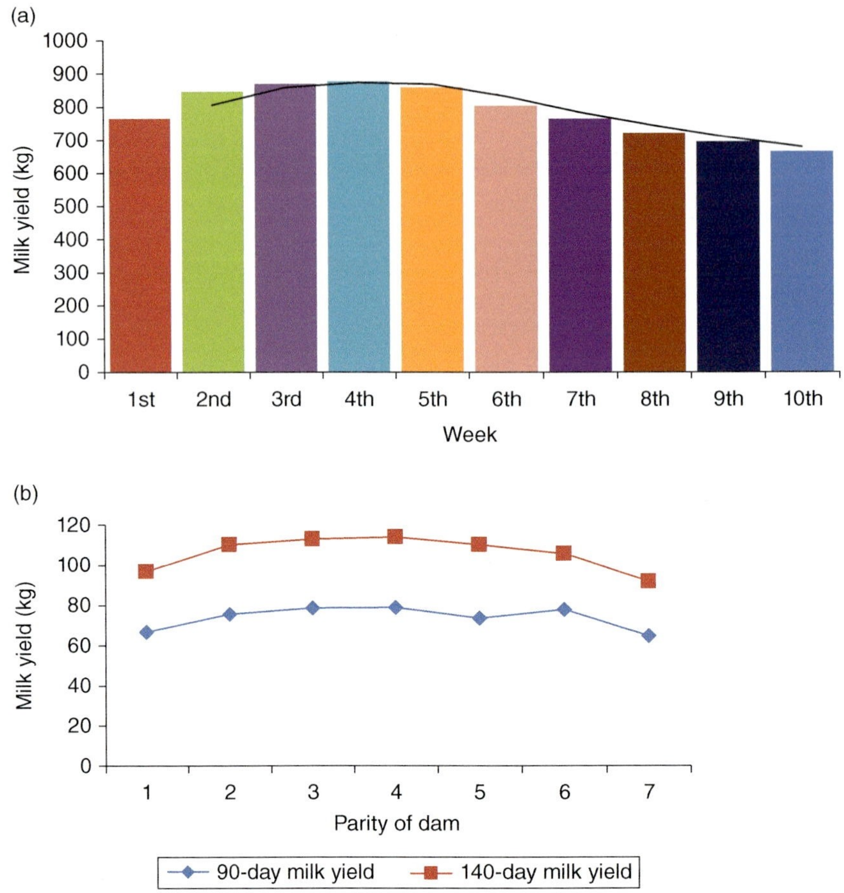

Fig. 2.12. Milk yield of Jamunapari goat during different periods (a) and parity (b) in semi-arid regions.

for milk yield of South African Saanen was 0.23 (Muller *et al.*, 2002). The heritability estimates for milk yield was 0.19 in Alpine, Toggenburg, Saanen and Nubian breeds (Weppert and Hayes, 2004). The estimated heritability for first-parity milk yield of US dairy goats was 0.36 across breeds and varied from 0.35 to 0.38 within breed (Montaldo *et al.*, 2010). Similarly, low-to-moderate estimates of heritability (0.17 to 0.30) for milk yield in Saanen goats were reported in central Mexico (Valencia *et al.*, 2007; Torres-Vázquez *et al.*, 2009, 2010). The heritability of 0.30 for US Alpine and Saanen goats for first lactation yield was observed by Rupp *et al.* (2011). Low heritability estimates of 0.04 for lactation length was reported by Montaldo *et al.* (2010) in Saanen goats reared in Mexico. The direct additive heritability estimates were 0.15 ±0.05, 0.26 ±0.07, 0.25 ±0.08 for milk yield at 90 days, milk yield at 140 days and total milk yield (TMY), respectively in Jamunapari goats. The additive heritability estimate for lactation length was 0.02 ±0.03. The repeatability was moderate to high and ranged from 0.68 to 0.73 for milk yield traits in Jamunapari goats (Rout *et al.*, 2017).

The genetic trend of milk production in Alpine and Saanen goats was 13.65 L/year and 12.53 L/year in France during 1990–2000. Similarly, the genetic trends in Alpine and Saanen goats were 8.63 L/year and 6.99 L/year in the US during 1995–2000 (Clément *et al.*, 2002). Also, a positive genetic trend of 7.0 kg/year was obtained for 305 days mature equivalent milk yield in the female Saanen goat population in the US between 1995 and 2000 (Wiggans *et al.*, 2003). This trend is equivalent to an annual genetic progress of 0.79% in average milk production (Montaldo and

Manfredi, 2002). There is no report available on genetic trends of milk yield traits for long-term selection experiments in the tropics. A positive genetic trend was observed for milk yield at 90 days, 140 days and TMY in Jamunapari goats. There was an increase in mean milk yield of 0.25 kg/year, 0.70 kg/year and 0.72 kg/year at 90 days, 140 days and TMY, respectively, in Jamunapari goats (Rout *et al.*, 2017).

Fibre production

MOHAIR FIBRE Mohair is only produced by Angora goats (*Capra hircus aegagrus*), but cashmere is produced by different breeds (Hunter and Hunter, 2001). Angora goats originated from the Asian Himalayas and highlands of Tibet and migrated to Turkey with pastoral tribes during the eleventh century. The goats are maintained and developed in the Turkish planes near Ankara. From Turkey, mohair goats were supplied to the US, South Africa and Australia (Shelton, 1993). The original Turkish Angora goats are small, delicate and produce about 1–2 kg of mohair. The Angora goats maintained in Texas are large in size and produce 1.5–4 of kg mohair in every 6 months. Mohair production declined significantly after the year 1980 as the demand for fibre in the international market decreased.

The traits considered for the selection programme are a clean fleece weight, average daily body weight gain, mature body weight, average staple length; average fibre diameter; and face and neck cover score. High heritable traits in Texas goats include staple length, mature body weight and body weight at 12 months of age. Moderately heritable traits include fleece weight, fleece density, average fibre diameter, medullation and weaning weight (Shelton, 1993).

CASHMERE FIBRE Cashmere is produced by approximately 150 breeds of goats. China has five breeds that are superior producers of cashmere fibre (Teh and Gipson, 1992). India has two breeds of goat that produce superior quality of fibre known as pashmina.

The heritability value of grease fleece weight was 0.25–0.45, average fibre diameter 0.47–0.83 and body weight was 0.22–0.29. The heritability of average staple length, secondary follicle density, secondary follicle:primary follicle ratio and change in average fibre diameter with age was 0.70, 0.17, 0.29, and 0.21, respectively (Pattie and Restall, 1992).

2.2 Performance of Valuable Traits in Tropical Regions

Many genetic improvement programmes have been carried out in tropical environments and production performances have been improved. The local breeds have been evaluated for growth, milk yield and fibre production in their own habitat. Genetic improvement on growth traits have been reported in different breeds and locations (Snyman and Olivier, 1996; Khan *et al.*, 2008; Roy *et al.*, 2008, 2011; Hermiz *et al.*, 2009; Thiruvenkadan *et al.*, 2009; Alade *et al.*, 2010; Ekambaram *et al.*, 2010). Genetic improvement of milk yield traits have been carried out in different local goats (Mavrogenis *et al.*, 1989; Kala and Prakash., 1990; Rabasco *et al.*, 1993; Muller *et al.*, 2002; Valencia *et al.*, 2007; Brito *et al.*, 2011; Selvaggi and Dario, 2015; Rout *et al.*, 2017).

Similarly, crossbreeding programmes have been carried out to evaluate the performance of different crossbreeds in tropical environments. In India, a systematic effort was made under the All India Coordinated Research Project (AICRP) on Goat Improvement, ISGP and PL-480 projects. Similarly, the performance evaluation in Africa has also been carried out in different local breeds by FARM-AFRICA. Crossbreeding has been undertaken in India with exotic breeds to increase milk, meat and mohair production. Exotic breeds such as Alpine and Saanen have been used for milk production and Angora for mohair production. Alpine and Toggenburg exotic breeds have been crossed with Sirohi for improving milk and reproduction performance of the local breeds (Groot *et al.*, 1992).

2.2.1 Milk production

Crossbreeding of Beetal and Malabari with exotic breeds Alpine and Saanen was carried out to increase milk production. Milk production and lactation length improved in both the half-breeds. The half-breeds were better than 75% inheritance in both the crosses. The Saanen breed showed better performance with both the indigenous breeds than Alpine. Saanen and Malabari crosses produced average milk yield of 127.3 kg per lactation with an improvement of 128% over the native Malabari. The Saanen × Beetal crosses with 75% of exotic inheritance were the best, showing 103.4% improvement over purebred Beetal (AICRP, 1985). The crosses of Alpine with Beetal and Malabari showed an improvement of 64% and 50% in lactation yields, respectively. However, the improvement in lactation length was relatively small (8% to 29%) in these crosses (Acharya, 1992). Alpine × Beetal crossbreeds had higher age at first kidding, kidding efficiency and multiple births than Saanen × Beetal crossbreds. Similarly, the Saanen and Beetal crossbreeds also had lower kid mortality rates than Alpine and Beetal crosses (AICRP, 1985).

Crossbreeding for improvement of milk production in Sirohi goats with Alpine and Toggenburg was implemented by Indo-Swiss Goat Production (Groot *et al.*, 1992). The lactation yield of 180 days was 130.1 kg for Sirohi, whereas it was 160.3 kg for Alpine × Sirohi and 160.9 kg for Toggenburg × Sirohi. The differences in milk yield were significant between crossbreeds and pure breeds; however, the observed increase in milk yield was lower than expected. Subsequently, it was also observed that selection in pure breeds was more effective and profitable than crossbreeding.

Similarly, a project called Caprin Ngozi was started in Burundi in 1980. The aim of the project was to improve milk production of local goats by crossbreeding with Alpine goats from France and Germany. Pure Alpine bucks were produced in different buck stations

2.2.2 Meat production

Jamunapari and Beetal as the improver breeds were used to cross with Black Bengal, Assam Local, Sirohi and Sangamneri to improve the meat production. Beetal × Black Bengal cross performed better than Jamunapari × Black Bengal cross. Beetal × Sirohi crosses attained about 32 kg of bodyweight at 6 months age under intensive feeding and 26.1 kg in semi-intensive system of management. The crossbreds of Beetal × Black Bengal exhibited 50.3% superiority at 3 months age over local Black Bengal. Beetal × Assam local crosses attained about 11.5 kg body weights at 12 months age, which showed an improvement of 31.5% over Assam local at 12 months of age (AICRP, 1985).

The combining ability of four important indigenous breeds was analysed in the PL-480 project at RBS College, Agra (India) involving Jamunapari, Black Bengal, Barbari and Beetal indigenous breeds. Jamunapari × Black Bengal crossbreeds performed better with respect to meat production while Beetal × Barbari crossbreeds performed better with respect to milk production.

Crossbreeding was carried out in East Africa with local the Galla breed and improver breeds Toggenburg and Anglo-Nubian (Ruvuna *et al.*, 1988). It was reported that the Anglo-Nubian × Galla crossbreed had the highest body weight at 9 months of age. Galla dams produced kids that were heavier than those from contemporary East African dams (Ruvuna *et al.*, 1992).

In Egypt, crossbreeding was carried out with Zaraibi and Damascus with Barki does. The Zaraibi × Barki cross had a 12–24 % higher body weight at all ages with a mortality rate of 9%. Similarly, Damascus × Barki showed 0–4% higher body weight at all ages with a mortality rate of 12% (Abdelsalam *et al.*, 1994).

2.2.3 Mohair and pashmina production

Indian goat breeds do not produce mohair; however, crossbreeding was carried out in Sangamneri and Gaddi with an exotic goat breed, Angora. The crossbreeds with 87.5% exotic inheritances were found to be the best among the crosses for mohair production under Indian environmental conditions. These crossbreeds produced about half of the mohair produced by purebred Angora contemporaries but with shorter staple length. The mohair yield of 87.5 Angora crossbreeds was 0.85 ±0.05 kg in comparison to 1.45 kg yield of the purebred Angora. However, the crosses had higher mortality rates and the mohair production appeared to be uneconomical (AICRP, 1985).

The pashmina yield varies from 78 g in first clip to 227 g in the third clip. Chegu and Changthangi goats showed an annual pashmina production of 132 g and 214 g, respectively, with an average fibre diameter of 12.4 microns. Selection based on greasy fibre weight and fibre diameter combined into an index is likely to provide maximum genetic progress in improving pashmina production.

2.2.4 Synthetic breed development

The USAID Project on small ruminant research was initiated in Kenya in 1980. The project aimed to develop a synthetic breed of goat as it was observed that crossbreeding may not be feasible in that area. Four breeds were selected: two breeds from a temperate climate (Toggenburg and Anglo-Nubian) and two tropical breeds from Africa such as Small East African and Somali. Each temperate breed was mated with each tropical breed to produce four types of F1 crossbreeds. Each crossbreed was mated with each of three other crossbreeds to produce the F2 generation. These four–way crosses were then mated together to produce large numbers of the synthetic breed. Selection was undertaken to produce a superior breed known as Kenya Dual Purpose Goat. The breed has since been developed and has taken more than 15 years to be distributed to farmers.

2.3 Behavioural Analysis

Dogs are considered to be a human's best friend as they share complex communication with them. Similarly, horses can also share emotional communicative behaviour with humans. It has also been shown that goats can share a similar affinity with humans. Experiments have shown that goats are highly intelligent, capable of complex communication with humans, and behave as potential partners to help in problem-solving situations. Goats are smarter than sheep and effectively communicate and interact with humans and attempt to provide relief in targeted work. Domesticated goats in grassy, temperate climates do not climb trees because their food is available in the grazing fields. But in hot regions where grass availability is patchy, like Asia, Africa, Mexico and some parts of Europe, goats leap upon green shrubs to forage. Tree-climbing goats play a crucial role dispersing nuts from argan trees, ensuring the success of future generations of this valuable resource.

Goats were probably the first ruminant livestock species and were domesticated almost 10,000 years ago (Mason, 1984; Zeder and Hesse, 2000; Hatziminaoglou and Boyazoglu, 2004). Behaviour is the adaptation to the environment and reflects the immediate response to the interaction between animal and its environment (Metz and Wierega, 1997). Social behaviour in a group is defined as group cohesion and ecological integration. Social behaviour is the interaction between two or more individuals in a group and group behaviour is modified in a group.

2.3.1 Communication

Communication plays a key role in the development and maintenance of social behaviours. Communication is essential for the recognition of individuals within the group. Olfactory signals are important for sexual and maternal behaviour. Urine is one of the main vectors for signals of individual identification, and goats urinate frequently during the oestrus to express their reproductive status. An olfactory learning process is the primary basis for the development of a relationship between a dam and her kids (Poindron *et al.*, 2007a,b). Kids castrated at a young age are rejected by their mother because they have a different odour; it is therefore not advisable to go for early age castration and medication. Vocal signals are important to establish the relationship between a mother and her kids and the ability to locate her kids in the open rangeland.

2.3.2 Social organization

Goats prefer to stay close together and do not like to be alone. If a goat is isolated from their group, it increases cortisol concentration and emotional stress (Kannan *et al.*, 2002). Pregnant goats should not be isolated. Isolation from the group is very stressful for goats and can affect the hypothalamic–pituitary–adrenocortical system and also affects the offspring severely (Roussel *et al.*, 2005). Therefore isolation of goats from their group for management purposes should not be routinely practiced. Once the social structure of the group is established, it continues for a long period (Barroso *et al.*, 2000; Rout *et al.*, 2002). It has been established that maternal selectivity in goats is developed within 4 hours after parturition (Bordi *et al.*, 1994). The first 2–4 hours after parturition is essential for the establishment of a relationship between mother and kid (Delgadillo *et al.*, 1997). Mother and kids should not be separated immediately. Goats remember separation for long periods, which can result in a reduction in growth rate. Psychological disturbance and abnormal behaviour are common in goats. Weaning should not be practiced at 6–7 weeks of age as kids become independent after 5 weeks of age. If there is frequent regrouping in a herd or an intensive system, there is an increase in aggression and reduction in production performance, and milk production decreases after the first regrouping (Fernandez *et al.*, 2007). Goats are less aggressive in extensive production systems compared to intensive production systems. The separation of bucks from does outside the breeding season is advisable; however, the introduction of bucks into a group of females is useful as a stimulus for synchronized ovulation (i.e. the buck effect).

2.3.3 Social dominance

Social dominance is clearly established and quite stable in goats and there is a linear hierarchical order in a herd (Barroso *et al.*, 2000; Rout *et al.*, 2002; Miranda-de la Lama, 2005).

2.4 Breeds in the Tropics

Goats have fulfilled agricultural, economic, cultural and even religious roles from very early times in human civilization. Goats are the most adaptable and geographically widespread livestock species, ranging from the mountains of Siberia to the deserts and tropics of Africa. Archaeological evidence indicates that the goat was one of the first animals to be domesticated by humans around 10,000 years ago at the dawn of the Neolithic period in the Fertile Crescent (Porter, 1996; Pringle, 1998). Goats played a central role in the Neolithic agricultural revolution and the spread of human civilization around the globe (Porter, 1996; Legge, 1996; Pringle, 1998; Zeder and Hesse, 2000). It has been suggested that at least two wild species of the genus *Capra* have contributed to the gene pool of domestic goats (Mannen *et al.*, 2001), while some studies have suggested that an independent domestication in Pakistan gave rise to the cashmere breeds (Porter 1996; Meadow 1996). It has been suggested that Mediterranean breeds derive some of their distinct characteristics from the extinct *Capra prissca*. Similarly, the Markhor and Bezoar are the progenitor of most Indian and Central Asian breeds.

Currently, nine wild species are recognized based on coat colour, horn morphology and facial pattern (Shackleton, 1997; Pidancier *et al.*, 2006). They are mostly found in mountain regions and are adaptable to extreme climatic conditions:

- *Capra aegarus* (Bezoar or wild goats) (found in Central and North East Asia);
- *Capra ibex* (alpine ibex) (found in the Alps);
- *Capra caucasica* (West Caucasian tur) (West Caucasus);
- *Capra pyrenaica* (Spanish ibex) (found in Iberian peninsula);
- *Capra falconeri* (Markhor) (found in Western Himalayas);
- *Capra cylindricornis* (East Caucasian tur) (found in East and Central Caucasus);
- *Capra nubiana* (Nubian ibex) (found in North-East Africa and parts of Arabia);
- *Capra sibirica* (Siberia ibex) (found in northern mountains of Ethiopia); and
- *Capra hircus* (domestic goat) (found world-wide).

Compared to other livestock (cattle, sheep and pigs), since the year 2000 goats have recorded the highest population growth rate. Asia and Africa harbour more than 90% of goats followed by America, Europe and Oceania. The zoological classification of goats is as follows:

Order:	Artiodactotyla
Sub-order:	Ruminantia
Family:	Bovidae
Genus:	*Capra*
Species:	*hircus*

2.4.1 Breeds and local genotypes from the tropics

The majority of the goat breeds are multipurpose animals (Mason, 1996), providing meat, milk, fibre and hides. However, the breeds have been classified according to meat, milk or fibre types. The dwarf goat breeds of Asia and Africa are specifically known for meat quality (Gall, 1981). Meat goats are selected on the basis of growth rate, feed conversion efficiency and carcass quality. Moreover, higher reproductive rates and lower generation intervals are of economic importance for meat goat breeds in the tropics. Similarly, milk-producing breeds in the tropics produce higher protein content and fat percentage and are more efficient milk producers compared to sheep and cattle in the tropics. The most common meat, dairy and fibre goats are described here, along with a summary of their specific use.

Asian subcontinent

ASSAM HILL The genotype is known locally as khasi. The population is distributed in Assam, India, and the neighbouring states of Meghalaya and Manipur. This is a small-sized breed and coat colour varies from white, white and black mixed, and black and brown mixed. The horns are flat at the base and point backwards (Fig. 2.13). The horn length of the male is 12.53 ± 2.65 cm and the female's is 4.2 ± 0.58 cm. The ears are flat and leaf-like, medium-sized and drooping; ear length is 2.36 ± 0.03 cm. Its muzzle is black and grey to light black colour. The adult body weight at 12 months

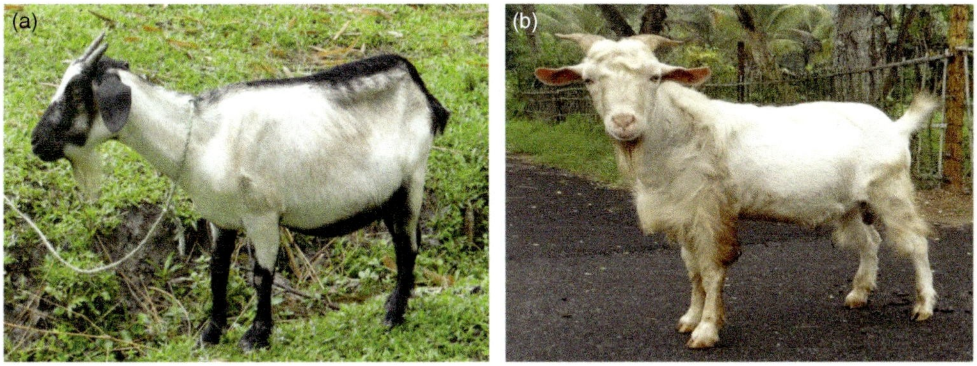

Fig. 2.13. (a) Male and (b) female Assam hill goat.

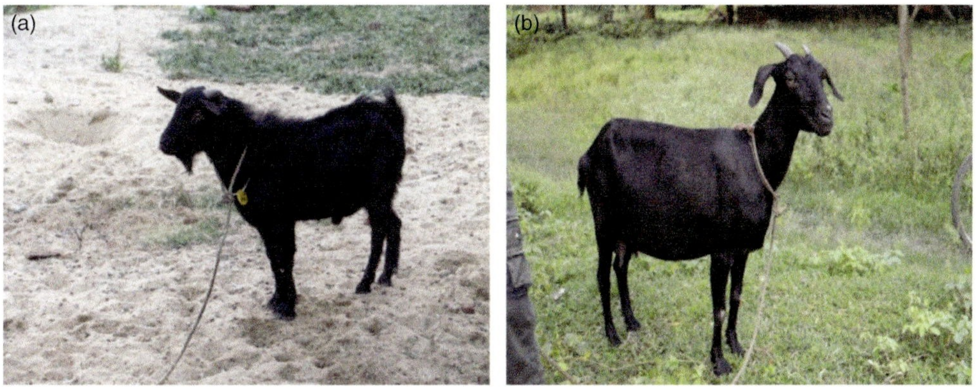

Fig. 2.14. (a) Male and (b) female Andamani goat.

of age is 13.94 ± 0.09 kg. The tail is medium in length and curved upward; tail length is 5.35 cm. The kidding rate is 1.58. The milk yield is 11.45 kg during 60 days of lactation.

ANDAMANI This goat is also known as the Andaman local goat. The horns are flat at the base and point backwards. The body colour varies from black and grey to light black (Fig. 2.14). The tail is medium in length and curved upward. The body weight at 12 months of age is 16.06 ± 0.04 kg. The kidding rate is 1.54. The milk yield is 0.308 kg per day.

BLACK BENGAL This goat is known locally as the deshi goat. The breed is mainly distributed in the West Bengal, Bihar, Odisha and Jharkhand regions of India. It is a small-sized and short legged goat. The chest is wide with a deep body. The body appears triangular from the side in females, with a heavier posterior region. The body colour is mainly black; however white, brown and grey colour goats are also observed in the breeding tract. The breed possesses a smooth, shiny, thin coat. The majority of the animals have a straight head, but convex and concave types have also been described. The horns are straight or curved. The ears are either erect or pendulous and vary from 10–15 cm in length at 12 months of age. Bengal goats may also have a beard, particularly in old age. The goats have a black nose and muzzle (Fig. 2.15). The average body weight at 12 months of age is 13.488 ± 0.108 kg. The breed possesses a short tail (5–15 cm). The average kidding rate is 1.89. The breed produces about 10.335 kg of milk during 60.80 days of lactation.

GADDI This breed is also known as the white Himalayan goat. The breed is distributed in Kangra, the Kullu Valley, Chamba, Lahul Spiti and Mandi

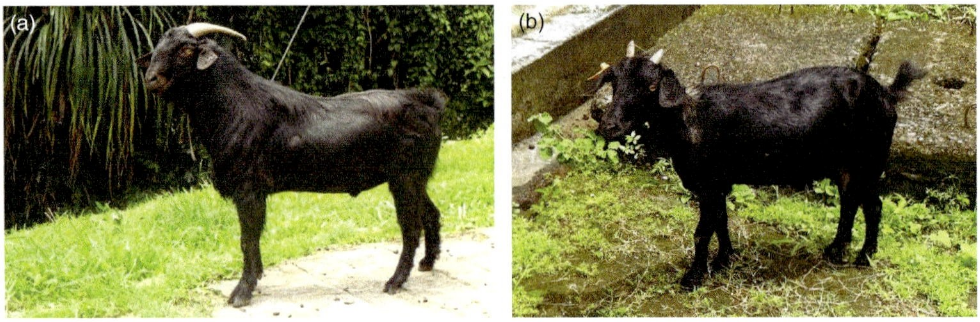

Fig. 2.15. (a) Male and (b) female Black Bengal goat.

Fig. 2.16. (a) Male and (b) female Gaddi goat.

districts of Himachal Pradesh and also in some parts of the Chamoli district of Uttarakhand, India. It is a medium-sized goat with long hair. The colour varies from pure white to mixtures of black and brown (in a small proportion of goats). The horns are long, pointed, bending upwards and backwards. The ears are drooping and pointed (Fig. 2.16). The hind quarters contain short tufts of hair. The body weight at 12 months of age is 27.55 ± 0.24 kg. Gaddi goats produce adequate milk for kids. The milk yield is about 64.78 kg at 90 days of lactation. The kidding rate is 1.2.

GANJAM The breed is known locally as Golla, Lanka or Dalua goats. The goats are distributed in the Ganjam and Gajapati districts, the adjoining areas of Nayagarh and Khurda districts of Odisha and parts of Srikakulam district of Andhra Pradesh, India. These animals are tall and leggy with a slender look and high ground clearance. This is a medium-sized goat. The predominant coat colours are reddish-brown and black, although goats with mixed or spotted colours have also been observed. The horns are long, twisted and directed upwards and backwards and may be curved both clockwise and anti-clockwise. The average horn length is 32 cm in adult males and 23 cm in adult females. Ears are mostly drooping, medium in size and mostly pendulous (Fig. 2.17). Ear length varies from 16–18 cm. The body weight at 12 months of age is 18.47 ± 0.04 kg. Wattles are present in both the sexes but are more common in females. The udder and teats are of medium size. The average milk yield is 15.54 kg at 60 days of lactation. The kidding rate is 1.09.

MALABARI This breed is also known as the Tellicherry goat. The breed is distributed in the Kozhikode or Calicut, Kannur and Malapuram districts of North Kerala, India. This breed is also found throughout Kerala, and parts of Tamil Nadu and Karnataka, India. The coat colour is most commonly white, but other colours such as black, brown or a combination of all these have also been

Fig. 2.17. (a) Male and (b) female Ganjam goat.

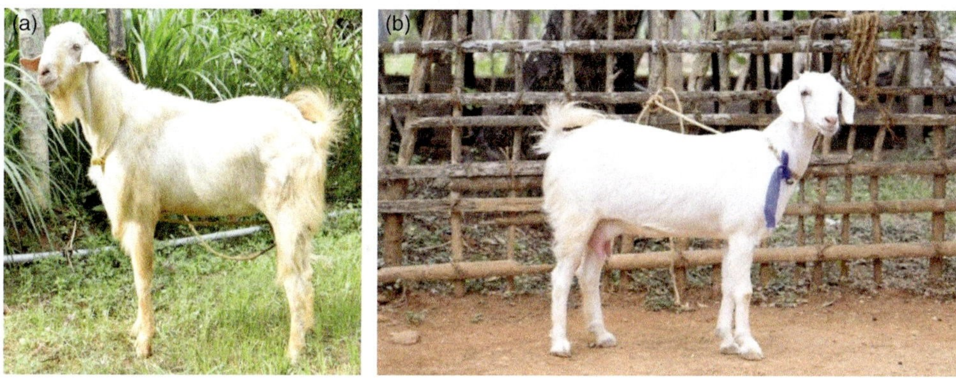

Fig. 2.18. (a) Male and (b) female Malabari goat.

observed. Both sexes have small, slightly twisted horns directed outwards and upwards. Ears are medium sized, directed outwards and downwards (Fig. 2.18). The average ear length is 16.30 ± 0.20 cm. The body weight at 12 months of age is 23.94 ± 1.75 kg. Males are bearded. The udder is small and round with medium-sized, small, thin teats. The tail length is 13.10 ± 0.05 cm. The goat produces 0.5 to 1.8 litres of milk per day with an average yield of 0.890 litres per day. The milk yield is 75.50 kg with a lactation length of 84.70 days. The kidding rate is 1.62.

SANGAMNERI The goat is found in the Ahmednagar, Pune and Nashik districts of Maharashtra, India. It is a medium-sized goat with a glossy, white coat. The ears are pendulous and horizontal, about 16 cm long. The tail is curved upwards and is about 18.45 cm long (Fig. 2.19). The average body weight at 12 months of age is 22.50 ± 0.33 kg. The breed is known for milk production. The milk yield is 99.03 ± 1.83 litres for the lactation period of 90 days. The kidding rate is 1.58.

CHANGTHANGI This breed is known locally as the Changra goat. The goats are distributed in the Leh and Ladakh regions of Jammu and Kashmir State. The goats are mainly found in the Changthang (meaning southern plains) region of Leh. However, they are also found in other parts of the Ladakh region (non-traditional belt) such as Nubra, Kargil and Drass. The breed is small to medium sized with a strong barrel-shaped body and powerful legs. The prevalent colour is white, but black, grey, brown, fawn and a mixture of these colours are also found. Their ears are small, pricked and point outwards. The horns are long and corkscrew shaped (directed backwards and upwards) (Fig. 2.20) and

Fig. 2.19. (a) Male and (b) female Sangamneri goat.

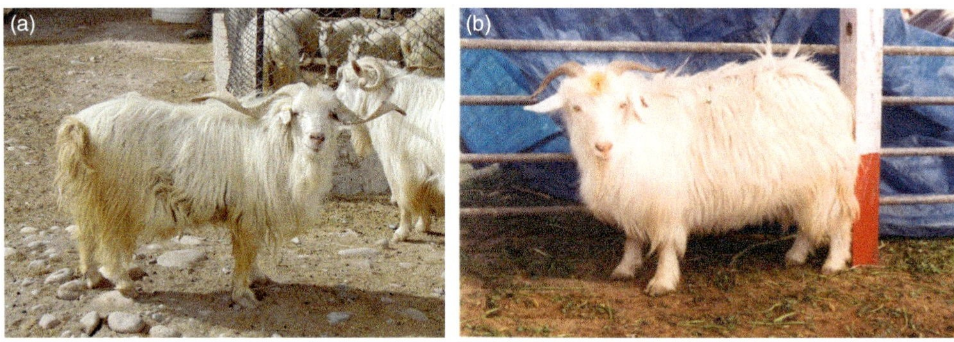

Fig. 2.20. (a) Male and (b) female Changthangi goat.

are classified into two main groups, resembling those of markhors in the majority of animals and the ibex in a few animals. The ear length is 12.50 ± 0.46 cm. Body weight at 12 months of age is 16.08 ± 0.64 kg. The breed is known for pashmina production; the average pashmina production is 269.66 ± 13.0 g. The breed also produces a good amount of milk for kids.

SURTI These goats are distributed in the Navsari, Surat region of Gujarat, India. It is a medium-sized goat. The predominant colour is white and sometimes brown hairs are found in different parts of body. The horns are small, pointing backwards and slightly curved. The ears are medium sized and pendulous (Fig. 2.21). The horn length is about 12 cm. The tail is short and erect. The body weight at 12 months of age is 21.99 ± 0.151 kg. The breed is known for milk production and milk yield at 90 days is 80.22 ± 1.94 litres during 90 days of lactation. The kidding rate is 1.68, indicating a high prolificacy breed.

MARWARI Marwari goats are distributed in an extensive area in Western Rajasthan. The breed is distributed in Barmer, Bikaner, Jalore, Jodhpur, Nagaur and Pali districts. The breed can also be found in the adjoining area of Sikar, Jhunjhunu, Churu, Sriganganagar, Hanumangarh districts of Rajasthan and in the adjoining area of northern Gujarat, especially Mehsana district. This is a medium-sized breed with a small muzzle. The body colour is mostly black, with a long, lustrous shaggy coat with a growth rate of 10–12 cm/year. The horn is short, twisted and pointed, running upwards and backwards. The breed has medium-sized drooping ears. The body weight at 12 months of age is 24.14 ± 0.03kg. The milk yield at 90 days is 83.03 ± 0.33 kg. The kidding rate is 1.08 (Fig. 2.22).

PANTJA These goats are distributed in the Tarai region of Udham Singh Nagar and Nainital districts of Uttarakhand, India. Their colour varies from fawn to brown (Fig. 2.23). The horns

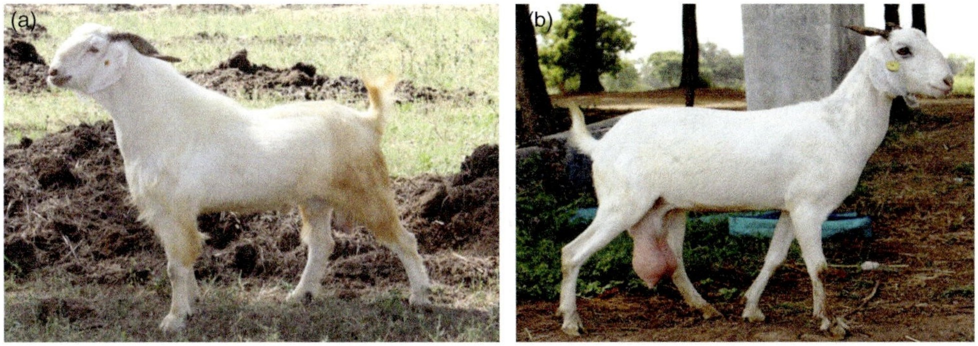

Fig. 2.21. (a) Male and (b) female Surti goat.

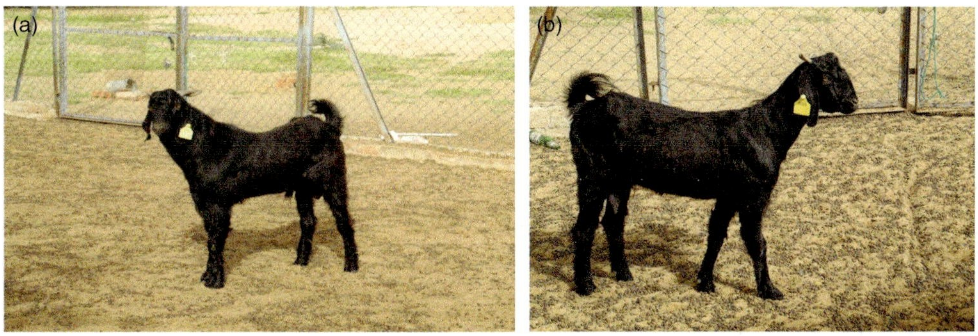

Fig. 2.22. (a) Male and (b) female Marwari goat.

Fig. 2.23. (a) Male and (b) female Pantja goat.

have a slight upwards and backwards orientation and are small (<15 cm) in size. The ears are drooping. The average body weight at 12 months of age is 18.91 ± 0.78 kg. The milk yield at 90 days is 46.34 ± 0.34 kg. The kidding rate is 1.53.

OSMANABADI Osmanabadi goats are distributed in the Latur, Tuljapur and Udgir Taluks of the Osmanabad district of Maharashtra, India. The body colour is black with white spots on the ears (Fig. 2.24). Polled bucks are also found occasionally in the herd. The kidding rate is 1.62. The body weight at 9 months of age is 20.5 kg. The average lactation milk yield is 108 kg.

JAKHRANA Jakhrana goats are distributed in the Jakhrana village of Alwar district of Rajasthan, India. The body colour is predominantly black, with spotty white characteristic ears and a white speckle in the muzzle. It is a large-sized goat. The breed has horizontal twisting horns (Fig. 2.25). The body weight at 12 months of age is 24.89 kg and the milk yield at 90 days is 124.32 litres. The kidding rate is 1.43. The average lactation length is 158.35 days. The gestation period, kidding interval and dry period is 152.47 days, 289.23 days, 142.64 days, respectively.

KUTCHI This breed is distributed in the Kutch region of Gujarat, India. The body colour is black, with white or speckled ears, neck and face. It is a medium- to large-sized breed. The nose is slightly Roman in shape. The average body length, height and heart girth in bucks is 77 cm, 86 cm and 75 cm, respectively, and 75 cm, 82 cm and 76 cm, respectively, in does. The ears are broad, long and drooping. The horns are corkscrew shaped, pointing upwards (Fig. 2.26). The average weight of an adult buck is 44 kg and 39 kg in does. The average lactation yield is 80.90 kg at 140 days of lactation.

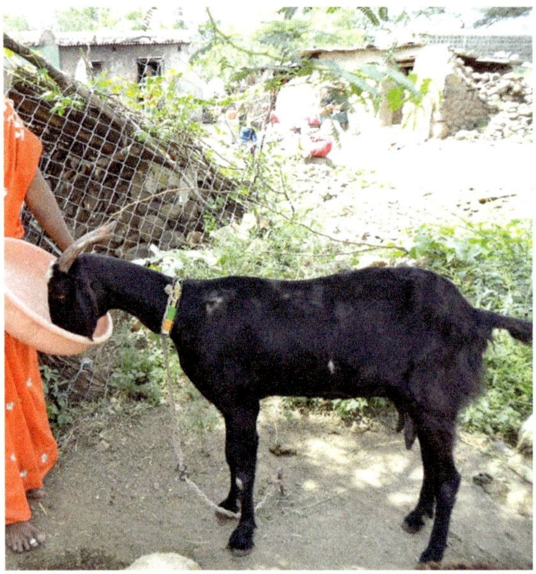

Fig. 2.24. Osmanabadi goat.

BARBARI This breed is distributed in Agra, Mathura, Etah and Aligarh districts of Uttar Pradesh, India. The colour varies greatly, but white with red or tan spots is considered typical. It is a medium-sized breed, with short legs and a wedge-shaped body. The body length, height and heart girth is 70 cm, 71 cm and 76 cm, respectively, in bucks and 59 cm, 56 cm and 64 cm, respectively, in does. The ears are small and tubular, oriented upwards and outwards. The horn is slightly twisted, directed upwards and backwards (Fig. 2.27). The average body weights are 38 kg and 23 kg for bucks and does, respectively. The average body weight at 12 months of age is 21.25 ± 0.18 kg. The milk yield at 90 days is 65.4 ± 3.7 kg. The kidding rate is 1.58.

JAMUNAPARI The Jamunapari goat is a milk-producing breed with an average body weight of 28.0 kg at 12 months of age and a kidding rate of 1.46 (Rout *et al.*, 2000, 2016). The Jamunapari goat has been used extensively for upgrading local goats in neighbouring South-East Asian countries (Rout *et al.*, 2004). It has a majestic white colour coat with long, pendulous ears and a Roman nose (Rout *et al.*, 2000). The ear length ranges from 26–28 cm. The horns are short and flat (10–15 cm),

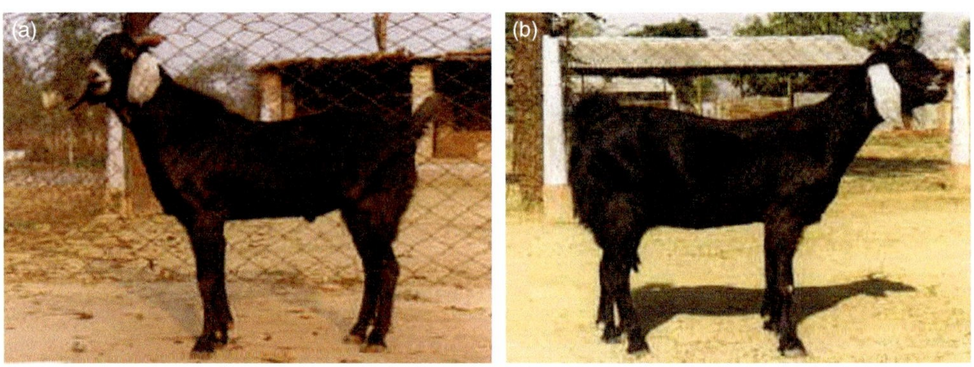

Fig. 2.25. (a) Male and (b) female Jakhrana goat.

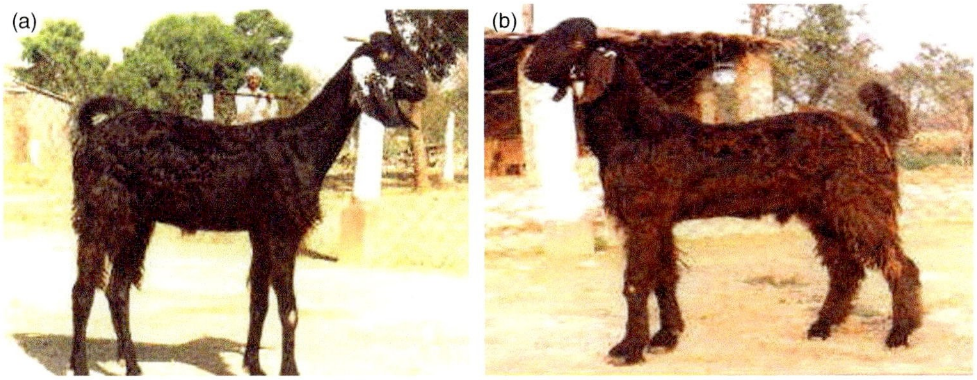

Fig. 2.26. (a) Male and (b) female Kutchi goat.

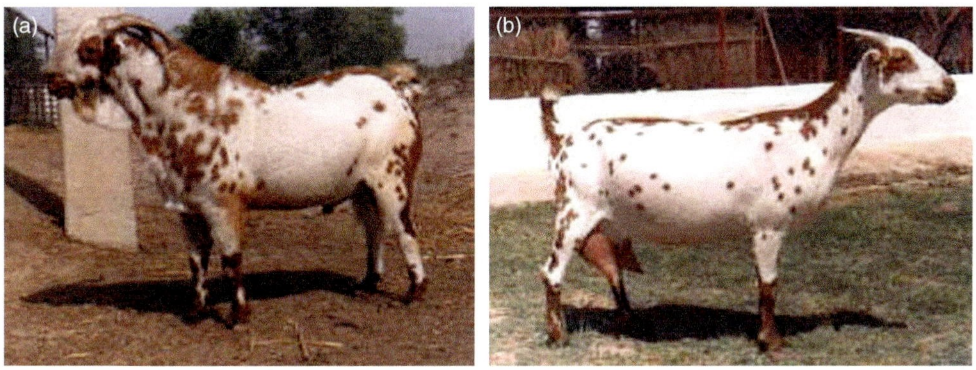

Fig. 2.27. (a) Male and (b) female Barbari goat.

horizontal and twisting backwards (Fig. 2.28). The body weight at 12 months of age is 28.30 ± 0.8 kg. The milk yield at 90 days of lactation is 90.23 kg. The kidding rate is 1.46.

BEETAL The breed is distributed in the Gurdaspur area of Punjab, India. The body colour is black, and males have a beard (Fig. 2.29). The face is convex, with black lips, blue-black eyes with white or brownish corneal surroundings. The average body length, height and heart girth are 86 cm, 92 cm and 86 cm, respectively, in bucks and 70 cm, 77 cm and 74 cm, respectively, in does. The ears are long, pendulous and drooping. The pinna is similar to the betel leaf in shape. The horns are horizontal with backwards and outwards twisting. The mature buck is 70 kg and doe is 46 kg. The body weight at 12 months of age is 27.16 kg. The average lactation yield is 150 kg.

MEHSANA These goats are found in the Mehsana, Gandhi Nagar and Ahmedabad districts of Gujarat, India. It is a medium-sized breed with a Roman nose and broad muzzle (Fig. 2.30). The coat is black with white spots at the base of the ear; both bucks and does have small beards and grey-black skin. The ears are white, leaf-like and drooping. The horns are slightly twisted and curved upwards and backwards. The body length, height and heart girth are 71 cm, 80 cm and 77 cm, respectively, in bucks and 68 cm, 74 cm and 73 cm, respectively in does. The body average weights of adult bucks and does are 37 kg and 32 kg, respectively. The milk yield is about 1–1.5 kg per day.

GOHILWADI This breed is distributed in the north-west arid and semi-arid region in the Bhavnagar, Amreli and Junagarh districts of Gujarat, India. It is a medium-sized goat. The body colour is black with coarse, long hairs. The body length, height and heart girth are 73 cm, 81 cm, 75 cm, respectively, in bucks and 72 cm, 80 cm and 75 cm, respectively,

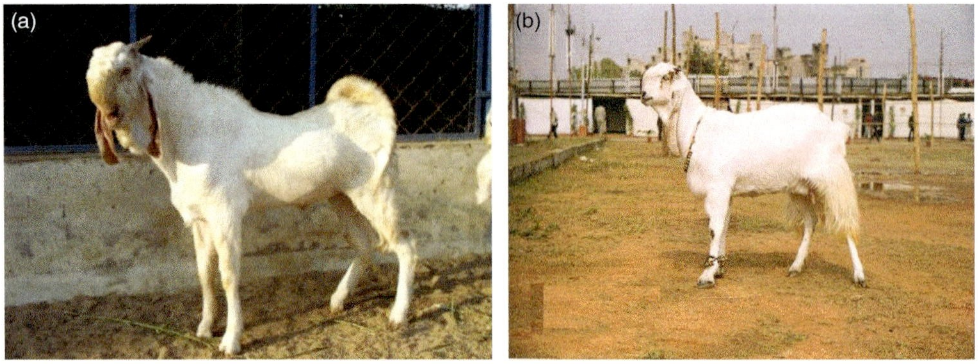

Fig. 2.28. (a) Male and (b) female Jamunapari goat.

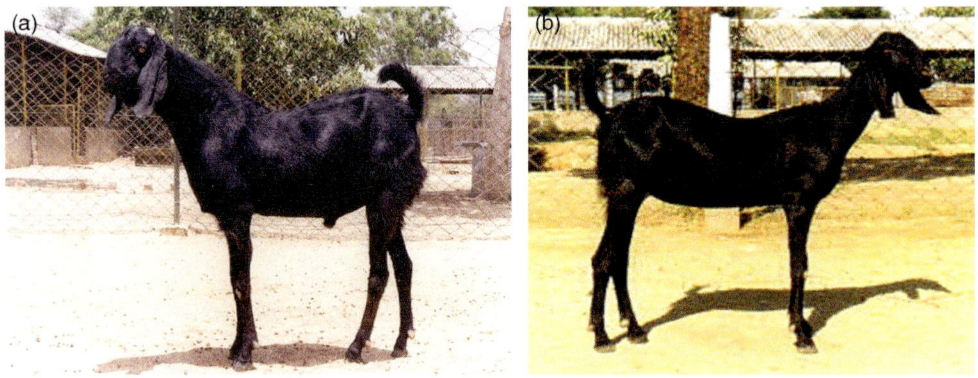

Fig. 2.29. (a) Male and (b) female Beetal goat.

Fig. 2.30. (a) Male and (b) female Mehsana goat.

in does. The ears are tubular and drooping. The horns are slightly twisted and turned backwards (Fig. 2.31). The average body weights are 52 kg and 42 kg for an adult buck and adult doe, respectively. The lactation milk yield is 80 kg. The milk yield is about 1.7 kg per day and the average lactation period is 227 days.

KANNI-ADU OR KANNIADU These goats are tall and predominantly black or black with white spots.

Fig. 2.31. (a) Male and (b) female Gohilwadi goat.

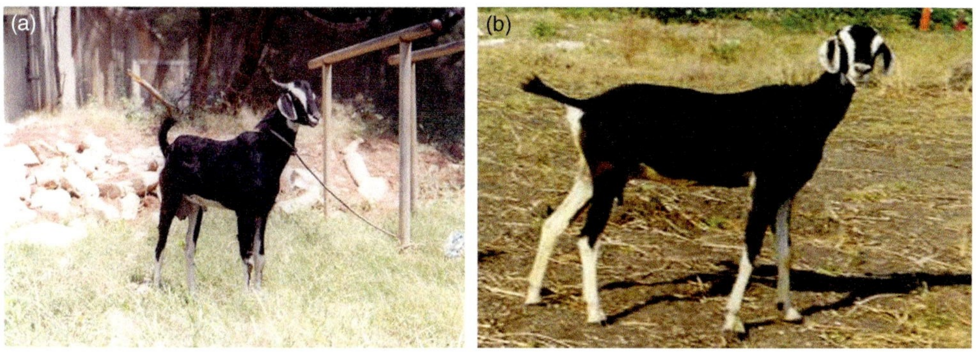

Fig. 2.32. (a) Male and (b) female Kanniadu goat.

Body is stout with long legs. The body length, height and heart girth are 71 cm, 84 cm and 78 cm, respectively, in bucks and 67 cm, 76 cm and 71 cm, respectively, in does. The breed has two strains in its native tract. Pai-kanni goats are the most common, with black hair and two parallel white stripes on either side of the face (Fig. 2.32). The body weight at 12 months of age is 16.86 ± 0.52 kg. The milk yield varies between 60 ml and 80 ml per day.

BERARI This breed is distributed in the Nagpur, Wardha and Akola districts of the Vidarbha region of Maharashtra, India. It is a tall animal with a dark brown body, black stripes on either side of the face and a completely black-brown colour on the underbelly and inner legs. It also has prominent black ring-like shading around the neck (Fig. 2.33). The average body weights of adult male and female goats are 33.95 kg and 28.58 kg, respectively.

KODI ADU This breed is distributed in the south-eastern coastal region of Tamil Nadu, India. The breed has two colour varieties – blackish-brown and reddish-brown. Their ears are medium in size, curved forward and slightly drooping but alert. Horns are positioned upwards, backwards and curve down or up sharply at the tip. The tail is thin and small to medium sized. The teats are placed laterally in does. The average weight for adult males is 39.50 kg and for females is 32.17 kg.

ZALAWADI This breed is distributed in the Surendra Nagar district of Gujarat, India. The body colour is black and skin pinkish-blue with long and lustrous hair. The neck is long and the throat has lobular appendages. The tail is short and curved; the body length, height and heart girth 76 cm, 83 cm and 77 cm, respectively, in bucks and 72 cm, 79 cm and 74 cm, respectively, in does. The ears are white, long, wide, leaf-like and drooping. The horns are long and straight with screw-like twisting (Fig. 2.34). The

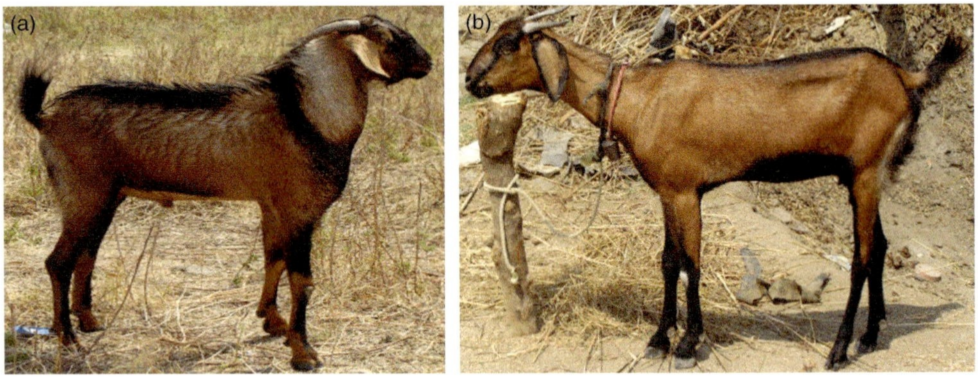

Fig. 2.33. (a) Male and (b) female Berari goat.

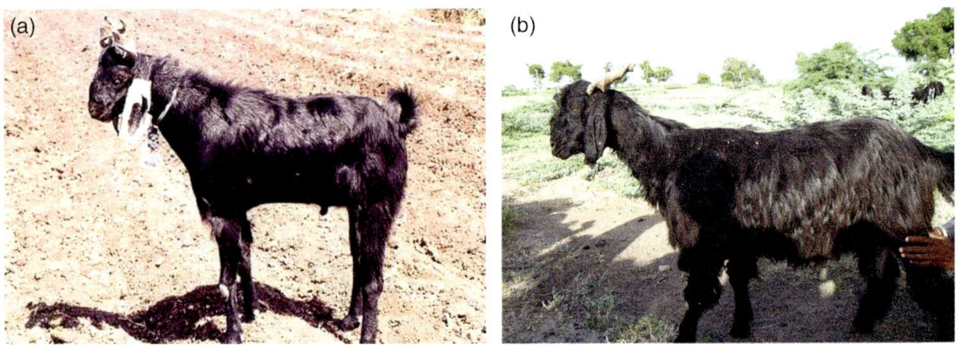

Fig. 2.34. (a) Male and (b) female Zalawadi goat.

average weight for bucks is 39 kg and does is 33 kg; the birth weight is 3 kg.

CHAPPAR This breed, also known as Jabli, Jablu, Kohistani or Takru, and originated in the Sind and Baluchistan area of Pakistan. It is a multipurpose breed and is known for meat, cashmere, hair and milk production. The coat colour is black, white or pied. The breed is of medium size with adult males weighing about 26 kg and adult females weighing about 20 kg (Devendra, 2007). The average daily milk yield is between 0.8 kg and 1.0 kg.

KAMBING KATJANG OR PEA This breed is native to Malaysia and Indonesia. It is also called Kambing, which means goat, and Katjang, which means beans. The breed resembles the goats found in Burma, Thailand, the Philippines and Taiwan. It is a medium-sized breed and adult males weigh about 25 kg and females weigh about 20 kg (Devendra, 2007). The wither height is 60–65 cm in males and 56 cm in females. The hair colour is black and coarse. It is a prolific breed.

SYRIAN MOUNTAIN This breed is distributed in Syria, Lebanon, and northern Israel (Mason, 1981). The breed is also known as Black Bedouin, Djelab, Jabali, Jabel or Jebel, Mamber Mambrine or Membrine, Northern Mountain (Israel), and Palestinian or Syrian black. This is a multipurpose breed and is reared for meat, milk and hair. The hair colour is usually black, and their ears are long and pendulous. This is a medium-sized breed and the adult males weigh 60 kg, and females weigh about 40 kg. The breed is of medium prolificacy with about 130% kidding rate.

DAMASCUS This breed is distributed in the Middle East, especially Syria and Lebanon. They are also known as Baladi or Shami, and Aleppo in Turkey (OSU, 2004). This is a dual-purpose breed but mostly reared for milk purposes. The face is straight and convex and the ears are short. The coat colour is usually red, brown or white. The hair is long, and males have tassels. Both horned and polled types are produced.

ABAZA These goats are distributed in north-eastern Anatolia, Turkey. The coat is made up of pinkish-white short and soft fibres, with coloured markings around the mouth, eyes and legs. The bucks have long, flat and sword-like horns. The does are usually polled. The breed is known for its milk production and average lactation yield is about 200 kg (Yalcin, 1986).

MALTA (MALTESE) These goats are distributed in the Aegean region. The animal has a Roman head with long, wide and drooping ears. The colour is usually brown and white spotted, but black or white animals may also be observed. The animals are usually polled or have small horns.

NACHI This breed is unique due to its natural dancing gait (it is a genetic trait). When animals walk, the feet and pastern move in a partially revolving motion and, with the head held high, the animals exhibit a dancing walk. A normal goat walk (i.e. not dancing) will disqualify an animal in show conditions. Small steps and a semi-circular movement of the front feet are the two most important attributes. Main districts inhabiting Nachi breed are the Jhang, Multan Muzaffargarh, Layyah, Bahawalpur and Bahawalnagar districts in the Punjab province of Pakistan. Nachi goats are tall in stature (almost as tall as Beetal goats). The coat colour is black (Fig. 2.35), but it may also be black with white spots. It has a medium head, Roman nose, small and thin horns and medium ears. The udder in the does is well developed. The adult body weight of does is about 60 kg and breeding males weigh more than 80 kg. The height at the withers average 88 cm and 98 cm in does and bucks, respectively.

Fig. 2.35. Nachi goat.

DAMANI The Damani is a milk goat found in the Bannu and Dera Ismail Khan districts of the North-West Frontier Province, Pakistan. They are medium in size with a mature weight of approximately 35 kg. They typically have a black hair coat with tan head and legs. The daily average milk production from the does is 1.8 litres. Characteristics of the breed include a long-haired coat and well-developed udders.

KAMORI The Kamori goat is a popular and beautiful breed of goat. It is a milk goat breed that is found in the Sindh province of Pakistan. The body of the purebred Kamori goat is dark brown in colour with small coffee-coloured or dark patches over their entire body. They have a long and well-developed body. The head is also very large with a Roman nose and long, wide, drooping ears. The tail is small. Both male and female goats have horns. The udder and teats in the Kamori does are fairly well developed. It is a medium to large goat breed. On average, the bucks weigh about 60 kg and the does weigh about 50 kg. The average daily milk production is approximately 1.5 litres. The Kamori has been exported to different countries.

KAGHANI The home tract of the breed is Kaghan valley (Abbotabad, Kohistan, Mansehra and Swat) in Pakistan. The Kaghani goat is mainly used for meat production, but also good for cashmere fibre production. It is also known as the Bakerwali goat in some areas. Kaghani goats are medium-sized animals. Their body colour is white, brown, grey or black. They have long hair throughout their body. Both bucks and does usually have horns. The average body weight of a Kaghani buck is up to 37 kg, and the does on average weigh up to 32 kg. On average, a Kaghani goat produces about 1–2 kg of long hair per year. The does are not good milk producers. On average they produce about 0.4–0.7 kg milk daily.

LEHRI This is mainly a meat breed. It is found in Lehri town in Sibi district of the Balochistan province, Pakistan, hence the name 'Lehri' (Pakistan). Black

Fig. 2.36. Lehri goat.

is the most common colour (Fig. 2.36), but white or grey animals may also be seen. The body is covered in long hair. It is a small-sized breed, with an average body weight of 30–35 kg. The average wool yield/annum is 0.8 kg. Phenotypically, this breed has a medium-sized head with pliable, long, broad and pendulous ears. Both males and females have horns. The tail is about 20cm long, and the udder and teats in does are small. Twin births are common.

DERA DIN PANAH The Dera Din Panah is a breed of domestic goat named after Dera Din Panah town in the Multan and Muzaffargarh districts of the Punjab province of Pakistan. The breed is used primarily for milk production. These goats are black and hairy with a well-developed body, large head with Roman nose, hairs on the chin, and long and broad ears. Cartilaginous appendages on the side of the neck are found in some animals. The horns are thick and long with two to three spiralled curves. The tail is medium length, covered with short, rough hairs. The udder and teats in does are well developed; milk yield is 245 litres over the 135 days of lactation. Adult males and females weigh 45 kg and 40 kg, respectively, and yield 1200 g of hair per head/annum. The height of an adult male is 81 cm and adult female is 70 cm. Twin births are common. These goats are raised for milk, meat and hair production.

PATERI The Pateri is one of the heaviest breeds of goat found in Sindh (Pakistan). The average adult body weight ranges from 45–70 kg and the average height range is 48–55 cm. It is basically a meat goat but with good milking quality. It produces 1.0–2.0 kg of milk per day. Body colour is white and their face, neck and ears are reddish-brown. The ears are long and drooping and wide at the end. They also have loose skin under their chin. They have medium-sized horns.

BUGRI OR BUGI TOORI This breed is found in the districts of Hyderabad, Badin and Mirpurkhas of Pakistan. These goats are white coloured, have a medium head with spirally twisted horns rising in an upright position, and drooping medium ears. While walking their head and neck is held upward. The average adult weight of males is 30 kg and females is 25 kg. Hair yield is 600 g per head/annum. These goats are raised for meat and hair.

TAPRI OR LAPPI This breed is found in the Hyderabad, Mirpurkhas to Khipro and Khairpur districts of Sindh province of Pakistan. Tapri goats are camel coloured or reddish-brown, although occasionally a white or black body colour may also be seen. The head and ears are small and the neck is short. The udder and teats in does are small. Adult males and females weigh 33 kg and 25 kg, respectively. Twin births are common. This breed is early maturing and prolific. They are raised mainly for meat purposes, but they can also produce 0.5–1.0 litres of milk per day.

NEPALESE HILL This is a predominant breed found in Nepal. The breed is well distributed across the Siwalik and Mahabharat hill regions. The breed is used for meat purposes. The body colour is brown or black with white markings, or white with black markings. They are named by their coat colours including Khairi, Kali, Seti, Singari or Dhobini. The northern hill goats are heavier than the southern hill goats. The male northern hill goats weigh about 35–45 kg, while the females weigh about 24–30 kg. The southern hill goat weighs about 12–16 kg (Epstein, 1977).

KHERI (HILL) This breed is distributed across the hills and inner valleys of Nepal. They are quite widespread and can be found in the mid hills and the immediate south of high mountains. This is a common breed of Nepal. It is a hardy breed with six different colour types – Seti (white), Kali (black), Khairi (brown), Ghorli (brown mixed with other colours), Singari (black with white markings) and Dhobini (white with black markings) (Fig. 2.37). The kidding rate is 1.6 and the kidding interval is

Fig. 2.37. Kheri (Hill) goat.

Fig. 2.38. Chyngra goat.

283 days (Neopane, 2000). They have a relatively small body size that may vary widely in body weight from 20 kg to 40 kg

SINGHALA This breed is spread across the southern flank of the high Himalayan mountain region of Nepal. It comprises about 35% of the total goat population of the country. The breed is heaviest in Nepal. The average adult body weight of male and female Singhala goats is 42 kg and 35 kg, respectively, with the average height at wither being 67 cm. The goats are black, light brown and white in colour with coarse hairs covering the body. The breed has long horns. The breed is late maturing with a low twinning rate. The average birth weight of the kids is 2.1 kg and yearling body weight is about 18.5 kg.

CHYNGRA These goats are migratory in nature, spread across the trans-Himalayan region of Nepal. This well-known breed produces the world's most expensive and finest quality of luxurious cashmere wool. The goats have spirally coiled twisted horns (17–28 cm). They also have coarse but silky long hair, but the under-coat is of pashmina or cashmere. Most of the Chyngra goats are white or black with white stripes in the face (Fig. 2.38), though various colours (ash, brown) are also present. The average adult body weight for bucks and does is 35–40 kg and 27–30 kg, respectively. The average wither height is 62 cm in female goats. The average birth weight of the kids is 1.71 kg, with average daily growth rate of 47 ± 10.04 g and 42 ± 8.02 g for male and female kids, respectively, from birth to 30 weeks.

TERAI Terai goats are found in the southern most plains (Terai) of Nepal. This breed appears to be recently developed from the population originating from Jamunapari and native breeds. Coat colours vary from black, white, brown, brown mixed with red, black with white marking, ash colour, and black and white ears. The average adult body weight of a buck is 30–35 kg and that of a doe is 18–32 kg. The average age at first kidding (15–16 months) and kidding interval (7–8 months) are similar to Kheri hill goats. The average birth weight of kids is 1.5 kg and yearling weight of male and female goats are 14.2 and 11.9 kg, respectively.

KURDI The Kurdi goat is distributed in Kurdistan, Iraq and Iran. The breed is a cashmere-producing breed. Their colour is usually white, black or brown. They are also known as Karadi, Kurdish, Marghaz, Markhaz, Morghose or Morghoz (Iran). The face is straight and concave and the ears are twisted forward (Mason, 1981). The horns are twisted. Males weigh about 50–60 kg.

CASHMERE This breed is distributed in the mountainous region of north-west India, particularly in the Ladakh region. The breed is also known as Pashmina, Changthangi (Kashmir), Mongolian Alashan Down, Albas Down, Hexi Down and Tibetan and Xinjian (China). The fibre (luxurious 'fibre of kings') is a secondary hair shed by goats during the spring and is very fine in diameter. The breed is usually white with long hair. The horns are twisted and the ears are either erect or horizontal. This breed is easy to raise and needs minimal management. They are sheared once a year, and an adult buck will shear about 1.1 kg of fleece, which contains both guard hair and cashmere.

China

CHENGDU BROWN Chengdu brown goats are distributed in the Sichuan province of China. The goats are a prolific breed and reared for both meat and milk production. The body colour is brown with a dark face and back stripes.

DUAN Duan goats are from the Guangxi province in China and are used for meat production. They are black, white or pied.

LAOSHAN The Laoshan goat is a dairy goat breed and distributed throughout the Shandong province of China. The Laoshan goat was developed from selective breeding of local goats with Saanen goats, first introduced to Loushan by German preachers early in 1904. Laoshan goats are usually white in colour with soft and short hairs (Fig. 2.39). A large percentage of the breed is hornless. The does have well-developed udders. As a medium-sized dairy goat, the average body height of the Laoshan bucks is about 85 cm and about 71 cm for does. A mature buck weight is about 76 kg and a mature doe about 48 kg. The does are pretty good milkers, and they produce about 557–870 kg of milk per lactation.

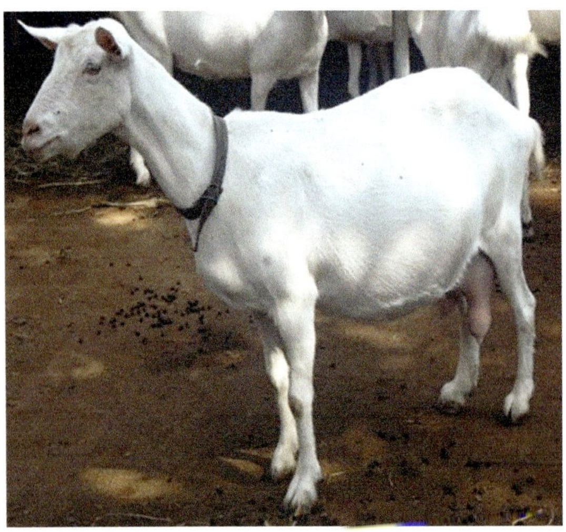

Fig. 2.39. Laoshan goat.

HAIMEN The Haimen breed is found in Zhejiang, China. It is a white-coated meat breed known for its prolific reproduction.

ZHONGWEI The Zhongwei (Chung-wei) is a breed of goat from the Ningxia Hui Autonomous Region and Gansu province of China (Cheng, 1984). Its natural habitat is the arid desert Steppes, and it is adapted to a diet of salty and sandy plants and shrubs. It is used primarily for the production of kid pelts and secondarily for cashmere fibre. The pelts are white with lustrous staples and attractive curls The breed has low genetic variability, likely due to the historic selection of pelt production traits. It is closely related to the Funiu white, Hexi cashmere, Luliang black, and Taihang breeds. Males weigh 39 kg on average and females weigh 24.5 kg on average. The breed has spiral horns and the fur colour is white, and occasionally black (Fig. 2.40).

JINING GREY This goat is reared for kid pelt and cashmere fibre production. The breed is spread across the Shandong province of China. The breed is known for producing attractive wavy patterns on the kid pelt. Jining Grey goats are small-sized animals and have small bodies. Their coat colour patterns vary between black, white or black and white (Fig. 2.41) . Both bucks and does have horns and have a forelock. The average body weight of the

Fig. 2.40. Zhongwei goat.

Fig. 2.41. Jining Grey goat.

Jining Grey buck is about 33.2 kg and the does weigh about 25.4 kg on average. Bucks will produce about 50–150 g of fibre, with cashmere fibre comprising between 18% and 30% of the total fleece.

African continent

MUBENDE The Mubende goat is an indigenous breed from the Kabale and Bundibugyo districts of Uganda. This breed has shiny straight hair and the colour is a mix of black and white. The meat is very good quality and the skin is used as leather in the tanning industry. Males have manes and are usually hornless. Adult males weight 25–35 kg and adult females weigh 22–28 kg. The breed also adapts well during drought and can survive for a few days without drinking water. These goats are also resistant to heart water (a tick-borne disease), worms and other diseases such as mange.

MARADI OR RED SOKOTO This breed is spread across the Sokoto area of Nigeria. The coat colour is red (Fig. 2.42), and the skin is valued for leather production. The strain of the goat is also known as Kano Brown and Boronu White. The goats have horns and horizontally positioned short ears. Mature animals weigh 20–30 kg.

NIGERIAN DWARF GOAT The Nigerian Dwarf goat is a small goat native to West Africa that is usually reared as a pet animal. It is growing in popularity, and is sometimes used for show purposes (OSU, 2004). Original dwarf goats were transported to West Africa after being captured for food for large cats when they were being delivered to zoos. The Nigerian Dwarf Goat breed is docile, playful and very fertile – triplets or quadruplets are common. It has the ability to produce an average of 1.2 kg of milk per day. It is a small breed with does measuring 43–48 cm in height while bucks measure 48–53 cm in height. The goats have a straight nose and erect ears. The coat is soft with short to medium hair and varies in colour from white, gold, and red to black and cream (Fig. 2.43).

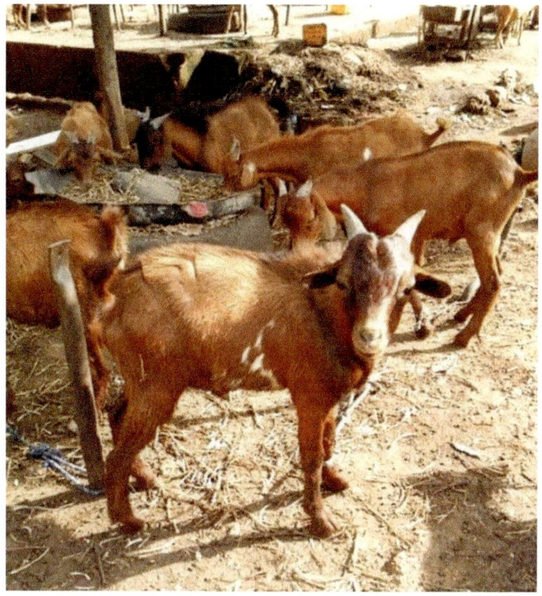

Fig. 2.42. Maradi or Red Sokoto goat.

SMALL EAST AFRICAN This breed is distributed in Kenya, Uganda and Tanzania (Devendra and Burns, 1983). It is a small-sized goat known for meat and skin production. It is an early-maturing breed, but twinning rates are low. The coat colour varies from pure white to pure black with different shades. The hair is short and fine. The ears are short and erect and males have beards. The animals have horns that sweep backwards, curving upwards at the tip. The live weight of the adult male and female are 32.5 kg and 26.5 kg, respectively.

GALLA The Galla goat is a large-sized breed found in Kenya, Ethiopia and Somalia. The breed is also known as the Boran or Somali goat (Somalia) and Benadir goat (Ethiopia). There are two strains known as Degyir and Degeun. Degyir goats have white hair all over their bodies with a black skin. Degeun goats are coloured around the head and lower legs with a black stripe along the spine. This is primarily a meat breed but it also produces

Fig. 2.43. Nigerian Dwarf goat.

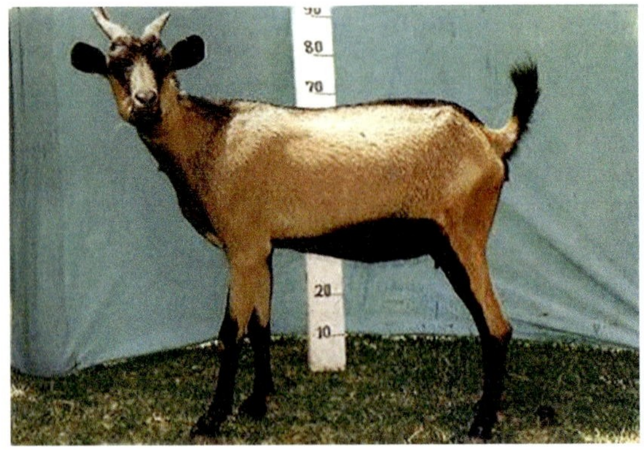

Fig. 2.44. Woyto-Guji goat.

an average of 500 ml of milk per day. The adult male body weight is about 70 kg and female body weight is about 50 kg.

AFAR This breed is distributed in Ethiopia. The coat colour varies from white to light brown; some animals may have black coats. The face is concave and narrow, and there are long, thin upward-pointing horns. The animals are leggy and a beard is present in males. Wattles are relatively common.

ABERGELLE This breed is distributed in Ethiopia. The goats are mostly reddish-brown in colour and males have spiral horns directed backwards. The coat colour is plain and patchy. Hair is short and smooth.

ARSI-BALE The breed is distributed in Ethiopia. The goats are medium to large in size. The coat colours have a patchy pattern. The face is straight and horns are straight and pointing backwards.

WOYTO-GUJI This breed is distributed in Ethiopia. The coat colour varies from brown to black or red. Black or brown stripes are observed along the back (Fig. 2.44). The horns are straight and point backwards. A beard is present on all males.

HARARGHE HIGHLAND This breed is distributed in Ethiopia. It is a small-sized goat. The coat colour varies from white, brown or black (Fig. 2.45). The face is straight and concave and mostly polled animals are observed in flock.

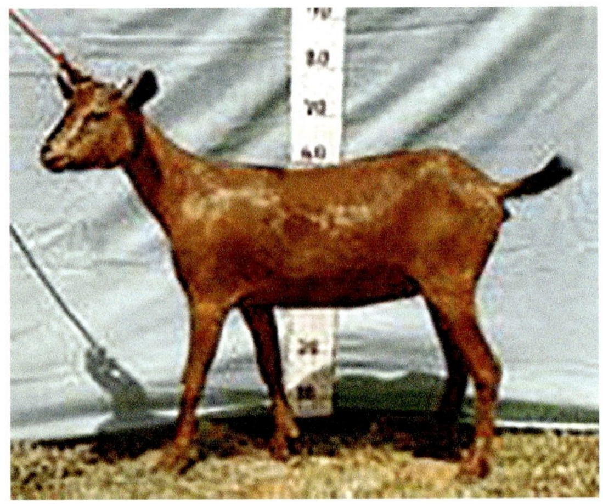

Fig. 2.45. Hararghe Highland goat.

SHORT-EARED SOMALI This breed is distributed in Ethiopia. It is a medium-sized goat with a white coat. The face is straight and horns are straight or curved (Fig. 2.46).

LONG-EARED SOMALI This breed is distributed in Ethiopia. It is a large-sized goat with a white coat. The breed has a straight face and the horns are curved and pointed backwards (Fig. 2.47).

CENTRAL HIGHLAND This breed is distributed in Ethiopia. This is a medium-sized breed with a broad face. The coat colour is reddish-brown with straight hair. The horns are straight and pointed backwards.

WESTERN HIGHLAND This breed is distributed in Ethiopia. It is a tall breed with a coat colour that is

Fig. 2.46. Short-eared Somali goat.

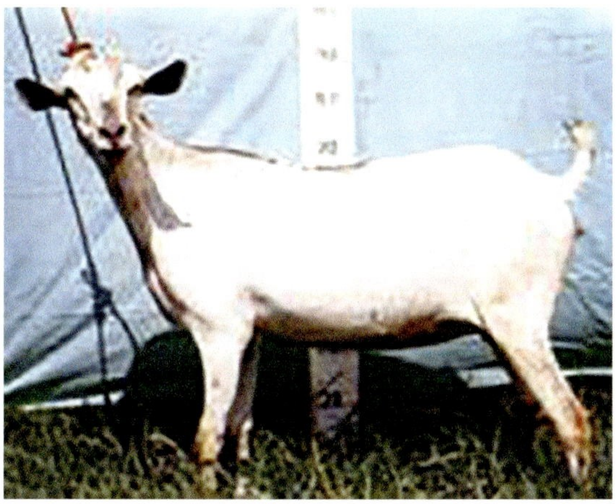

Fig. 2.47. Long-eared Somali goat.

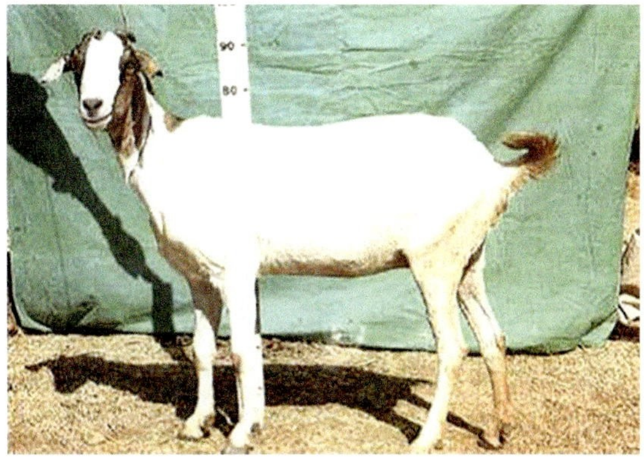

Fig. 2.48. Western Highland goat.

Fig. 2.49. Western Lowland goat.

mainly a white or fawn colour. The face is concave and animals have hair on their thighs (Fig. 2.48).

WESTERN LOWLAND This breed is distributed in Ethiopia. It is a small-sized breed with a fawn or white colour coat. The face is straight. Most males have straight horns oriented backwards (Fig. 2.49).

KAFFA This breed is distributed in Ethiopia. It is a small-sized breed with a red or black coat colour. The animals have a short neck and short prick ears. The face is straight and the horns point backwards (Fig. 2.50).

KALAHARI RED This breed was developed from crossbreeding between Red Boer and Nubian goats. The Kalahari Red breed is robust and thrives well in harsh desert conditions. Kalahari Red goats are very productive with good mothering abilities. They have strong teeth and excellent growth rates. The coat colour is reddish-black (Fig. 2.51). They have drooping ears and backwards-pointing horns (Campbell, 2003)

BOER This breed is distributed in South Africa. It is specifically a meat-purpose breed. The breed has some genes from goats of Europe and India (Casey and van Niekerk, 1988). The coat colour is white and the head is brown in colour (Fig. 2.52). The breed has long and pendulous ears and is known for high fertility and fast growth. The kidding rate is 2.0. It is a highly prolific breed with excellent growth rate and carcass quality. The

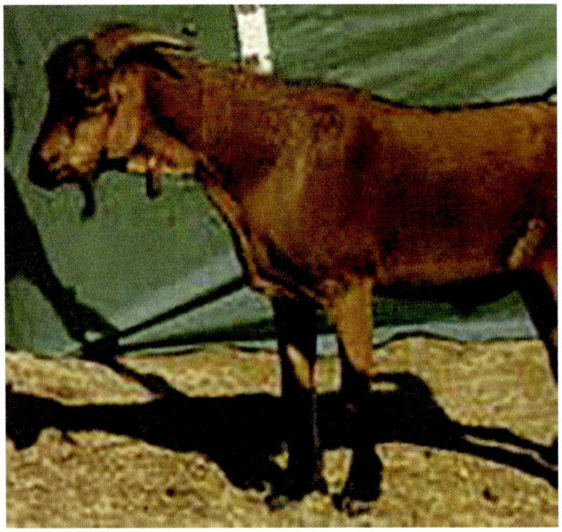

Fig. 2.50. Kaffa goat.

Fig. 2.51. Kalahari Red goat.

Fig. 2.52. Boer goat.

breed is being used as an improver breed in 48 countries. The adult body weight of the male is 110–135 kg and the female is 90–100 kg.

SAVANNA This breed is raised by pastoralists for both milk and meat products (Shrestha and Fahmy, 2007). It was also developed in South Africa from the crossbreeding of indigenous goats with white Boer goats. The white coat colour is a dominant trait over other colours in goats. The breed is preferred for religious ceremonies and to celebrate the birth of a son. It has high productivity, high muscular development, and strong bones, legs and hooves (Campbell, 2003).

WEST AFRICAN LONG-LEGGED This breed is a dual-purpose goat that is reared for meat and milk products. The face is straight and convex and the ears are short. The coat colour is white, black or brown. This breed resembles the West African Dwarf goat except that they are larger in size, and their horns are long and twisted (Mason, 1981). They also are known as Sahelian, Sudan or Fulani goats. Local names in Mali include Gorane, Niafounke and Nioro. The breed is found in northern Africa spreading between Chad and Senegal, and in the tropical forest and desert areas.

PYGMY The Pygmy goat was derived from the Cameroon Dwarf or the West African Dwarf goat. Northern, south-western, and eastern African countries also have similar forms of Pygmy goats. This breed is small with short legs, a large torso and broad head. Besides being used as pet animals, the breed is also used for meat. They are very popular as exotic animals. The hair is straight and medium to long. Males have beards, but beards are non-existent, sparse or trimmed in females. All body colours are acceptable, the predominant colour being a grizzled (agouti) pattern produced by the intermingling of light and dark hairs of any colour (National Pygmy Goat Association, 1932)

WEST AFRICAN DWARF (WAD) This is one of the major goat breeds in humid West and Central Africa and is distributed in different countries. The breed is known for milk and meat production. Multiple births are common and the kidding interval is about 220 days (Wilson, 1991).

Other regions

CREOLE This breed, also known as Reunion Creole, is known for its meat quality. The breed is widely distributed in the West Indies and has spread to 18 different Caribbean and South American countries as a transborder breed (FAO, 2004). The goats have short ears and horns. The colour varies between black, brown and pied. The adult males weigh 25 kg, and females weigh about 20 kg. Wither heights are 60 cm and 50 cm for adult males and females, respectively (Alexandre *et al.*, 1999).

CRIOLLO Criollo goats are distributed in Mexico, Argentina Bolivia, Peru, and Venezuela (Mason, 1980). They originated during the Spanish conquest and colonization (Lanari *et al.*, 2003). The most common colour is black, and their horns are short. The breed is known for meat production but milk production is also valued. The average adult weighs 39 kg. The breed is adapted for grazing in harsh environmental conditions.

ANGORA Angora goats are distributed in Turkey. It is a small and very delicate goat. The coat colour varies from red to tan and brown. Some goats have grey or black coats. The breed produces 1–2 kg of fibre known as mohair. The adult Angora goat has a pair of horns that curve backwards. Their ears are pendulous, and their legs are short.

2.5 Breeds and their Utility

2.5.1 Improver breeds for milk, meat and fibre

Goat breeds in the tropics are classed based on their origin, adaptability and speciality in the production environment (Table 2.10). A summary of the diversity in production performance is presented in Table 2.11 (Shrestha and Fahmy, 2007).

2.5.2 Conservation of genetic resources

Planned conservation programmes of goats in the tropics has largely been neglected. Goat breeds are adapted for different characteristics such as adaptability, fecundity, disease resistance, grazing habit, cheese quality, meat quality and skin characteristics. The special characteristics of goat breeds should be conserved for future use in their home tract (Baumung *et al.*, 2004). The wide range of production environment requires an equally diverse range of biological types to enable sustainable production. Tropical countries have diverse environment and climatic conditions, diverse types of feed resources, and various levels of management and disease occurrence. The genetic material required to meet these challenges must be determined by the nature of the production environments. It has been emphasized that animals are fundamental to most agro-ecosystem and production systems as they are critical for system resilience and flexibility, and are responsible for increasing the productivity in different production systems. Food security will only be achieved and maintained by utilizing appropriate genetic resources (Hammond and Leitch, 1995). Therefore, the conservation of animal genetic resources has a role in sustaining production systems. The planned exposures in different ecosystems have taken place and breeds have naturally evolved to fulfil the needs of farmers in specific environments. It is therefore necessary to conserve different type of breeds that have evolved through natural selection historically or due to geographical isolation. It is impossible to determine which characteristics of any breed have potential value in the future to cover all eventualities. Genetic variation is the base material for animal breeders. It is this variation that is represented by differences between the breeds that cannot be readily regenerated, and this irreversible loss reduces the diversity required for future selection.

Threatened populations

There is no simple numerical level at which a population is defined as being endangered or eligible for consideration as a candidate for conservation. It is dependent upon a number of factors such as the actual number of animals, the rate of decline in the population size, the closeness of the relationship between individuals within the population, the geographical range and the rate of reduction of that range, special threats from introduced species, and rapid changes in environmental conditions including climate, predators and parasites.

The International Union for the Conservation of Nature (IUCN) provides clear definitions for the

Table 2.10. Goat breeds with potential for genetic improvement.

Speciality	Breed	Country of origin	Climate	Environment
Meat	Bugri	Pakistan	Tropical	Dry
Meat	Damani	Pakistan	Tropical	Dry
Meat	Fijian	Figi	Tropical	Humid
Meat	Kaghani	Pakistan	Tropical	Dry
Meat	Kail	Pakistan	Tropical	Dry
Meat	Katjang	Indonesia	Tropical	Humid
Meat	Katukachchiya	Sri Lanka	Tropical	Dry
Meat	Lehri	Pakistan	Tropical	Dry
Meat	Sudan desert	Sudan	Tropical	Very dry
Meat	Tapri	Pakistan	Tropical	Dry
Meat	Terai	Nepal	subtropical	Humid
Milk	Black Bedouin	Israel and Egypt	Tropical	Very dry
Milk	Dera din Panah	Pakistan	Tropical	Dry
Milk	Kamori	Pakistan	Tropical	Dry
Milk	Nubian	Sudan	Tropical	Dry
Milk	Zaraiby	Egypt	Tropical	Dry
Prolificacy	Criollo	South America	Tropical and subtropical	Dry
Prolificacy	Sudan desert	Sudan	Tropical	Very dry
Prolificacy	West African dwarf	West Africa	Tropical	Humid
Skin	Maradi (Red Sokoto)	Nigeria	Tropical	Humid
Skin and meat	Mubende	Uganda	Tropical	Humid
Pashmina	Changthangi	India	Tropical	Dry
Meat	Singhala	Nepal	High mountains	Cold
Meat and skin	The Small East African Goat	Kenya	Tropical	Dry
Meat	Galla goats	Kenya, Somalia	Tropical	Dry
Fibre	Angora Goat	Turkey	Tropical	Dry cold
Meat	Kalahari Red	South Africa	Tropical	Semi-arid and arid
Meat	Boer	South Africa	Tropical	Dry
Meat and prolificacy	Nigerian Dwarf	West Africa	Tropical	Dry
Meat	Chappar	Baluchistan(Pakistan)	Tropical	Dry
Meat	Creole	West Indies	Tropical	Dry
Meat	Criollo	Mexico	Tropical	Dry
Meat	Kambing Katajang or pea	Malaysia and Indonesia	Tropical	Dry
Meat	Nepalese Hill	Nepal	Tropical	Dry
Meat	Savana	South Africa	Tropical	Dry
Meat, milk and hair	Syrian Mountain	Syria	Tropical	Dry
Meat and Milk	West African Long-Legged	Senegal	Tropical	Dry
Milk	Damascus	Syria and Lebanon	Tropical	Dry
Fibre	Kurdi	Kurdistan, Iraq, Iran	Tropical	Dry
Meat and Milk	Barbari	India	Tropical	Semi-arid to arid
Meat	Black Bengal	India	Tropical	Hot humid
Meat	Black Bengal	Bangladesh	Tropical	Hot humid
Meat and Milk	Kutchi	India	Tropical	Dry hot
Meat	Ganjam	India	Tropical	Hot humid
Meat	Khasi (Assam local)	India	Tropical mountain	Humid
Meat	Kheri	Nepal	Subtropical	Humid
Meat and Milk	Marwari	India	Tropical	Dry
Meat and milk	Osmanabadi	India	Tropical	Semi-arid

Continued

Table 2.10. Continued.

Speciality	Breed	Country of origin	Climate	Environment
Meat and milk	Sirohi	India	Tropical	Dry
Milk	Sangamneri	India	Tropical	Semi-arid
Milk and meat	Beetal	India	Tropical	Arid
Milk	Damascus	Syria and Lebanon	Subtropical	Arid
Milk	Jamunapari	India	Tropical	Dry
Milk	Jakhrana	India	Tropical	Semi-arid
Milk and meat	Malabari	India	Tropical	Humid
Milk	Surti	India	Tropical	Semi-arid and Humid
Milk and meat	Zaraibi	Egypt	Tropical	Dry

Table 2.11. Diversity in the production performance of goats by region of the world.

Region/performance	Number of observations	Mean ± S.D	Minimum	Maximum
Africa				
Body weight (kg)				
Male	32	53 ± 37	20	130
Female	32	39 ± 22	20	94
Litter size (kid)	17	1.4 ± 0.33	1	2.1
Milk yield (kg/lactation)	3	126 ± 116	50	500
Asia and Pacific				
Body weight (kg)				
Male	106	43 ± 17	16	130
Female	106	32 ± 12	14	100
Litter size (kid)	79	1.4 ± 0.37	1	2.9
Milk yield (kg/lactation)	63	136 ± 109	16	550
Europe				
Body weight (kg)				
Male	123	66 ± 13	35	120
Female	124	49 ± 12	24	120
Litter size (kid)	28	1.6 ± 0.35	1	2.2
Milk yield (kg/lactation)	41	299 ± 225	40	775
Latin America and Caribbean				
Body weight (kg)				
Male	10	40 ± 14	15	70
Female	11	30 ± 13	13	50
Litter size (kid)	11	1.4 ± 0.3	1.1	2.0
Milk yield (kg/lactation)	2	63 ± 4	60	65
Near East				
Body weight (kg)				
Male	48	44 ± 14	17	75
Female	47	33 ± 8	15	50
Litter size (kid)	27	1.6 ± 0.43	1.1	2.5
Milk yield (kg/lactation)	32	150 ± 97	35	460
North America				
Body weight (kg)				
Male	5	41 ± 21	22	67
Female	4	35 ± 26	13	60
Litter size (kid)	4	1.0	1	1

Source: Galal (2005)

rarity of species in its international Red Data Books. Generally, the population size is used to determine the point at which a species is considered for conservation. In addition to this, it is necessary to examine the crossbreeding or grading up programmes at the population level to ensure that future selection and improvement can be carried out. For goats, a breed will be considered as endangered if there are less than 500 breeding females and four or less distinct male lines. Breeds can also be listed if there are between 500 and 1000 breeding females and the numbers are declining, or less than 20 males (Quartermain, 1991). All the breeds mentioned are vulnerable to the extent that either the populations are declining numerically or their security is not yet secure, or the populations are of adequate size but security is dependent upon special intervention programmes (Maijala *et al.*, 1984; Quartermain, 1991; Rout *et al.*, 2012).

The goat population is facing the threat of being endangered due to various problems. The lack of proper breeding policy is one of the contributing factors for the endangerment of tropical goats. The indiscriminate crossbreeding and intermixing of the blood among breeds with local animals are the major cause for putting tropical goats in danger of extinction. It is not possible to determine which characteristic has potential value in the future because it is impossible to envisage all future eventualities that may include climate change, mutations in disease or parasite population, the effects of political change, wars and the availability of energy (Office of Technology Assessment, 1987; Wilson, 1988; FAO, 1989; Weiner, 1989; National Academy of Sciences, 1992). Therefore, any population that exhibits unusual characteristics or traits, is geographically isolated or has some genetic influence from another breed would be considered a unique population.

India is facing major environmental problems due to increases in agricultural land, human population, rangeland management and protected forests for wildlife, and so on. The livestock sector is facing problems due to a shortage of grazing land, availability of feed and fodder and loss of habitat. The pure Jamunapari population has declined in its habitat for different reasons. Beetal goats are becoming the focus of conservation because their number is declining as a result of changing agricultural patterns in the Punjab region. Barbari goats are facing dilution in breed characteristics due to crossbreeding with Sirohi and Jamunapari goats in their home tract. Due to a decrease in natural browsing and geographical restrictions, Surti and Jakhrana goats are decreasing in number. Moreover Epstein (1974), Acharaya *et al.* (1982) and Devendra and Burns (1983) have drawn attention to the conservation programme for specific goat populations in developing countries.

Identification of conservation stocks

If breeds are disappearing in their natural habitat then action to conserve the breeds should be taken immediately. A range of information is needed in order to identify the populations that are in danger of extinction and should be considered as candidates for conservation. The following information should be gathered and used when planning conservation strategy.

1. Descriptive information on breeds, geographical location and information on production characteristics.
2. Estimating the availability of the total number of animals in relation to sex and the population trend including selling pattern.
3. The proportion of the female population being used in crossbreeding. By estimating the number of purebred males and females along with the number of young stock with evidence of crossbreeding, it is possible to estimate the rate of breed dilution. Even a 20% per year decrease in purebred young stock will result in a very dramatic crash in population size over a relatively short space of time.
4. The number of herds or breeding units available in their natural habitat, as large size herds may be more vulnerable to diseases and affected more seriously by economic or political changes than smaller herds.
5. Estimate the health risk: whether the population in the regions where lethal epidemics are endemic may be at greater risk than those in regions where such diseases are not present.
6. Estimate other risks, including political, climatic or economic. In particular the risk of drought, storms, flooding, war or rapid socio-economic change, which could result in the disappearance of indigenous populations.
7. Characterization of the breed, which includes the measurement and description of external

appearance, production characteristics, climatic adaptation, disease resistance, parasite tolerance, management and any other special features. It may also involve the collection of biochemical information from blood types, milk proteins and the comparative analysis of DNA fragments. All of this information is useful in determining the long-term conservation strategy with respect to a breed, but is not essential in establishing an initial programme to prevent the early loss of a breed or population.

Mechanism of conservation

The mechanism for the conservation of individual genes within populations is closely linked to the conservation of species and breeds. The most important feature of a small population conservation programme is the rate of genetic loss and the increase in homozygosity. The increase in homozygosity within a small population causes the loss of ability to adapt, inbreeding depression and extinction. Viable population size must therefore be linked to the ability to conserve genetic diversity within any conservation population. The maintenance of diversity in a population depends on founder population and effective population size. The population size, birth and survival rates, sex ratio and levels of variation must be taken into account before deciding the manner in which the breeds will be conserved.

The mechanism for conserving genetic resources is not an end in itself, but rather a means of ensuring that an animal's genetic resources are better understood and available, and more effectively used and developed by present and future generations. Generally two approaches (*ex situ* and *in situ*) have been used for conserving the goat population in India:

- *In situ:* The generation and loss of alleles is a dynamic process that should be maintained at close equilibrium through sound management. The maintenance of an animal in its natural habitat or in their as close as possible adaptive environment is called *in-situ* conservation.
- *Ex situ*: This conservation method uses the cryogenic storage of an animal's genetic resources (e.g. semen, ova or embryos, DNA segments in frozen blood or other tissues), along with the maintenance of breeds of domesticated animals in farms, zoos and other locations away from the animal's home tract. The CIRG in India has developed specific guidelines for the storage of genetic resources. The storage of germplasms as well as semen of proven bucks is in progress. Both *in-situ* and *ex-situ* methods are equally feasible for conserving Indian goats. Technological methods are available for conserving the germplasm in the form of gametes and embryos with liquid nitrogen, thereby creating a gene bank to address the problem of extinction.

2.6 Breeding Strategy and Breed Development

2.6.1 Breeding policy

Breeding policy describes the breeding objectives for breed improvement of a specific population and is dependent on the biology of a specific population and the economic value of produce in the market. The heritability of traits, correlations between traits and marketable products and reproductive techniques are major factors affecting the breeding programme. Economic factors are additional points to consider in the improvement of traits, along with the cost of implementing the breeding programme.

It is necessary to consult with farmers to get their input with regard to the purpose of goats (meat, milk, fibre) and preference in selection for keeping males and females. Secondly, it is necessary to analyse the genetic potential of the animal and what it is feasible to attain through genetic improvement with respect to economic profitability. The outcome is subject to discussion in relation to the overall aim of the executing agency, government policies and other social and economic factors. The policy should be finalized with the participation of farming households and should be implemented in a cooperative manner.

The traits that have economic importance should be given first priority in a successful breeding programme. As there are lots of variations within breeds or within populations, it is necessary to standardize the production trait, method of data recording and analysis format. Breeding programmes should incorporate the criteria relevant to existing field situations and farmer preferences to better develop the breed standard for a particular area of operation. Standard operating procedures need to be developed for precision recording and decision making for improving the traits in a given region.

2.6.2 Breeding strategy

After the breeding policy has been formulated, it is necessary to define a breeding strategy; that is, how the traits will be transmitted from one generation to the next and the choice of selection method. In a breeding programme a population is divided into a base population and one or more subgroups consisting of animals producing improved breeding stock. Different methods of selection can be adapted for males and females within each group. For production of the next generation of breeding bucks only a few top bucks are selected, which are then mated to selected top-producing mothers.

2.6.3 Breeding plan

The breeding plan includes the total number of animals in the programme and the number of animals to be selected for each selection plan. The breeding plan also shows the feasibility of implementing a particular breeding scheme in terms of the number of animals involved. This is initially based on assumptions for the breed parameters such as kidding rate, mortality rate in kids and buck-to-doe ratio. It is important to define the proportion of high-producing animals to be selected and the proportion of animals to be culled (based on culling criteria).

Suppose our programme requirement is 250 bucks to be distributed across farms and 25 bucks to be kept in the nucleus herd. The breed parameters are:

- fertility rate: 90%;
- kidding rate: 1.1; and
- mortality rate: 15%.

Field recording was carried out in 2000 adult does and nucleus flock recording was carried out with 150 adult does. We selected 25% of the best does from the field recording; therefore we have 500 selected does. As we have 90% fertility, we have 450 does from the village herd and 135 does from the nucleus herd; we have 585 does for producing kids. Total kids born from 585 does is 644 (kidding rate 1.1). The mortality in the flock till 3 months of age is 97 (15% mortality rate). At present we have the 546 kids and we may have 273 male and females in the flock. Therefore we can easily distribute 250 does to farmers every year.

The strategy may be different with different breed parameters:

- fertility rate: 90%,
- kidding rate: 1.6; and
- mortality rate: 15%;

Now we need to select 300 does from the village flock and we have 150 does in the nucleus flock. We have 450 does in total and the available does for kidding will be 405 (based on fertility rate). Total kids born was 648 and mortality of kids up to 3 months of age was 97. The total kids available in the flock will be 508 and we will have 254 males and 254 females. We can easily distribute 250 females to farmers.

2.7 Breeding Operations and Community Breeding Programmes

Breeding operations include the whole range of activities necessary to implement a breeding programme.

2.7.1 Data format and recording of data in the farmer's field

The format for data recording needs to be developed and field data should be collected regularly and then checked and analysed. A herd card maintained at the goat-keeper's home is essential to ensure effective data recording. Monitoring of data recording is required to analyse deviations in recording, age-wise variation in weight and morning and evening milk yield.

2.7.2 Record keeping and database management

Performance recording is the first step of effective implementation of a breeding programme. Recording data should be carried out in order to identify the best-performing animals in a herd. Performance recording is useful not only for breed-improvement programmes, but also for effective support for decision making for managing the farmer's flock with respect to feed supplement, reduction of mortality, kidding management and so on. The parameters to be recorded should be relevant to the breeding objectives.

For implementing breed-improvement programmes in field conditions at farm level, it is necessary to record specific traits at different ages. Generally, animals are selected after 3 months of age; therefore the body weight at 3 months old and the dam's milk yield at 60 and 90 days should be recorded for the selection process. If we select from a larger population, then higher selection intensity leads to increased genetic progress.

2.7.3 Management of breeding, reproduction and production performance recording

The first step is the selection of the best bucks and does as the major aim of the programme is the balanced dissemination of identified superior genes in a population. The best animals at 3 months of age are selected and reared under similar village conditions. During the selection process, it is necessary to screen for breed characteristics, anatomical features, behavioural traits and veterinary health status. Screening tests for brucellosis and *Mycobacterium paratuberculosis* should be carried out before including an animal in a breeding programme. Bucks with physical abnormalities, poor mating behaviour, low fertility or those having a low breeding value must be culled. Breeding bucks must be distributed in the field well in advance of the beginning of the mating season as they need time to adjust to local differences in feeding and management patterns, and should be used to the maximum extent possible once the goats come into heat.

2.7.4 Genetic defects and inbreeding

Bucks with visible physical defects should not be used for breeding purposes. Physical defects such as dwarfism, parrot mouth and an absence of hair should be identified. Goats are known to show a higher frequency of intersexuality or masculinization of females when breeding with polled (hornless) animals. This condition is caused by a linkage of genes controlling the absence of horns to a defective development of sexual organs. Offspring of polled animals should not be used for producing breeding bucks. Breeding bucks are preferably kept for 1 year in the same herd, but never more than 2 years in order to avoid mating of daughters. For the same reason, using breeding bucks in the herds of origin must also be avoided.

2.7.5 Monitoring and evaluation

The programme needs to be monitored to analyse its impact. It is necessary to observe the performance of breeding bucks, production of viable offspring and the response of farmers regarding the benefits of the programme. The evaluation of the breeding programme can be achieved through monitoring of field activities, analysis of production performance and genetic progress.

2.8 Breed Improvement in Village Conditions

Genetic improvement is carried out through selection in village conditions. The farmers must be clear about their objectives for the programme. The farmers should select the traits with the best market potential, and the traits should be easily measured in field conditions.

2.8.1 Breeding operation

Farmers should be clear about the objectives of the programme. The farmer should know the production potential of each individual in his flock. The recording of traits such as body weight and milk yield should be done with utmost care. The farmers should know about selection criteria and the trait quality should be discussed. Genetic potential of the individual or breed should be discussed in relation to anatomical, physiological and production performance. Adequate knowledge regarding deformities and the selection of healthy animals should be held by the farmers. The flock should be free from genetically abnormal individuals. Moreover, the knowledge of inbreeding and its control should be provided to farmers. Selection of the best males and females should be taken up by farmers. The farmers should be aware of age, health status, physical characteristics, body features, structural soundness, genetic defects, production efficiency (6-month and 12-month weight), and reproduction efficiency (producing twins or singles).

Similarly, the animals should be strong and healthy. The behaviour of animals should be normal, not aggressive. The sexual organs should be well developed. Before starting the genetic improvement programme, it is necessary to create the environment for high-producing or improved animals. If the improved animals are kept in the same environment, then it is a waste of effort producing them. The improvement in management practices is required before putting improved or graded animals in field conditions. After standardizing environmental input, it is necessary to put selection pressure on flocks to obtain the desired improvement. Improved bucks can be maintained by the same farmers and resources can be shared. Similarly, the necessary input from implementing agencies can be provided to farmers to maintain improved genetic diversity in their flock.

A national programme should consider the genetic diversity of the nation before taking up any programme for future use. The objectives of the programme should fulfil the interest of farmers and defined traits should be recorded for the improvement programme. The steps to be followed for breed development are described in Fig. 2.53.

There have been few selection programmes for goats in the tropics. Selection is a long-term programme and needs commitment from the implementing agencies. Identification of a population and accurate recording of performance traits in the population are also necessary for a successful programme. Selection is an appropriate strategy for goat breeds in harsh and extreme climates in the tropics. As it is very difficult to improve the management conditions of a field, it is important that the production performance and adaptability characteristics of the improver breeds need to be taken in to account in order to ensure that they perform well. The selection programme starts by establishing a nucleus/multiplier flock with specific intervention and appropriate technologies in different locations. The selection programme in a farmer's flock (i.e. village

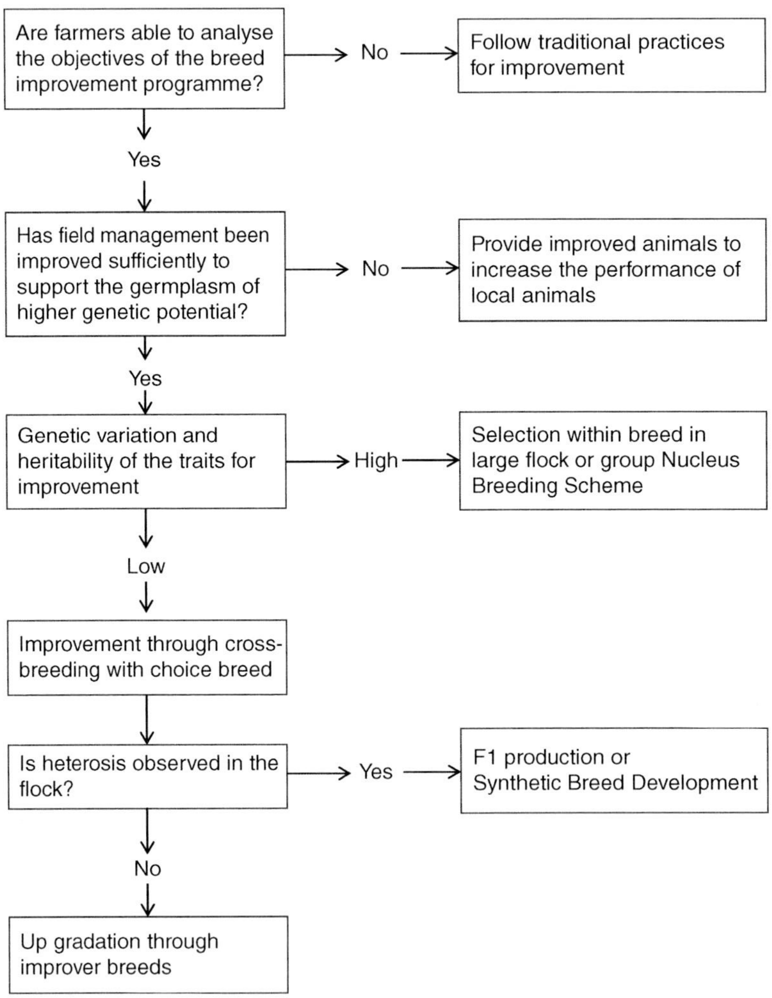

Fig. 2.53. How to decide the method of breed improvement.

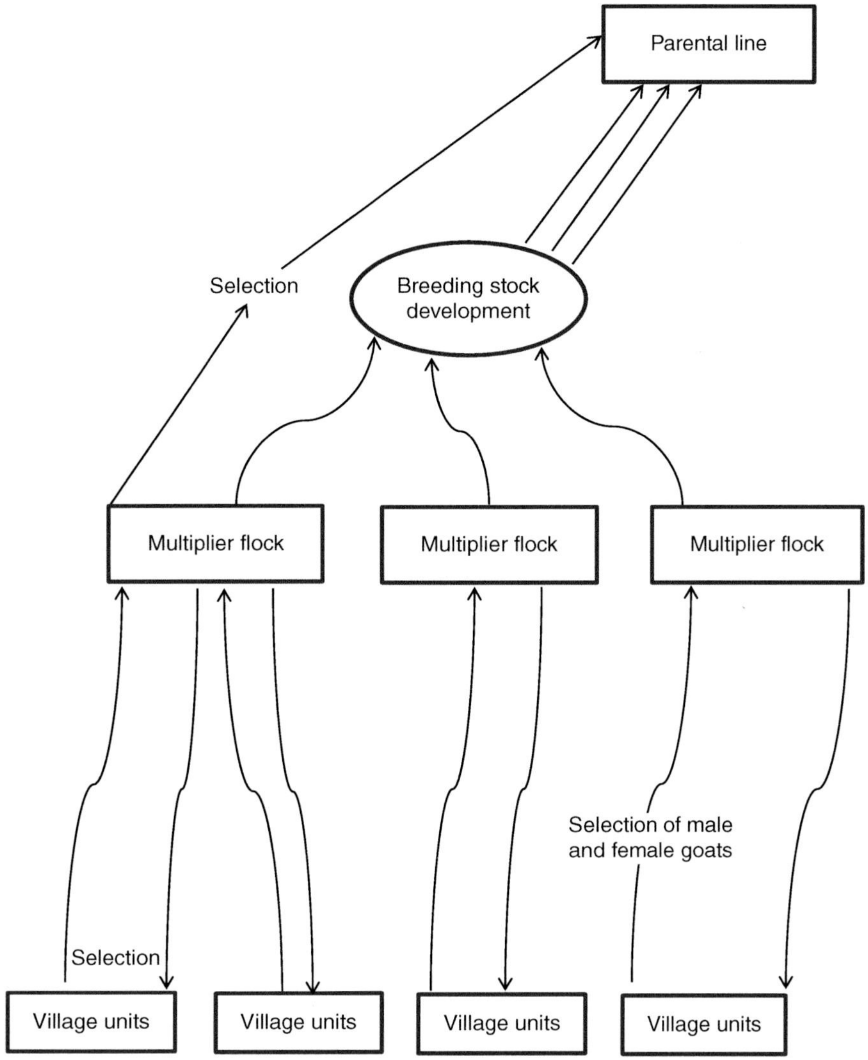

Fig. 2.54. Structure of developing a breeding stock in its natural habitat.

units) will continue, where the best individuals will be selected based on morphology, structural soundness and production performance to develop parental lines. A detailed structure of breeding stock development is described in Fig. 2.54.

2.8.2 Mating

After selecting the best animals, it is important to decide on a mating plan. If it is for pure breeding, then selected males will be mated to females based on a certain mating plan.

Crossbreeding should be considered if the desired trait has a low degree of heritability (e.g. reproductive traits, disease resistance traits). The crossbreeding should be taken up only when the levels of management have been improved to maintain the crossbreed's potential. It is important to implement an effective extension programme to improve management conditions, including feed and fodder availability. Crossbreeding should be allowed if the crossbreeds are provided with better environmental conditions to express their genetic potential and perform well. The F1, a new crossbreed, needs better management for optimum expression of genes. Once the adaptation of F1 is established in the environment, then further upgrading is possible for testing the combining ability. Subsequently, it is very important to follow a specific mating plan to maintain the blood level of exotic germplasm for optimum production. After grading up to a desired level, mating between the populations will be followed by selection and *inter se* mating.

Genetic gain from crossbreeding is based on two types of gene action, i.e. additive gene action and heterosis. Additive gene action plays a major role whereby the production trait is improved due to the additive effect. Heterosis is the effect of crossbreeding when the F1 is better than the mean of its parents. It becomes difficult to maintain hybrid vigour in a subsequent generation. The effect of heterosis can be higher in a harsher environment than in more favourable environments. The improver breed may not perform very well in an unfavourable environment. However, crossbreeds perform well in harsh and stressful environments and perform better than both the parents.

After the crossbreeding programme, it is important to maintain heterosis or the additive effect. It is possible to produce F1 stock continuously, which is known as terminal crossing. It is an expressive approach.

The second method is alternate mating of tropical and improver breeds, termed criss-cross mating. The local breed is first mated with an improver breed and then mated with a local breed. The heterosis effect is maintained up to 66.66%.

The third method is the production of a synthetic breed; that is, the reciprocal crossing of three or more breeds and selection of their offspring with good management over a long period of time. This method is very difficult to achieve in tropical breeds.

2.9 Breed Registration, Judging and Breed Promotion Activities

2.9.1 Breed registration

The population within species can be classified as domesticated and wild/feral breeds, strains or varieties. Breeds have evolved in different regions with specific attributes. It is essential to explore the biodiversity of a species in different regions for future conservation and use. Breed registration is a process to protect local animal genetic diversity as per the requirement of convention on biological diversity and to provide recognition to livestock keepers or to breed developers. The registration process mainly revolves around the concept of breed and depends on guidelines issued. In India, a breed registration guideline has been developed by the National Bureau of Animal Genetic Resource under the auspices of the Indian Council of Agricultural Research. The guidelines and format for goat breed registration have been developed and are available in the public domain (www.nbagr.res.in). The format is provided in the protocol section at the end of this chapter to develop breed registration guidelines as per the requirement of each country.

2.9.2 The show-ring judging and goat shows

The show ring has been very useful for breeding programmes of livestock species all over the world. The owner gets an opportunity to showcase their best animals and the animals are evaluated by professional judges. Shows generally take place during livestock fairs or goat fairs where the breeders and farmers exhibit their animals. Goat shows have different components such as breeding animals, dairy animals and show-purpose animals. The judges use specific score cards to evaluate animals (see Protocol 5 at the end of this chapter) (Table 2.12).

Table 2.12. Scorecard for evaluating goats for different traits.

Goat type trait	Dairy adult doe	Dairy young doe	Dairy buck	Market	Fibre
General appearance	35	55	55	75	40
Dairy character	20	30	30	–	–
Body capacity	10	15	15	25	–
Mammary system	35	–	–	–	–
Fleece	–	–	–	–	60
Total	100	100	100	100	100

Source: Ebert and Solaiman (2010).

Goats such as breeding bucks, milking does and fibre-producing goats attract public attention as evaluated by professional judges. Some goat shows are also markets where the animals are for sale. The judges also interact with exhibitors/farmers to ascertain the degree of involvement the farmer has in goat rearing and their contributions to the present best form.

2.10 Application of Biotechnological and Genetic Tools for Enhancing Goat Productivity

Molecular genetics promises to have a great impact on goat production in the future. DNA from the individual organism is unique. The application of molecular tools has been recognized in the field of parent identification, identification and traceability of individual animals, genetic disorder screening, genetic disease monitoring and sex determination. Marker-assisted selection and gene introgression have been applied in livestock selection programmes. The practical applications of transgenesis in livestock production have been identified in areas such as enhancing prolificacy and reproductive performance, increased feed utilization and growth rate, improved carcass composition, improved milk production and composition, and increased resistance to disease. Application of genomic information in livestock has been reviewed by Sellner *et al.* (2007). Animal pharming, the process of using transgenic animals to produce human drugs, is becoming more common and goats have been flagged a suitable choice. Transgenic goats are producing (in their milk) human protein C and antithrombin-III (AT-III) for the treatment of thrombosis, glutamic acid decarboxylase for the treatment of diabetes and human immunodeficiency virus. Spliced spider genes have been inserted into the mammary gland of goats to manufacture silk in milk.

2.10.1 The Online Mendelian Inheritance in Animals (OMIA) database

The Online Mendelian Inheritance in Animals database (http://omia.org) is a catalogue of inherited disorders and genes in different animal species. Disorders in goats and their genetic basis have been described. The trait disorders need to be known by researchers and veterinarians for control of the specific conditions (Table 2.13).

Table 2.13. The details of disorders in goats in OMIA (http://omia.org).

Total trait disorder	84
Mendelian trait/disorder	17
Mendelian trait/disorder likes causal variants known	614
Potential model for human disorder	37

2.10.2 Innovative phenotyping

Phenotyping is the key to successful implementation of a breeding programme. First, automatic trait measurement and recording need to be carried out for genetic improvement programmes (Berry *et al.*, 2012). Second, it is necessary to identify new traits or new phenotypes for efficient selection. Moreover, indicator traits need to be defined for phenotyping of the individual for different disease resistance and functional traits. There are a number of new traits that can be included in selection programmes, for example:

- milk proteins, fat content and sub-components such as amino acids, fatty acids and lactoferrin; indirect measure of somatic cell, milk electrical conductivity, milk mineral content;
- health-related traits recording such as veterinary records and other indicator traits to analyse genetic variation to disease susceptibility;
- behaviour traits like temperature sensitivity, mothering ability, etc.;
- monitoring of time related to rumination and metabolic activity;
- automated measurement of behavioural traits; and
- meat quality traits.

2.10.3 DNA markers

Since the Mendelian era, single-locus genetic markers such as eye colour in *Drosophila* or haemoglobin polymorphism have been described. The advancement of the polymerase chain reaction (PCR) technique has revolutionized the field and different markers have been developed.

Multi-locus markers

MINISATELLITE OR VNTR (VARIABLE NUMBER TANDEM REPEAT) MARKERS Minisatellites consist of regions of DNA where a short sequence (usually less than 20 base pairs) is tandemly repeated. Their variability stems from the fact that the number of repeats at a particular locus varies from individual to individual. The minisatellites, discovered by Jeffreys *et al.* (1985), were the first tandem repeat markers identified and are unique to individuals. Most of the simple repetitive DNA motifs are composed of two to seven nucleotides and can be found in the eukaryotic genomes among both transcribed and non-transcribed sequences. When used as probes against genomic DNA under low stringent conditions, the cloned minisatellites generate a complete pattern of hypervariable bands, that are individually specific (Jeffreys *et al.*, 1985; Singh, 1991; Lang *et al.*, 1993). They were mostly dominant markers with only one identifiable allele. Minisatellites are located mainly in telomeric regions, therefore good coverage of genome using minisatellite markers could not be achieved. The callipyge gene in sheep has been identified by single-locus minisatellite (Cockett *et al.*, 1994).

RAPD (RANDOM AMPLIFIED POLYMORPHIC DNA) MARKERS Primers of eight to ten base pairs are used to amplify a mixture of random fragments of DNA from the genome and each reaction produces multiple bands by PCR amplification. Some of these bands are polymorphic and can be used as genetic markers. RAPD markers have been used in plants to develop linkage maps. RAPD is the dominant marker and not useful to identify heterozygous and homozygous individuals. Moreover, the reproducibility is not easy and it is necessary to generate bands for each pedigree in the population.

AFLP (AMPLIFIED FRAGMENT LENGTH POLYMORPHISM) MARKERS These are dominant markers and require the generation of a new map for each population. AFLP markers are the multi-locus markers of choice (Vos *et al.*, 1995). DNA is digested with restriction endonucleases and linkers are ligated to each end of the restriction fragments. Selective PCR primers are then used to amplify subsets of fragments from the mixture of genomic restriction fragments. The selective primers encompass the linkers added on to the end of the restriction fragments. The amplified fragments are separated according to size. The bands that are present in some individuals but absent in others can be used as genetic markers.

Single-locus markers

RFLP (RESTRICTION FRAGMENT LENGTH POLYMORPHISM) MARKERS. RFLP markers are codominant markers and easily reproducible. The disadvantage is that the process requires large quantities of DNA. Genomic DNA is digested with restriction enzymes and the bands are separated by agarose gel electrophoresis and detection of desired bands by Southern blotting.

MICROSATELLITE MARKERS Short tandem repeats (STR) or microsatellite markers are present in the genome and the number of repeats within many of them is highly variable in a population of a particular species. Microsatellite polymorphism is now used for a wide range of applications in genetics, including construction of genetic linkage maps, linkage mapping of quantitative traits and disease genes, diagnosis of genetic disorders, paternity testing and forensic studies, and also for selective breeding in livestock. Microsatellites are becoming increasingly used in population genetic studies in humans, cattle, fish and *Drosophila*. The markers are also used to estimate relationship among breeds.

STR markers are polymorphic DNA loci that contain repeated nucleotide sequences and a repeat unit can be two to seven nucleotides in length. The number of nucleotides per repeat unit is the same for a majority of repeats within a STR locus. The number of repeat units at an STR locus may differ, so alleles of many different lengths are possible. PCR-based STR analysis has the following advantages over conventional methods of DNA analysis such as RFLP: the small size of STR loci improves the chance of obtaining a result for samples containing less amounts of DNA; many STR loci can be typed from a single PCR reaction; STR alleles have discrete sizes making it easier to interpret results; and tests are easy to standardize and are reproducible. Microsatellites are single locus, codominant and highly reproducible and easy to genotype. Microsatellites cover the genome and variation is due to simple sequence repeats. Microsatellites are typed by designing primers to the unique DNA sequences on the either side of the repeat and amplifying by PCR. The size of the PCR product is ascertained by electrophoresis on a DNA sequencing gel. The International Society for Animal Genetics (ISAG) recommends the use of a validated panel of microsatellites for characterization and other analysis (ISAG, 2010)

SNPS (SINGLE NUCLEOTIDE POLYMORPHISMS) SNPs are distributed in the genome and it has been estimated that an SNP occurs once in every kilo base of unique sequence in humans and livestock. The unique region of DNA can be amplified and analysed for variation in DNA sequence. SNP typing can be carried out for genetic diversity, signature by selection process and trait association studies. Goats showed the highest population growth rate when compared to cattle, pigs and sheep and other economically important species. High-throughput genomic resources were not available in goats until 2011, when the International Goat Genome Consortium developed and released the first high-throughput SNP chip with more than 50 000 SNPs (Illumina CaprineSNP50 Beadchip) using ten biologically and geographically different breeds (Tosser-Klopp *et al.*, 2014). The first complete assembly of the goat genome was released in 2013 by Dong *et al.* (2013) and a second version of the reference genome was released during 2017 (Bickhart *et al.*, 2017).

2.10.4 Milk protein diversity

The discovery of milk protein polymorphism (Aschaffenburg and Drewry, 1955) and its impact on milk production and technological properties has promoted characterization of milk protein variants in different species (Grosclaude *et al.*, 1987; Moioli *et al.*, 1998). Casein, the main protein fraction of ruminant milk, is encoded by four tightly linked genes (Grosclaude *et al.*, 1978; Grosclaude and Martin, 1997) clustered in a 250-kb genomic DNA segment (Ferretti *et al.*, 1990) in the following order: aS1-casein (CSN1S1), b-casein (CSN2), aS2-casein (CSN1S2) and j-casein (CSN3) (Threadgill and Womack, 1990; Jansa Perez *et al.*, 1994; Grosclaude and Martin, 1997). These genes have been mapped on chromosome 6 in cattle and goats (Hayes *et al.*, 1993; Popescu *et al.*, 1996).

Milk protein polymorphism has been analysed in different breeds of goats (Feligini *et al.*, 2005) and has been used to establish relationships between different species or breeds, for establishing geographical diversity and conservation decisions (Beja-Pereira *et al.*, 2003), and for use as a marker to find linkage association with production traits (Leroux *et al.*, 1992; Martin, 1993; Martin and Leroux, 1994; Ikonen *et al.*, 2001). The milk casein locus has been associated with milk production traits as well as manufacturing properties (Leroux *et al.*, 1992; Martin, 1993; Martin and Leroux, 1994; Ikonen *et al.*, 2001). Polymorphisms in casein clusters have been intensively investigated both at protein and DNA level. The analysis of goat casein complex has led to 16 alleles associated with the CSN1S1 locus (Rammuno *et al.*, 2005; Caroli *et al.*, 2006; Kumar *et al.*, 2007), 3 alleles associated with the CSN2 gene (Chessa *et al.*, 2005), 7 alleles associated with the CSN1S2 gene (Sacchi *et al.*, 2005; Caroli *et al.*, 2006), and 16 alleles with CSN3 gene (Prinzenberg *et al.*, 2005).

Milk composition traits such as protein, fat, lactose and SNF have been analysed in relation to different breeds and geographical regions (Grosclaude and Martin, 1997). The CSN1S1 gene is the best natural model and different genotypes occur due to interallelic combinations (Bevilacqua *et al.*, 2002; Feligini *et al.*, 2005). The caprine CSN1S1 locus has been characterized by at least 16 alleles that have been associated with various levels of protein synthesis (Grosclaude *et al.*, 1994; Rammunno *et al.*, 2000; Bevilacqua *et al.*, 2002; Sacchi *et al.*, 2005). CSN1S1 is related to the cheese production and milk-processing properties of milk. The amount of total casein in caprine milk was positively correlated with the presence of the CSN1S1 allele and was highest in cases of A, B and C alleles (Grosclaude *et al.*, 1987; Grosclaude *et al.*, 1994). The effects of CSN1S1 polymorphism on milk yield and composition have been thoroughly studied in different breeds (Grosclaude *et al.*, 1994; Vassal *et al.*, 1994; Barbieri *et al.*, 1995). It has been reported that goats associated with the high content (A or B allele) of CSN1S1 casein produce milk characterized by a significantly higher percentage of protein, fat, total calcium, better curd-firming time, curd firmness and cheese yield compared with goats homozygous for alleles associated with a low or intermediate content (E or F allele). The benefit of using the information on the CSN1S1 genotype in a selection programme for dairy goats will improve the protein content (Sanchez *et al.*, 2005). The application of gene-assisted selection has been applied to the CSN1S1 gene in goats. The polymorphism for the CSN1S1 in goats has a strong effect on protein content, total protein output and other qualitative traits. The quantitative effects of CSN1S1 variants on protein content have shown a significant difference between the extreme genotypes AA and FF. The Indian goats exhibited CSN1S1 A allele in higher frequency in all the breeds, indicating the higher casein yield in their milk (Rout *et al.*, 2010).

2.10.5 mtDNA diversity and Y-chromosome haplotype

Mitochondrial DNA (mtDNA) contains highly informative polymorphic sites and its simple maternal inheritance without recombination makes it useful for population studies in many organisms (Brown *et al.*, 1986; Loftus *et al.*, 1994; Bradley *et al.*, 1996; Luikart *et al.*, 2001). Luikart *et al.* (2001) carried out a worldwide survey of domestic goat mtDNA diversity and identified three major mtDNA lineages. Lineage A was the most common in all continents. Haplogroup B1 is distributed in Asia, sub-Saharan Africa and Greece, and haplogroup B2 is distributed in China and Mongolia. Lineage C was observed in a few samples from Mongolia, Switzerland and Slovenia. Haplogroup D and E is distributed in Asia and Central Europe. Haplogroup F was distributed in Sicily and haplogroup G was distributed in North-east and Northern Africa. These lineages were judged to have diverged over 200 000 years ago; this ancient divergence time and the different geographical localizations of the lineages suggested the likelihood of either multiple domestication events or introgression of additional lineages after the original domestication. A more complex origin for domestic goats has been proposed by analysing mitochondrial HVRI (hypervariable) regions in Indian goat breeds from different agro-climatic regions (Joshi *et al.*, 2004).

Y-chromosome haplotype analysis showed the distribution of Y1A haplotype worldwide. However, Y2A and Y1B are distributed in Europe, North Africa and the Near East. Similarly haplotype Y2B was found in Asia and haplotype Y2C in Turkey (Pereira *et al.*, 2009, Cinar Kul *et al.*, 2015)

2.10.6 Genomics in animal breeding

Genetic improvement in livestock mainly involves selection of the best males and females and production of superior progeny through mating of these improved animals. The performance of progeny is mainly dependent on the expression of different traits, and traits are mostly quantitative in nature. Quantitative traits are governed by multiple or many genes along with the environment. The traits include growth rate, feed efficiency, milk yield, protein yield, fat yield and meat quality. Breeding value for the traits of interest of the individuals is the major criteria to select the animal for the improvement programme. The breeding value of an individual is defined as the sum of additive effects of all the loci that contribute to the trait (quantitative trait loci), deviated from the population mean. Similarly, selection based on estimated breeding value from phenotype has been effective in different species for different traits. The major requirement is to record the phenotypic data on individuals or relatives during a specific time period. It is necessary to select an individual at an early age to reduce the generation interval. Again, traits such as milk production and meat quality can be measured late in life, and as a result these phenotyping constraints limit the amount of genetic progress. Animal geneticists have searched for some markers to measure the traits at a young age for increasing accuracy of selection. Initially, blood groups, blood protein polymorphism and physiological measurements were recorded to select individuals for particular traits. Blood group was used to select individuals for disease resistance in chickens (Hansen *et al.*, 1967). Serum insulin-like growth factor–I has been used for selection of growth parameters.

DNA markers have been used to identify genes or genomic regions that control traits of interest. The first application was carried out in different species to develop genetic tests for single-gene defects. Selection was carried out based on genetic markers associated with quantitative trait loci (QTL) and phenotypic traits. Selection was carried out based on candidate genes and its association with traits of interest. Again, QTL mapping studies were also carried out in different populations in cattle, sheep and pigs (Soller, 1978; Smith and Simpson, 1986; Lande and Thompson, 1990). High-density SNP genotyping was then used for whole-genome selection. The first high-density SNP genotyping platform was the 50 K Bovine illumina SNP Panel (Matukumalli *et al.*, 2009). Similar SNP panels are available for other livestock species including poultry, pigs, sheep and horses. The first high-throughput SNP chip with more than 50,000 SNPs (Illumina CaprineSNP50 Beadchip) is also available for goats (Tosser-Klopp et al., 2014). The high-density SNP genotyping has been used to implement genomic selection (Meuwissen *et al.*, 2001; Hayes *et al.*, 2009). Genomic selection is a process that uses a large panel of dense genome-wide markers to predict the breeding value (Meuwissen *et al.*, 2001). Genomic selection has been implemented in cattle in different countries. Milk production cannot be observed in bucks, meat tenderness cannot be measured in the live animal and fibre production cannot be observed at a young age. Similarly, feed conversion efficiency is not measured in a traditional selection method in a flock. To get more accuracy in genomic selection, the size of the reference population (i.e. the number of animals with phenotypes and genotypes from which the prediction regulation is estimated) is critical (Goddard, 2009). Genetic improvement during a short period of time, breed selection, mating plan and individual measurement will help the goat industry to cater for the future needs of market demand.

Demands for animal proteins such as milk and meat will increase in future to fulfil the requirements of the world's growing population (Delgado *et al.*, 1999; Alexandre *et al.*, 1999). Simultaneously, there will be increased competition for land, water and feed resources by which the input for livestock industry will increase. Again, climate change directly affects the production performance and increases the recurring cost. It is necessary to manage the animal during heat stress and onset of different tropical diseases. This will lead to increased input costs, therefore increasing the price of livestock products. Therefore, livestock production systems need to increase production system efficiency. At present, the goats are managed in extensive systems and farmers are able to get better return per unit expenditure; however, the gain in production systems needs to accelerate as the future input for maintaining the flock will be more competitive.

2.10.7 Genetics of adaptation

In the future, livestock production is going to face heat stress, water scarcity and coastal flooding due to climate change. Therefore, it is necessary to evaluate different genotypes in location-specific areas for national and global application in the future. Goats have evolved and adapted to many diverse environments and carry unique combinations of genes that define productive and adaptive capabilities. Goats have a wider genetic base as they are distributed over varied geographical and climatic conditions with differential production capability. Due to this wider genetic variability, goats can be manipulated to develop as suitable adaptable animals in relation to different ecotypes.

Heat stress affects animal bioenergetics, and has a negative impact on animal performance and well-being. Goats are distributed in different ecological conditions and are supposed to be more tolerant to extreme weather conditions because of their metabolic size and water conservation capacity (Silanikove, 2000). Short term variations in weather patterns affect the performance of goats and are exhibited in physiological and behavioural changes in them (Appleman and Delouche, 1958; Solanki, 1989; Kannan *et al.*, 2000; Hamzaoui *et al.*, 2013; Silanikove and Koluman, 2015). Goats are highly adaptable to heat stress due to specific anatomical and physiological characteristics. The morphological characteristics that result in better adaptation are large salivary glands, a higher surface area of absorptive mucosa and the capacity to increase the volume of the foregut with high fibrous foods (Silanikove and Koluman, 2015). The anatomical structure of goats favours better regulation of body temperature during periods of extreme heat. Short hair, abundant sweat glands and minimal subcutaneous fat allow goats to regulate body temperature in an efficient manner.

Exposure of goats to ambient temperatures above the upper critical limit results in heat stress (Yousef, 1985; Lu, 1989). The suggested upper limit of heat tolerance for goats was above 40°C (Appleman and Delouche, 1958; Solanki, 1989). The temperature–humidity index (THI) has been used to define the environmental effect in different livestock with respect to climatic conditions (McDowell *et al.*, 1976; Collier *et al.*, 2006). THI is the most prevalent index to evaluate heat stress as it combines the effect of heat and humidity. This index has been developed as a weather safety index to monitor and reduce heat-stress–related losses. THI values for Indian goats in semi-arid regions can be classified as 74 or less comfortable; 80–85 stressful; 86–88 more stressful; more than 88 extreme distress. A similar weather heat-stress risk class for dairy goats has been described by Silanikove and Koluman (2015).

Phenotyping to heat stress

Goats are more resilient and adjust to different higher environmental temperatures by expressing different adaptive strategies (Silanikove and Koluman, 2015). Goats generally utilize their thermoregulatory mechanism to relieve stress. Rectal temperature (RT), respiration rate (RR) and heart rate (HR) are commonly used to assess heat tolerance. Rectal temperature is considered a good index of thermoregulatory capacity (Yousef, 1985). Due to specific anatomical and physiological adaptation, goats can efficiently manage the heat in desert conditions. The goats did not exhibit wide variation during heat-stress periods; however, significant variation was observed in RR and HR.

The physiological response of goats such as RT, RR and HR has been evaluated in different environmental conditions. The range of variability in RR and HR was 24–108 and 80–160, respectively, during heat-stress periods in Jamunapari and Barbari goats. The range of variability in response to heat stress was more in RR and HR during a heat-stress period; therefore, HR and RR were considered to be good markers to define heat-stress susceptibility and tolerance. There was significant difference ($p < 0.05$) observed in RR and HR between heat-stress and thermoneutral (comfortable) periods. During thermoneutral conditions, the HR and RR were in the normal range, with only 3.19% of adult animals and 2.47% of kids being observed in higher ranges (>50 for RR and >130 for HR). In contrast, 21.57% of adult animals and 28.97% of kids had higher RR and HR during heat-stress periods. There is a direct shift of HR and RR (more than 25%) to the higher side due to heat stress in Jamunapari and Barbari goats. The histogram distribution of HR and RR during heat-stress and thermoneutral periods showed the presence of two contrasting classes in the population. Based on the distribution of RR and HR, it was also observed that RR and HR could be used as stress indicators for identifying contrasting phenotypes. Phenotyping for heat-stress susceptibility was proposed based on RR and HR. Based on the distribution of RR and HR over the breed in the population, it was observed that individuals having RR ≥50 and HR ≥130 are recognized as a heat-stress susceptible (HSS) phenotype and RR ≤30 and HR ≤100 are recognized as a heat-stress tolerant (HST) phenotype in adult goats. There is a slight deviation for kids. Kids with a RR ≥55 and HR ≥145 are recognized as HSS individuals and those with RR ≤30 and HR ≤100 are recognized as HST individuals.

The variation in rectal temperature between HST and HSS in Jamunapari goats was 0.709°C, while the variation in RR was 15.728 breath/min and HR was 18.173 beats/min. The rectal temperature variation was 0.361°C between HST and HSS phenotypes in Barbari goats, and the variation in RR was 24.051 breath/min and HR was 19.133 beats/min. The combination of RR and HR can be utilized to identify heat-tolerant animals. There is a direct shift of HR and RR to the higher side during heat-stress periods in the goats from semi-arid regions. It has been also reported that the change in environmental temperatures from 20°C to 40°C resulted in the respiration rate increasing from 30 to over 200 breaths/min in East African goats (Maloiy and Taylor, 1971) and domestic Swedish goats (Olsson *et al.*, 1995). Reference respiratory rates for adult goats range between 15 and 30 breaths/min according to Pugh and Baird (2012). Increased respiration rate under heat-stress conditions is a known mechanism for dissipating heat load by evaporation. Rectal temperature and respiration rate values peaked in heat-stress goats during the first week and then gradually decreased, which indicates a partial adaptation to the heat-stress conditions. The general homeostatic responses in ruminants include reduction in feed intake and increases in RR, HR and sweating rate. Similarly, the exposure of goats to heat-stress conditions depressed their milk yield (Lallo *et al.*, 2012). Therefore physical, physiological and biochemical processes play an important role in negating the effect of heat stress to maintain thermal balance.

2.11 Protocols

Protocol 1: Recording body measurement

Body height

Body height is measured as the vertical distance from the point of the withers to the base of the hoof (Fig. 2.55). The distance is measured in centimetres.

Body length

Body length is measured as the distance between the point of the shoulder to the point of the pin bone or tuberischii (Fig. 2.56). The distance is measured in centimetres.

Body/heart girth

This measurement is the circumference of the thoracic cavity immediately behind the forelimbs and parallel to the body axis (Fig. 2.57). The distance is measured in centimetres.

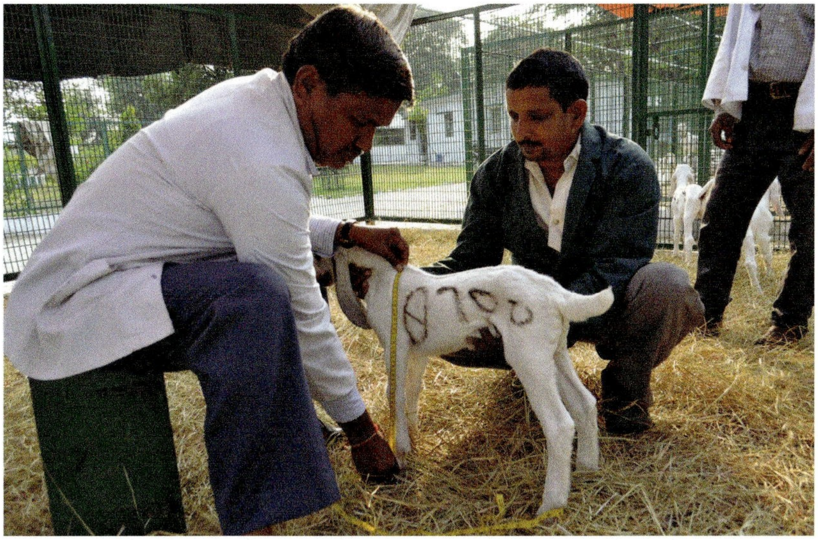

Fig. 2.55. Recording of body height.

Fig. 2.56. Recording of body length.

Fig. 2.57. Recording of body girth.

Protocol 2: Data recording sheets

a) Livestock recording format

Name of cluster: .... Name of village: ....
Name of livestock owner: .... House no.: ....
No. of livestock: .... Identification tag no.: ....

	Males	Females	Kids
No. of goats:	....	....	....
No. of goats sold:	....	....	....

Date: ....

Amount: ....

b) Growth recording format

Animal No.: .... Name of village: .... Sire No.: ....
Breed: .... Name of owner: .... Dam No.: ....
Date of birth: .... Wt. of dam at kidding: ....(kg)
Type of birth: .... Parity of dam: ....
Sex: ....

Age	Date	Weight (kg)	BL (cm)	BH (cm)	HG (cm)	Remarks
Birth						
1 Month						
2 Months						
3 Months						
6 Months						
9 Months						
12 Months						

BL, body length; BH, body height; HG, heart girth.

c) Milk recording format

Animal No.: .... Name of village: .... Date of kidding: ....
Breed: .... Type of birth: ....
Sire No.: .... Name of owner: .... Kid no.: ....
Dam No.: .... Sex of kids: ....
Parity: .... Date of drying: ....
Lactation length: ....

Week No.	Date	Milk yield (kg)			Weekly Yield (kg.)	Remarks
		Morning	Evening	Total		
1.						
2.						
3.						

d) Buck distribution format

Name of cluster: .... Identification/tag no.: ....
Name of village: .... Criteria: ....
Name of farmer: .... Mode of distribution: ....
Date of buck provided: ....

Total number sold during last season: ....

e) Mortality register

Animal No.: .... Name of village: .... Name of owner: ....
Date of birth: .... Date of death: .... Symptoms: ....
Cured/died: .... PM findings: .... Diagnosis: ....
System affected: .... Remarks: ....

Protocol 3: Correction factors

For calculating weaning weight or average daily gain, it is necessary to express the 90 days weight as adjusted values. Adjustments are made to calculate the 90-day body weight because litter size and age of doe can affect weaning weight. Weaning weight decreases as litter size increases, and weaning weight of kids of first parity are lighter when compared to third parity. Therefore, adjusted 90-day body weight can be calculated by multiplying correction factors with 90-day weight.

Table 2.14. Correction factors for 90-day weaning weight due to litter size born and weaned, dam age, and sex of kid.

Number born	Number weaned	Correction factor
Effect of litter size		
1	1	1.00
2	1	1.04
2	2	1.18
3	1	1.08
3	2	1.23
3	3	1.27
Effect of dam age, years		
1		1.10
2		1.09
3 and older		1.00
Sex of kid goat		
Buck		**1.00**
Doe		**1.11**
Wether		**1.08**

Source: Adapted from Browning *et al.* (2007).

Similarly, correction factors are used for adjusting lactation yield as milk yield is affected by litter size, parity, season and month of kidding. Adjusted lactation yield can therefore be calculated by multiplying

Table 2.15. Correction factor milk yield due to litter size, parity and month of kidding.

Parameter	Correction factor
Litter size	
1	1.00
2 or more	0.96
Parity	1.05
1	1.05
2	1.01
3 or more	1.00
Month of kidding	
September	1.00
October	1.00
November	1.04
December	1.08

Table 2.16. Example of adjusted 90-day yield calculations.

Goat A	Goat B
90-day milk yield: 92 kg	90-day milk yield: 99 kg
Litter size: single (1.00)	Litter size: twin (0.96)
Parity: first (1.05)	Parity: fourth (1.00)
Month of kidding: December (1.08)	Month of kidding: September (1.00)
Adjusted 90-day yield: 104.32 kg	Adjusted 90-day yield: 95.04 kg

appropriate correction factors with milk yield for 90 days or 140 days for lactation yield.

Protocol 4: Goat productivity indices

The production performance of goats reared in different production systems can be carried out after standardizing performances such as productivity indices (Peacock, 1987; Bosman *et al.*, 1997). These indices can also be used to compare the productivity efficiencies of different breeds. Moreover, the 3mw doe productivity is expressed in terms of metabolic body weight and the Klieber ratio to measure the efficiency of production indices.

Productivity analyses

The modified indices by Bosman *et al.* (1997) have limitations related to the number of parameters required. Age at first kidding and flock mean weight or metabolic weights are required parameters in the modified indices, but they are cumbersome to generate in field conditions. For these reasons, flock mean weight has been replaced by postpartum weight (ppw) in specific cases.

Individual level productivity has been assessed using productivity indices, including productivity expressed as the body weight of 90-day old kids produced per doe per year (Index I) and per kg ppw per year (Index II) (Peacock, 1987; Bosman *et al.*, 1997). In this case, live weight at 3 months (LW3M) is calculated as the sum of weights per parturition per doe at 3 months from the collected data for the respective breeds.

$$\text{Index I}: y = \frac{\text{LW3M}}{\text{KI}} \times 365$$

$$\text{Index II}: y = \frac{\text{Index I}}{\text{weight of dam at kidding}}$$

where y = live weight production per parturition in kg, standardized per annum; LW3M = litter weight at three months after birth; KI = kidding interval.

Analyses of overall flock productivity

Overall productivity was assessed and compared across the three indigenous goat breeds using the following index (Peacock, 1987; Bosman *et al.*, 1997):

$$y = \frac{N \times \frac{\text{LSB}}{1} \times \frac{\text{S3M}}{1} \times \frac{\text{L3MW}}{1}}{\text{weight of dam at kidding (postpartum weight)}}$$

where y = productivity in kg live weight per kg postpartum weight per year; N = number of parturitions per year; LSB = litter size at birth; S3M = survival rate at 3 months; L3mw = live weight at 3 months (3mw = 3mwc for AB); PPWm = mean postpartum weight of does.

In the calculation of overall flock productivity, measurements were lacking on bucks, fraction of females not actually reproducing and goats entering/exiting the flock. For this reason, PPWm was used instead of overall flock mean weight. Overall mean values of N, 3mw (3mwc for AB), LSB, S3M, KI and PPWm, were used in the calculation of indices. N was calculated based on overall mean KI values. When KI is less than 365 days, N is definitely more than one times and when the KI is more than 365, N is less than one times.

$$\text{Productivity index} = \frac{\text{conception rate} \times \text{litter size} \times \text{survivability to} \times 365}{\text{kidding interval}}$$

Protocol 5: Score card for judging

Doe score card

1. Identification of the goat:
   a. Breed
   b. Breeder's name
   c. Village
   d. Category of goat
2. General appearance: (35 points)
   a. Breed characteristics
   b. Head
   c. Shoulder blades
   d. Back
   e. Loin
   f. Rump
   g. Hips
   h. Pin bones
   i. Tail head
   j. Tail
   k. Forelegs
   l. Hindlegs
   m. Feet
   n. Horns
3. **Dairy character:** (20 points)
   a. Neck
   b. Withers
   c. Ribs
   d. Flanks
   e. Thighs
   f. Coat colour

4. **Body capacity:** (10 points)
    a. Barrel
    b. Heart girth
    c. Length
    d. Height
    e. Back curvature
5. **Mammary system:** (35 Points)
    a. Udder capacity and shape
    b. Texture
    c. Teats
    d. Mammary veins

Buck score card

1. **Identification of the goat:**
    a. Breed
    b. Breeder's name
    c. Village
    d. Category of goat
2. **General appearance:** (40 points)
    a. Breed characteristics
    b. Head
    c. Shoulder blades
    d. Back
    e. Rump
    f. Legs and Feet
    g. Horn characteristic
    h. Scrotum
    i. Coat colour
3. **General behaviour:** (15 points)
    a. General vigour/activeness/alertness
    b. Adjustment to new environment
4. **Dairy character:** (30 points)
    a. Neck
    b. Withers
    c. Ribs
    d. Flanks
    e. Thighs
    f. Coat colour
5. **Body capacity:** (15 points)
    a. Barrel
    b. Heart girth
    c. Length
    d. Height
    e. Back surface

Final score

Goat no.	Name of the breeder and address	Scores – sections 2 to 5					Rank
		2	3	4	5	Total	
1							
2							
3							

Protocol 6: Breed registration

Breed descriptor for registration (www.nbagr.res.in)

I. GENERAL DESCRIPTION

1. Name of the breed
2. Synonyms
3. Background for such a name / origin
4. Since when the breed is known
5. Strains (or within breed types)
6. Most closely related breeds (in appearance)
7. Classification
    a. Size (small/medium/large)
    b. Utility (mutton/milk/dual/fibre/hair or any other)
8. Distribution
    a. Native tract of distribution in terms of longitude and latitude
    b. Approximate area of distribution (sq. km)
    c. Place(s):
    State:
    District:
9. Estimated population
    a. Year of estimation
    b. Population
    c. Source/reference
10. Community
    a. Communities responsible for developing the breed
    b. Description of community (Farmers/nomads/isolated/tribal)

11. Flock
    a. Average size
    b. Composition
       Doe ... %   Bucks.... %   Kids ...... %
12. Any other information

II PHYSICAL CHARACTERS

1. Colour
   Distinctive colour markings, if any:
2. Head profile (straight/convex/slightly convex)
3. Ears (erect/pendulous/horizontal)
4. Wattles (present/absent)
5. Horns
   a. Number
   b. Shape
   c. Orientation
   d. Size (small < 15 cm/medium 15–25 cm/large >25 cm)
6. Coat Type (hair/cashmere/pashmina/mohair)
7. Beard (present/absent)
8. Any other information

III PERFORMANCE

Body weight (kg)

	Male			Female		
Weight at	Average	Range	*N*	Average	Range	*N*
Birth						
Weaning/3 months						
6 months						
1 year						
First kidding						
6 teeth (adult)						

Body measurements (cm)

	Male			Female		
Body measurement	Average	Range	*N*	Average	Range	*N*
Chest girth						
Body length						
Height at withers						

Carcass characters

	Male			Female		
	Average	Range	*N*	Average	Range	*N*
Age at slaughter (days)						
Weight at slaughter (kg)						
Dressing % (hot)						
Dressing % (cold)						

Dairy performance

	Average	Range	*N*
Daily milk yield (g)			
Total lactation milk yield (kg)			
Lactation length (days)			
Fat %			
SNF %			

Reproduction

Reproduction	Average	Range	*N*
Age at first mating in males (days)			
Age at first mating in females (days)			
Age at first oestrus (days)			
Oestrus cycle duration(days)			
Age at first kidding (days)			
Kidding interval (days)			
Service period (days)			
Litter size			
Lifetime number of kiddings			

Fibre characteristics

Trait	Average	Range	*N*
Greasy fleece weight (kg)			
Clean fleece weight (kg)			
Staple length			
Fibre diameter			

1. Age at shearing/combing/collection/clipping (months)
2. Type of fibre [mohair (true/heterotypes/kemps)/Cashmere/Pashmina/Hair]
3. Fleece colour

Any other information specific to the breed

Details added according to the breed being registered.

2.12 References

Abdelsalam, M.M., Haider, A.E., Aboul–Naga, A.M., Elkimary, I.S. and Eissa, M. (1994) Improving performance of desert barki kids by crossing with Zaraibi and Damascus goats. *Egyptian Journal of Animal Production* 31, 85–97.

Acharaya, R.M. (1982) Sheep and goat breeds of India. FAO Animal Production and Health Paper 30. Food and Agriculture Organization of the United Nations, Rome, Italy, pp. 190.

Acharya, R.M. (1992) Goat genetic resources and their management. *Research in Goats-Indian Experiences*, pp. 28–38.

Agazzi, A. Cigalino, G., Mancin, G., Savoini, G. and Dell'Orto, V. (2007) Effects of dietary humates on growth and an aspect of cell-mediated immune response in newborn kids. *Small Ruminant Research* 72, 242–245.

Alade, N.K., Dilala, M.A. and Abdulyekeen, A.O. (2010) Phenotypic and genetic paramotor estimates of litter size and body weights in goats. *International Journal of Science and Nature* 1(2), 262–266.

Alexandre, G., Aumont, G., Mandonnet, N. and Naves, M. (1999) The Creole goat of Guadeloupe (French West Indies): an important genetic resource for the humid tropics. *Animal Genetic Resources Information* 26, 45–55.

Ali, A. and Khan, M.S. (2008) Genetic trends in growth and reproductive traits of Beetle goats. *Indian Journal of Animal Breeding and Genetics* 27, 15–25.

All-India Coordinated Research Project on Goats (AICRP) (1985) *Terminal Report.* Central Institute for Research on Goats, Makhdoom, Mathura, India.

Appleman, R.D. and Delouche, J.C. (1958) Behavioral, physiological and biochemical responses of goats to temperature, 0 to 40°C. *Journal of Animal Science* 17, 326–335.

Aschaffenburg, R. and Drewry, J. (1955) Occurrence of different β-lactoglobulin in cow's milk. *Nature* 176, 218–221.

Barbieri, E., Manfredi, E., Elsen, J.M., Ricordeau, G., Bouillon, J. *et al.* (1995) Effects of the α_{S1}-casein locus on dairy performances and genetic parameters of Alpine goats. *Genetic Selection Evolution* 27, 437–450.

Barillet, F. (2007) Genetic improvement for dairy production in sheep and goats. *Small Ruminant Research* 70, 60–75.

Barroso, F.G., Alados, C.L. and Boza, J. (2000). Social hierarchy in the domestic goat: effect on food habits and production. *Applied Animal Behaviour Science* 69, 35–53.

Baumung, R., Simianer, H. and Hoffmann, I. (2004) Genetic diversity studies in farm animals- A survey. *Journal of Animal Breeding and Genetics* 121, 361–373.

Beja-Pereira, A., Luikart, G., England, P.R., Bradely, D.G., Jann, C.O. *et al.* (2003) Gene culture co-evolution between cattle and human lactase genes. *Nature Genetics* 35, 311–313.

Belichon, S., Manfredi, E. and Piacere, A. (1999) Genetic parameters of dairy traits in the Alpine and Saanen goat breeds. *Genetic Selection Evolution* 31, 529–534.

Berry, D.P., Cromie, A.R., Mc.Hugh, N., Burke, M., Pabiou, T. and MacCarthy, J. (2012) New traits for dairy cattle breeding. In: Proceedings of the 38th International Committee for Animal Recording (ICAR) Meeting, Cork, Ireland.

Bevilacqua, C., Ferranti, P., Garro, G., Veltri, R., Lagonigro, C. *et al.* (2002) Interallelic recombination is probably responsible for the occurrence of a new α s1-casein variant found in the goat species. *European Journal of Biochemistry* 269, 1293–1303.

Bickhart, D.M., Rosen, B.D., Koren, S., Sayre, B.L., Hastie, A.R. *et al.* (2017) Single-molecule sequencing and conformational capture enable de novo mammalian reference genomes. *Nature Genetics* 49, 643–650. DOI:10.1038/ng.3802.

Bishop, S.C. and Russel, A.J.F. (1996) The inheritance of fibre traits in a crossbred population of Cashmere goats. *Animal Science Journal* 63(3), 429–436.

Boichard, D., Bouloc, N., Ricordeau, G., Piacere, A. and Barillet, F. (1989) Genetic parameters for first lactation dairy traits in the Alpine and Saanen goat breeds. *Genetics Selection Evolution* 21, 205–215.

Bordi, A., De Rosa, G., Napolitano, F., Litterio, M., Marino, V. *et al.* (1994) Postpartum development of the mother–young relationship in goats. *Applied Animal Behaviour Science* 42, 145–152.

Bosman, H.G., Moll, H.A.J. and Udo, H.M.J. (1997) Measuring and interpreting the benefits of goat keeping in tropical farm systems. *Agricultural Systems* 53(4), 349–372.

Boyazoglu, J., Hatziminaoglu, I. and Morand-fehr, P. (2005) The role of the goat in society: past, present and perspectives for the future. *Small Ruminant Research* 60, 13–24.

Bradley, D.G., MacHugh, D.E., Cunningham, P. and Loftus, R.T. (1996) Mitochondrial diversity and the origin of the African and European cattle. *Proceedings of the National Academy of Science of the United States of America* 93, 5131–5135.

Brito, L.F., Silva, F.G., Melo, A.L.P., Caetano, G.C., Torres, R.A. *et al.* (2011) Genetic and environmental factors that influence production and quality of milk of Alpine and Saanen goats. *Genetics and Molecular Research* 10(4), 3794–3802.

Brown, G.G., Gadaleta, G., Pepe, G. and Saccone, C. (1986) Structural conservation and variation in the D loop containing region of vertebrate mitochondrial DNA. *Journal of Molecular Biology* 192, 503–511.

Browning, R., Donnely, Jr. B., Payton, T. and Byars, M. (2007) Preweaning body weights of meat goat kids produced in a three-breed diallel managed on south-eastern pastures. *Journal of Animal Science* 85, 40–41.

Campbell, Q.P. (2003) The origin and description of southern Africa's indigenous goats. *South African Society of Animal Science, Popular Scientific Papers* 4, 18–22.

Caroli, A., Chiatti, F., Chessa, S., Rignanese, E.M., Bolla, P. *et al.* (2006) Focusing on the Goat Casein Complex. *Journal of Dairy Science* 89, 3178–3187.

Casey, N.H. and van Niekerk, W.A. (1988) The Boer Goat Origin, adaptability, performance testing, reproduction and milk production. *Small Ruminant Research* 1, 291–302.

Chaudhary, U.B., Tripathi, M.K., Gupta, B., Dutta, T. and Sirohi, H.V. (2013) Effect of inorganic and organic zinc and copper supplementation on performance, nutriont utilization, rumen fermentation and blood biochemistry of kids. *The Indian Journal of Animal Sciences* 83(12), 1313–1322.

Chaudhary, U.B., Das, A.K., Tripathi, P. and Tripathi, M.K. (2015) Effect of concentrate supplementation on growth performance, carcass traits and meet composition of Sirohi kids under field condition. *Animal Nutrition and Feed Technology* 15, 251–260.

Cheng, P. (1984) *Livestock Breeds of China*. Animal Production and Health Paper no. 46. Food and Agriculture Organization of the United Nations, Rome, Italy.

Chessa, S.E., Budelli, F., Chiatti, A.M., Cito, P., Bolla, P. and Caroli, A. (2005) Predominance of β-casein (CSN2) C allele in goat breeds reared in Italy. *Journal of Dairy Science* 88, 1878–1881.

Cinar Kul, B., Bilgen, N., Lenstra, J.A., Korkmaz Agaoglu, O., Akyuz, B. and Ertugrul, O. (2015) Y-chromosomal variation of local goat breeds of Turkey close to the domestication centre. *Journal of Animal Breeding and Genetics* 132(6), 449–453.

Clément, V., Boichard, D., Piacere, A., Barbat, A. and Manfredi, E. (2002) Genetic evaluation of French goats for dairy and type traits. *Proceedings of the 7th World Congress on Genetics Applied to Livestock Production*, Montpellier, France, 19–23 August 2002, INRA 01-46.

Cockett, N.E., Jackson, S.P., Shay, T.L., Nielsen, D., Moore, S.S. *et al.* (1994) Chromosomal localization of the callipyge gene in sheep (*Ovis aries*) using bovine DNA markers. *Proceedings of the National Academy of Sciences USA* 91, 3019–3023.

Collier, R.J., Stiening, C.M., Pollard, B.C., Van Baale, M.J., Baumgard, L.H. *et al.* (2006) Use of gene expression microarrays for evaluating environmental stress tolerance at the cellular level in cattle. *Journal of Animal Science* 84, 1–13.

Constantineuscu, G.M. and Constantineuscu, I.A. (2010) Functional anatomy of goat. In: Solaiman, S.G. (ed.) *Goat Science and Production*, John Wiley & Sons, Inc., New York, pp. 89–138.

Darcan, N. and Cankaya, S. (2008) The effects of ventilation and showering on fattening performances and carcass traits of crossbred kids. *Small Ruminant Research* 75, 192–198.

Delgadillo, J.A., Poindron, P., Krehbiel, D., Duarte, G. and Rosales, E. (1997) Nursing, suckling and postpartum anoestrus of creole goats kidding in January in subtropical Mexico. *Applied Animal Behaviour Science* 55, 91–101.

Delgado, C., Rosegrant, M., Steinfeld, H., Ehui, S. and Courbois, C. (1999) Livestock to 2020 – The next food revolution. Food, Agriculture, and the Environment Discussion Paper. no. 28. IFPRI (International Food Policy Research Institute), Washington, DC, USA; FAO (Food and Agriculture Organization of the United Nations), Rome (Italy); ILRI, Nairobi (Kenya).

Devendra, C. (2007) *Goats: Biology, Production and Development in Asia*. Academy of Sciences Malaysia, Kuala Lampur, Malaysia.

Devendra, C. (2010) Concluding synthesis and the future for sustainable goat production. *Small Ruminant Research* 89, 125–130.

Devendra, C. (2012). Multifunctional relevance and contribution to food and nutrition security. In: Raseedee, A., Ariff Omar, A., Ali Rajion, M.A., Alimon, A.R., Liang, J.B. and Haw, A.K. (eds), *Proceedings of the First Asia Dairy Goat Conference*, Kuala Lumpur, Malaysia, pp. 1–6.

Devendra, C. and Burns, M. (1983) *Goat Production in the Tropics*. CABI, Wallingford, UK.

Devendra, C. and Haenlein, G.F.W. (2003) Goat breeds. In: Roginski, H., Fuquay, J.W. and Fox, P.F. (eds), *Encyclopedia of Dairy Sciences*, Academic Press, Amsterdam, The Netherlands, Vol. 2, pp. 585–598.

Devendra, C. and Liang, J.B. (2012) Conference summary of dairy goats in Asia: Current status, multifunctional contribution to food security and potential improvements. *Small Ruminant Research* 108(1–3), 1–11.

Devendra, C. and Nozawa, K. (1976) Goats in south East Asia–their status and production. *Z. Tierz, Zuchtbiol* 93, 101–120.

Dong, Y., Xie, M., Jiang, Y., Xiao, N., Du, X. *et al.* (2013) Sequencing and automated whole-genome optical mapping of the genome of a domestic goat (Capra hircus). *Nature Biotechnology* 31, 135–141.

Ebert, R.A. and Solaiman, S.G. (2010) Animal evaluation. In: Solaiman, S.G. (ed.) *Goat Science and Production*, John Wiley & Sons, Inc., New York, pp. 77–88.

Ekambaram, B., Gupta, R.B., Gnana, P.M., Sudhaker, K. and Reddy, V.R. (2010) A Study on the performance of Mahabubnagar goats. *Indian Journal of Animal Research* 44(1), 48–51.

Epstein, H. (1974) Vanishing livestock breeds in Africa and Asia. In: Proceedings of the 1st World Congress on Genetics Applied to Livestock Production, Madrid, Spain, Editorial Garsi, Madrid II, pp. 31–35.

Epstein, H. (1977). *Domestic Animals of Nepal*. Holmes and Meier Publishers, Inc., New York.

FAO (1989) FAO programmes for the Preservation of Animal Genetic Resources - A Report of the Expert Consultation, September 1989. FAO, Rome, Italy.

FAO (2004) Production Yearbook 2003. Food and Agricultural Organization of the United Nations, Rome, 57, 213–215, 241–242.

Faruque, S., Chowdhury, S.A., Siddiquee, N.U. and Afroz, M.A. (2010) Performance and genetic parameters of economically important traits of Black Bengal goat. *Journal of Bangladesh Agriculture University* 8(1), 67–78.

Feligini, M., Silvia, F., Cubric-Curik, V., Brambilla, A., Parma, P. *et al.* (2005) Caprine as1-casein polymorphism: characterization of A, B, E and F variants by means of various biochemical and molecular techniques. *Food Technology and Biotechnology* 43(2), 123–132.

Fernandez, M.A., Alvarez, L. and Zarco, L. (2007). Regrouping in lactating goats increases aggression and decreases milk production. *Small Ruminant Research* 70, 228–232.

Ferretti, L., Leon, P. and Sgaramell, V. (1990) Long range restriction analysis of the bovine casein genes. *Nucleic Acid Research* 18, 6829–6833.

Galal, S. (2005) Biodiversity in goats. *Small Ruminant Research* 60, 75–81.

Gall, C. (1981) Milk production. In: Gall, C. (ed.) *Goat Production*. Academic Press, London, U.K., pp. 309–344.

Gerstmayr, S. (1988) Estimating systematic effects and variance components in Turkish Angora goats for body weight and fleece weight. *Animal Breeding Abstracts* 56, 2773.

Gipson, T. (1996) Breeds and breeding plans. In: *Meat Goat Production Handbook*. Sponsored by Rural Economic Development Centre, Raleigh, NC, and Mid-Carolina Council of Governments, Fayetteville, North Carolina, USA.

Goddard, M.E. (2009) Genomic selection: Prediction of accuracy and maximisation of long term response. *Genetica* 136(2), 245–257.

Gowane, G.R., Chopra, A., Prakash, V. and Arora, A.L. (2011) Estimates of (co)variance components and genetic parameters for growth traits in Sirohi goat. *Tropical Animal Health Production* 43(1), 189–198.

Groot, B., De Narayan and Somi, R.A. (1992) Performance of Sirohi goats under village conditions in Rajasthan, India. V International Conference on Goats, 2–8 March, 1992, New Delhi, pp. 534–544.

Grosclaude, F. and Martin, P. (1997) Casein polymorphism in the goat. In: Proceedings of the IDF seminar held in Palmerston North, New Zealand, session IV, pp. 241–253.

Grosclaude, F., Joudrier, P. and Mahe, M.F. (1978) Polymorphisme de la casein a-S2 bovine: etroite liaison du locus a-S2-Cn avec les loci a-S1-Cn, b-Cnet j-Cn; mise en evidence d'une deletion dans le variant a-S2-CnD. *Annual Genetic Selection Animal* 10, 313–317.

Grosclaude, F., Mahe, M.F., Brignon, G., Di-Stasio, L. and Jeunet, R.A. (1987) Mendelian polymorphism underlying quantitative variations of goat aS1-casein. *Genetics Selection Evolution* 19, 399–412.

Grosclaude, F., Ricordeau, G., Martin, P., Remuef, F., Vassal, L. *et al*. (1994) Du gene au fromagee: le polymorphismie de la caseineα-S1 caprine, seseffets, son evolution. *INRA Productions Animales* 7, 3–19.

Haenlein, G.F.W. and Caccese, R. (1984) Goat milk versus cow milk. In: Haenlein, G.F.W. and Ace, D.L. (eds) *Extension Goat Handbook*. USDA Publication, Washington, DC, p. I, E-1.

Haenlein, G.F.W. (2004) Goat milk in human nutrition. *Small Ruminant Research* 51, 155–163.

Hammond, K. and Leitch, H.W. (1995) The FAO global programme for management of farm animal genetic resources. *World Animal Reviews*, 84/85.

Hamzaoui, S., Salama, A.A.K., Albanell, E., Such, X. and Caja, G. (2013) Physiological responses and lactational performances of late-lactation dairy goats under heat stress conditions. *Journal of Dairy Science* 96, 6355–6365.

Hansen, M.P., van Zandt, J.N. and Law, G.R.J. (1967) Differences in susceptibility to Marek's disease in chickens carrying two different B-blood group alleles. *Poultry Science* 46(Suppl.1), 1268.

Hatziminaoglou, Y. and Boyazoglu, J. (2004) The goat in ancient civilisations: from the Fertile Crescent to the Aegean Sea. *Small Ruminant Research* 51, 123–129.

Hayes, B.J., Bowman, P., Chamberlain, A.C., Verbyla, K. and Goddard, M.E. (2009) Accuracy of genomic breeding values in multi-breed dairy cattle populations. *Genetics Selection Evolution* 2009 41(1), 51. DOI: 10.1186/1297-9686-41-51.

Hayes, H., Petit, E., Bouniol, C. and Popescu, P. (1993) Localization and as2–343 casein gene (CASAS2) to the homologous cattle, sheep and goat chromosomes 4 by in-situ hybridization. *Cytogenetics and Cell Genetics* 64, 281–285.

Hermiz, H.N., Alkass, J.E., Hobi, A.A. and Asofi, M.K. (2009) Genetic and phenotypic parameters of body weights in Iraqi local goat and their crosses with Damascus. *Journal of Duhok University* 12, 189–194.

Hunter, L. and Hunter, E.L. (2001) In: Franck, R.R. (ed.), *Silk, Mohair, Cashmere and Other Luxury Fibres*. Woodhead Publishing Ltd., Cambridge, UK, pp. 68–132.

Ikonen, T., Bovenhuis, E.J., Ojala, E.J., Ruottinen, M. and Geogest, O. (2001) Associations between casein haplotypes and first lactation milk production traits in Finnish Ayrshire cows. *Journal of Dairy Science* 84, 507–514.

ISAG (2010) Applied Genetics in Sheep and Goats Workshop. In: Proceedings of the 32nd international conference on animal genetics, Edinburgh, UK, 26–30 July 2010. Available at: https://www.isag.us/Docs/Applied_GeneticsSheepGoats_CT.pdf (accessed 29 September 2017).

Jansa Perez, M., Leroux, C., Sanchez, A.B. and Martin, P. (1994) Occurrence of a LINE element in the 3' UTR of an allelic form of the goat as1-casein gene associated with a reduced level of protein synthesis. *Gene* 147, 179–187.

Jeffreys, J., Wilson, V. and Thein, S.L. (1985) Individual specific fingerprints of human DNA. *Nature* 316, 75.

Jembere, T., Dessie, T., Rischkowsky, B., Kebede, K., Mwai, A.O. *et al*. (2017) Meta-analysis of average estimates of genetic parameters for growth, reproduction and milk production traits in goats. *Small Ruminant Research* 153, 71–80.

Jin, H. and Zhang, B. (1995) Study of the skin follicle structure and selection methods in Cashmere goats. *Acta Agriculture University Pekin* 21, 94–99.

Joshi, M.B., Rout, P.K., Mandal, A., Tyler-Smith, C., Singh, L. *et al.* (2004) Phylogeography and origin of Indian domestic goats. *Molecular Biology Evolution* 21(3), 454– 462.

Kala, S.N. and Prakash, B. (1990) Genetic and phenotypic parameters of milk yield and milk composition in two Indian goat breeds. *Small Ruminant Research* 3, 475–484.

Kannan, G., Terrill, T.H., Kouakou, B., Gazal, O.S., Gelaye, S. *et al.* (2000) Transportation of goats: effects on physiological stress responses and live weight loss. *Journal of Animal Science* 78, 1450–1457.

Kannan, G., Terrill, T.H., Kouakou, B., Gelaye, S. and Amoah, E.A. (2002) Simulated preslaughter holding and isolation effects on stress responses and live weight shrinkage in meat goats. *Journal of Animal Science* 80, 1771–1780.

Khan, M.S., Khan, M.A. and Mahmood, S. (2008) Genetic resources and diversity in Pakistani goats. *International Journal of Agriculture Biology* 10, 227–231.

Kumar, A., Rout, P.K., Mandal, A. and Roy, R. (2007) Identification of CSN1S1 allele in Indian goats by PCR-RFLP method. *Animal* 1(8), 1099–1104.

Kumar, R., Tripathi, P., Chaudhary, U.B. and Tripathi, M.K. (2015) Development of Complete Feed for Economically and Environmentally Sustainable Goat Production. Annual Report CIRG 2015–16, pp. 44–45.

Lallo, C.H.O., Paul, I. and Bourne, G. (2012) Thermoregulation and performance of British Anglo-Nubian and Saanen goats reared in an intensive system in Trinidad. *Tropical Animal Health Production* 44, 491–496.

Lanari, M.R., Taddeo, H., Domingo, E., Perez Centeno, M. and Gallo, L. (2003) Phenotypic differentiation of exterior traits in local Criollo goat population in Patagonia (Argentina). *Archives Animal Breeding* 46(4), 347–356.

Lande, R. and Thompson, R. (1990) Efficiency of marker-assisted selection in the improvement of quantitative traits. *Genetics* 124, 743–756.

Lang, J.W., Aggarwal, R.K., Majumdar, K.C. and Singh, L. (1993) Individualization and estimation of relatedness in crocodilians by DNA fingerprinting with a Bkm-derived probe. *Molecular and General Genetics* 238, 49–58.

Legge, T. (1996) The beginning of caprine domestication in Southwest Asia. In: Harris, D.R. (ed.) *The Origins and Spread of Agriculture and Pastoralism in Eurasia*. Smithsonian Institute, Washington DC, pp. 236–262.

Leroux, C., Mazure, N. and Martin, P. (1992) Mutation away from splice site recognition sequences light cis-modulate alternative splicing of goat aS1-casein transcripts. Structural organization of the relevant gene. *The Journal of Biological Chemistry* 267, 6147–6157.

Loftus, R.T., Machugh, D.E., Bradley, D.G., Sharp, P.M. and Cunningham, P. (1994) Evidence for two independent domestications of cattle. *Proceedings of National Academy of Science USA* 91, 2757–2761.

Lu, C.D. (1989) Effects of heat stress on goat production. *Small Ruminant Research* 2, 151–162.

Luikart, G., Gielly, L., Excoffier, L., Vigne, J.D., Bouvet, J. *et al.* (2001) Multiple maternal origins and weak phylogeographic structure in domestic goats. *Proceedings of National Academy of Science, USA* 98, 5927–5932.

Madeli, U.C. and Patro, B.N. (1984) Heritability and correlation among body weights at different ages in Ganjam goats. *Indian Veterinary Journal* 61(3), 233–235.

Mahgoub, O., Lu1, C.D., Hameed, M.S., Richie, A., Al-Halhali, A.S. *et al.* (2005) Performance of Omani goats fed diets containing various metabolizable energy densities. *Small Ruminant Research* 58, 175–180.

Maijala, K., Cherekaev, A.V., Devillard J.M., Reklewski, Z., Rognoni, G. *et al.* (1984) Conservation of Animal genetic resources in Europe. Final Report of an EAAP working party. *Livestock Production Science* 11, 3–22.

Maloiy, G.M.O. and Taylor, C.R. (1971) Water requirements of African goats and haired-sheep. *Journal of Agricultural Science* 77, 203–208.

Mannen, H., Nagata, Y. and Tsuji, S. (2001) Mitochondrial DNA reveal that domestic goat (Capra hircus) are genetically affected by two subspecies of bezoar (Capra aegagrus). *Biochemical Genetics* 39(5–6), 145–154.

Martin, P. (1993) Polymorphisme des lactoproteines-caprines. (Polymorphism of goat milk proteins). *Le Lait* 73, 511–532.

Martin, P. and Leroux, C. (1994) Characterization of a further alpha-s1-casein variant generated by exon skipping. In: *XXIV International Conference of Animal Genetics*, July 23–29, ISAG, Prague, E., 43, 88.

Mason, I.L. (1980) Sheep and goat production in the drought polygon of northeast Brazil. *World Animal Review* 34, 23–28.

Mason, I.L. (1981) Breeds. In: Gall, C. (ed.) *Goat Production*, Chapter 3. Academic Press, London, UK.

Mason, I.L. (ed.) (1984) *Evolution of Domesticated Animals*. Longman, London, UK, 452 pp.

Mason, I.L. (1996) *A World Dictionary of Livestock Breeds, Types and Varieties*, 4th edn. CABI, Wallingford, UK.

Matukumalli, L.K., Lawley, C.T., Schnabel, R.D., Taylor, J.F., Allan, M.F. *et al.* (2009) Development and characterization of a high density SNP genotyping assay for cattle. *PLoS One* 4(4), e5350. DOI: 10.1371/journal.pone.0005350.

Mavrogenis, A.P., Papachristoforou, C., Lysandrides, P. and Roushias, A. (1989) Environmental and genetic effects of udder characteristics and milk production in Damascus goats. *Small Ruminant Research* 2, 333–343.

McDowell, R.E., Hooven, N.W. and Camoens, J.K. (1976) Effects of climate on performance of Holsteins in first lactation. *Journal of Dairy Science* 59, 965–973.

Meadow, A. (1996) Animal domestication in the Middle East: a revised view from the eastern margin. In: Possehl, G. (ed.) *Harappan Civilisation*. Oxford and IBH, New Delhi, pp. 295–320.

Metz, J. and Wierenga, H. (1997) Behavioural criteria for the design of housing systems for cattle. *Cattle Housing Systems, Lameness and Behaviour*, Martinus Nijhoff Publishers, Boston, MA, USA.

Meuwissen, T.H., Hayes, B.J. and Goddard, M.E. (2001) Prediction of total genetic value using genome-wide dense marker maps. *Genetics* 157, 1819–1829.

Miranda-de la Lama, G.C. (2005) Social strategy and the effect of environmental enrichment on the reactivity of handling and adrenocortical activity in dairy goats (Capra hircus). M.Sc. Thesis. Universidad Nacional Autónoma de México, México.

Mohammadi, H., Moradi, S.M. and Moradi, S.H. (2012) Genetic parameter estimates for growth traits and prolificacy in Raeini Cashmere goats. *Tropical Animal Health and Production* 44, 1213–1220. DOI: 10.1007/s11250-011-0059-z.

Moioli, B., Pilla, F. and Tripaldi, C. (1998) Detection of milk protein genetic polymorphism in order to improve dairy traits in sheep and goats: A review. *Small Ruminant Research* 27, 185–195.

Montaldo, H.H. and Manfredi, E. (2002) Organisation of selection programmes for dairy goats. 7th World Congress on Genetics Applied to Livestock Production, August 19–23, 2002, Montpellier, France. Communication no. 01–35, p. 9.

Montaldo, H.H., Torres-Hernández, G. and Valencia-Posadas, M. (2010) Goat breeding research in Mexico. *Small Ruminant Research* 89, 155–163.

Mukundan, G. and Bhat, P.N. (1978) Genetic parameters of production traits in Malabari goats and their Saanen and Alpine. 14th International Congress of Genetics Moscow, 21–30 August, Abstract Part I.

Muller, C.J.C., Cloet, S.W.P. and Schoeman. S.J. (2002) Estimation of genetic parameters for milk yield and milk composition of South African Saanen goats. Commun. No. 01–52 in Proceedings of 7th World Congress on Genetics Applied to Livestock Production, August 19–23, 2002, Montpellier, France. Institute National de la Recherche Agronomique (INRA).

National Academy of Sciences (1992) A Global Strategy for the Conservation of Animal Genetic Resources. National Academy of Sciences, Washington DC, USA.

National Pygmy Goat Association (1932) 149th Ave SE, Snohomish, WA, USA. Available at: http://www.npga-pygmy.com (accessed 22 May 2019).

Neopane, S. P (2000) Characterization of local breed of goat in Central Terai of Nepal. In: *Proceedings of Fourth National workshop on livestock and fisheries research in Nepal*, April 24–26, National Animal Science Research Institute (NASRI)/ NARC, Khumaltar, Nepal, pp. 214–217.

Office of Technology Assessment (1987) Technologies to Maintain Biological Diversity. Office of Technology Assessment, United States Congress, Washington DC, USA.

Oklahoma State University (OSU) (2004) Breeds of Livestock. Available at: http://www.ansi.okstate.edu/breeds/goats/ (accessed 22 May 2019).

Olsson, K., Josater-Hermelin, M., Hossaini-Hilali, J., Hydbring, E. and Dahlborn, K. (1995) Heat stress causes excessive drinking in fed and food deprived pregnant goats. *Comparative Biochemistry and Physiology* 110, 309–317.

Park, Y.W. and Haenlein, G.F.W. (2007) Goat milk, its products and nutrition. In: Hui, Y.H. (ed.) *Handbook of Food Products Manufacturing*, John Wiley & Sons, Inc., New York, USA, pp. 447–486.

Pattie, W.A. and Restall, B.J. (1992) Genetical and environmental factors affecting cashmere production. *Proceeding 5th International conference on Goats, New Delhi* 2(2), 502–512.

Peacock, C.P. (1987) Measures for assessing the productivity of sheep and goats. *Agricultural Systems* 23(3), 197–210.

Pereira, F., Queiros, S., Gusmao, L., Nijman, I.J., Cuppen, E *et al.* (2009) Tracing the history of goat pastoralism: new clues from mitochondrial and Y chromosome DNA in North Africa. *Molecular Biology and Evolution* 26(12), 2765–2773.

Phengvichith, V. and Ledin, I. (2007) Effect of feeding different levels of wilted cassava foliage (Manihot esculenta, Crantz) on the performance of growing goats. *Small Ruminant Research* 71(1–3), 109–116.

Pidancier, N., Jordan, S., Luikart, G. and Taberlet, P. (2006) Evolutionary history of the genus Capra (Mammalia, Artiodactyla): discordance between mitochondrial DNA and Y-chromosome phylogenies. *Molecular Phylogenetics and Evolution* 40(3), 739–749.

Poindron, P., Lévy, F. and Keller, M. (2007a) Maternal responsiveness and maternal selectivity in domestic sheep and goats: the two facets of maternal attachment. *Developmental Psychobiology* 49, 54–70.

Poindron, P., Terrazas, A., Navarro Montes de Oca, M.L., Serafín, N. and Hernández, H. (2007b) Sensory and physiological determinants of maternal behavior in the goat (*Capra hircus*). *Hormones Behavior* 52, 99–105.

Popescu, P., Long, S., Riggs, P., Womack, J., Schmutz, S *et al.* (1996) Standardization of cattle karyotype nomenclature: report of committee for standardization of the cattle karyotype. *Cytogenetics and Cell Genetics* 74, 259–261.

Porter, V. (1996) *Goats of the World*. Farming Press, Ipswich, UK.

Pringle, H. (1998) Neolithic agriculture: reading the signs of ancient animal domestication. *Science* 282, 1448.

Prinzenberg, E.M., Gutscher, K., Chessa, S., Caroli, A. and Erhardt, G. (2005) Caprine J-casein (CSN3) polymorphism: New developments of the molecular knowledge. *Journal of Dairy Science* 88, 1490–1498.

Pugh, D.G. and Baird, N. (2012) *Sheep and Goat Medicine*, 2nd edn. Elsevier Saunders, Maryland Heights, Missouri, USA.

Quartermain, A.R. (1991) Evaluation and utilization of goat breeds. In: Maijala, K. (ed.) *World Animal Science*. B8. Disciplinary. Approach. Elsevier, Amsterdam, The Netherlands, pp. 451–469.

Rabasco, A., Serradilla, J.M., Padilla, J.A. and Serrano, A. (1993). Genetic and non-genetic sources of variation in yield and composition of milk in Verata goats. *Small Ruminant Research* 11, 151–161.

Rammunno, L., Cosenza, G., Palppalardo, M., Pastore, N., Ramunno, D.L. *et al.* (2000) Identification on goat of the CSN1S1F allele by means of PCR-RFLP method. *Animal Genetics* 31, 333.

Rammuno, L., Cosenza, G., Rando, A., Pauciullo, A., Illario, R. *et al.* (2005) Comparative analysis of gene sequence of goat CSN1S1 F and N alleles and characterization of CSN1S1 transcript variants in mammary gland. *Gene* 345, 289–299.

Rashidi, A., Kashan, N.E., Miraei-Ashtiani, S.R., Rahimi, S. and Torshizy, R.W. (2000) Variance-covariance components and genetic parameters estimates for body weights in Markhoz goats. *Iranian Journal of Agricultural Science* 31, 455–462.

Rashidi, A., Sheikhahmadi, M., Rostamzadeh, J. and Shrestha, J.N.B. (2008) Genetic and Phenotypic parameter estimates of body weight at different ages and yearling fleece weight in Markhoz goats. *Asian-Australian Journal of Animal Science* 21(10), 1395–1403.

Roussel, S., Boissy, A., Montigny, D., Hemsworth, P.H. and Duvaux-Ponter, C. (2005) Gender-specific effects of prenatal stress on emotional reactivity and stress physiology of goat kids. *Hormones and Behavior* 47, 256–266.

Rout, P.K., Saxena, V.K., Khan, B.U., Roy, R., Mandal, A. *et al.* (2000) Characterization of Jamunapari goats in their home tract. *Animal Genetics Resource Information* 27, 43–52.

Rout, P.K., Mandal, A., Singh, L.B. and Roy, R. (2002) Studies on behavioral patterns in Jamunapari goats. *Small Ruminant Research* 43(2), 185–188.

Rout, P.K., Singh, M.K., Roy, R., Sharma, N. and Haenlein, G.F.W. (2004) Jamunapari-a dairy goat breed in India. *Dairy goat Journal (USA)* 82(3), 37–39.

Rout, P.K., Kumar, A., Mandal, A., Singh, S.K. and Roy, R. (2010) Characterization of casein gene complex and genetic diversity analysis in Indian goats. *Animal Biotechnology* 21, 122–134.

Rout, P.K., Thangraj, K., Mandal, A. and Roy, R. (2012) Genetic variation and population structure in Jamunapari goats using microsatellites, mitochondrial DNA, and milk protein genes. *The Scientific World Journal* 2012, 45–52.

Rout, P.K., Kaushik, R. and Ramachandran, N. (2016) Differential expression pattern of heat shock protein70 gene in tissues and heat stress phenotypes in goats during peak heat stress period. *Cell Stress and Chaperones* 21(4), 645–651.

Rout, P.K., Kaushik, R., Ramachandran, N. and Jindal, S.K. (2017) Identification of heat stress susceptible and tolerant phenotypes in goats in semi-arid tropics. *Animal Production Science* 58(7), 1349–1357.

Rout, P.K., Matika, O., Kaushik, R., Dass, G., Singh, M.K., Buhsan, S. (2018) Genetic analysis of growth parameters and survival potential of Jamunapari goats in semiarid tropics. *Small Ruminant Research* 165, 124–130.

Roy, R., Mandal, A. and Notter, D.R. (2008) Estimates of (co)variance components due to direct and maternal effects for body weights in Jamunapari goats. *Animal* 2(3), 354–359.

Roy, R., Dass, G.G. and Tiwari, H.A. (2011) Improvement and sire evaluation of Jamunapari goats for milk production. Annual Report (2010–2011) of Central Institute for Research on Goats Makhdoom, Mathura, (UP) India, pp. 19–22.

Rupp, R., Clément, V., Piacere, A., Robert-Granié, C. and Manfredi, E. (2011) Genetic parameters for milk somatic cell score and relationship with production and udder type traits in dairy Alpine and Saanen primiparous goats. *Journal of Dairy Science* 94, 3629–3634.

Ruvuna, F., Cartwright, T.C., Blackburn, H., Okeyo, M. and Chema, S. (1988) Gestation length, birth weight and growth rates of pure-bred indigenous goats and their crosses in Kenya. *Journal of Agricultural Science, Cambridge* 111, 363–368.

Ruvuna, F., Taylor, J.F., Okeyo, M., Wanyoike, M. and Ahuya, C. (1992) Effects of breed and castration on slaughter weight and carcass composition of goats. *Small Ruminant Research* 7, 175–183.

Sacchi, P., Chessa, S., Budelli, E., Bolla, P., Ceriotti, G. *et al.* (2005) Casein haplotype structure in five Italian goat breeds. *Journal of Dairy Science* 88, 1561–1568.

Sanchez, A., Ilahi, H., Manfredi, E. and Serradilla, J.M. (2005) Potential benefit from using the αs1-casein genotype information in a selection scheme for dairy goats. *Journal of Animal Breeding and Genetics* 122(9), 21–29.

Sawalha, R.M.A. and Tabbaa, M.J.A. (2004) Genetic and some environmental factors affecting body weights and dimensions of growing Damascus kids in Jordan. *Dirasat, Agricultural Sciences* 31(9), 74–87.

Schoeman, S.J., Els, J.F. and van Niekerk, M.M. (1997) Variance components of early growth traits in the Boer goat. *Small Ruminant Research* 26, 15–20.

Sellner, E.M., Kim, J.W. and McClure, M.C. (2007) Applications of genomic information in livestock. *Journal of Animal Science* 85, 3148–3158.

Selvaggi, M. and Dario, C. (2015) Genetic analysis of milk production traits in Jonica goats. *Small Ruminant Research* 126, 9–12.

Shackleton, D. (1997) Wild Sheep and Goats and their Relatives: Status Survey and Conservation Action Plan for Caprinae. IUCN, Gland, Switzerland.

Shafiq, M. and Sharif, M. (1996) Genetic evaluation of goats on productive traits by BLUP procedures. 17th Annual Report. Livestock Production Research Institute, Bahadurnagar Okara, Pakistan.

Shelton, M. (1993) *Angora Goat and Mohair Production*. Mohair Council of America, San Angelo, Texas.

Shrestha, J.N.B. and Fahmy, M.H. (2007) Breeding goats for meat production: a review. Selection and breeding strategies. *Small Ruminant Research* 67, 113–125.

Silanikove, N. (2000) The physiological basis of adaptation in goats to harsh environments. *Small Ruminant Research* 35, 181–193.

Silanikove, N. and Koluman, N. (2015) Impact of climate change on the dairy industry in temperate zones: Predications on the overall negative impact and on the positive role of dairy goats in adaptation to earth warming. *Small Ruminant Research* 123, 27–34.

Singh, L. (1991) DNA profiling and its applications. *Current Science* 60, 580–585.

Smith, C. and Simpson, S.P. (1986) The use of genetic polymorphisms in livestock improvement. *Journal of Animal Genetics and Breeding* 103(1–5), 205–217.

Snyman, M.A. (2012) Genetic analysis of body weight in South African Angora kids and young goats. *South African Journal of Animal Science* 42(2), 146–155.

Snyman, M. and Olivier, J.J. (1996) Genetic parameters for body weight, fleece weight and fiber diameter in South African Angora goats. *Livestock Production Science* 47(1), 1–6.

Solanki, G.S. (1989) Physiological response of goat under short term exposure of low ambient temperature in marginal and semi-arid region at Rajkot. *Proceedings of the Indian National Science Academy* B55(5 & 6), 361–368.

Soller, M. (1978) The use of loci associated with quantitative effects in dairy cattle improvement. *Animal Production* 27, 133–139.

Sullivan, B.P., Kennedy, B.W. and Schaeffer, L.R. (1986) Heritabilities, repeatabilities and correlations for milk, fat and protein yields in goats. *Journal of Dairy Science* 69(1), 100.

Teh, T.H. and Gipson, T.A. (1992) Cashmere production in the People's Republic of China. Proceedings of Cashmere Production in America, San Angelo, Texas, USA, pp. 1–8.

Thiruvenkadan, A.K., Murugan, M., Karunanithi, K., Muralidharan, J. and Chinnamani, K. (2009) Genetic and non-genetic factors affecting body weight in Tellicherry goats. *South African Journal of Animal Science* 39(5), 107–111.

Threadgill, D.W. and Womack, J.E. (1990) Genomic analysis of the major bovine protein genes. *Nucleic Acid Research* 18, 6935–6942.

Titi, H.H. (2003) Replacing soybean meal with sunflower meal with or without fibolytic enzymes in fattening diets of goat kids. *Small Ruminant Research* 48, 45–50.

Torres-Vázquez, J.A., Valencia-Posadas, M., Castillo-Juárez, H. and Montaldo, H.H. (2009) Genetic and phenotypic parameters of milk yield, milk composition and age at first kidding in Saanen goats from Mexico. *Livestock Science* 126, 147–153.

Torres-Vázquez, J.A., Valencia-Posadas, M., Castillo-Juárez, H. and Montaldo. H.H. (2010) Genetic and phenotypic trends for milk yield and milk composition traits of Saanen goats from Mexico. *Revista Mexicana de Ciencias Pecurias* 1, 337–348.

Tosser-Klopp, G., Bardou, P., Bouchez, O., Cabau, C., Crooijmans, R. *et al.* (2014) Design and characterization of a 52 K SNP chip for goats. *PLoS One* 9, e86227.

Valencia, M., Dobler, J. and Montaldo, H.H. (2007) Genetic and phenotypic parameters for lactation traits in a flock of Saanen goats in Mexico. *Small Ruminant Research* 68, 318–322.

Vassal, L., Delacroix-Buchet, A. and Bouillon, J. (1994) Influence des variants AA, EE, et FF de la caséine α αs1caprine sur le rendementfromager et les caractéristiques sensorielles des from ages traditionels: premières observations. *Lait* 74, 89–103.

Vos, P., Hogers, R., Bleeker, M., Reijans, M., van de Lee, T. *et al.* (1995) AFLP: a new technique for DNA fingerprinting. *Nucleic Acids Research* 23, 4407–4414.

Webb, E.C. (2014) Goat meat production, composition and quality. *Animal Frontiers* 4, 433–437.

Webb, E.C., Casey, N.H. and Simela, L. (2005) Goat meat quality. *Small Ruminant Research* 60, 153–166.

Weiner, G. (ed.) (1989) Animal Genetic Resources – A global programme for sustainable development. FAO Animal production and health paper 80, Food and Agriculture Organization of the United Nations, Rome, Italy.

Weppert, M. and Hayes, J.F. (2004) Direct genetic and maternal genetic influences on first lactation production in four breeds of dairy goats. *Small Ruminant Research* 52, 173–178.

Wiggans, G., Hubbard, S.M. and Wright, J.R. (2003) AIPL goat evaluation description. Genetic evaluation of dairy goats for yield and type. Available at: http://www.aipl.arsusda.gov/reference/goat/goatsfs.html (accessed 22 May 2019).

Wilson, E.O. (ed.) (1988) *Biodiversity*. National Academy Press, Washington DC, USA.

Wilson, R.T. (1991) *Small Ruminant Production and Small Ruminant Genetic Resources in Tropical Africa*. Food and Agriculture Organization of the United Nations, Rome, Italy, pp. 106–114.

Yalcin, B.C. (1986) *Sheep and Goats in Turkey* (FAO Animal Production and Protection Paper No. 60). Food and Agriculture Organization of the United Nations, Rome.

Yousef, M.K. (1985) Thermo neutral zone. In: Yousef, M.K. (ed.) *Stress Physiology in Livestock*, CRC Press, Boca Raton, Florida, pp. 67–74.

Zamiri, M.J., Eilami, B. and Kianzad, M.R. (2012) Effects of castration and fattening period on growth performance and carcass characteristics in Iranian goats. *Small Ruminant Research* 104, 55–61.

Zeder, M.A. and Hesse, B. (2000) The initial domestication of goats (*Capra hircus*) in the Zagros mountains 10,000 years ago. *Science* 28, 2254–2257.

Zeuner, F.E. (1963) A history of domesticated animals, *Harper and Row*, New York, USA, 560 pp.

Zhou, H.M., Allain, D., Li, J.Q., Zhang, W.G. and Yu, X.C. (2002) Genetic parameters of production traits of inner Mongolia Cashmere goats in China. *Journal of Animal Breeding and Genetics* 119(6), 385–390.

3 Reproduction Management

3.1 Introduction

The tropical countries of Asia, Africa and South America harbour about 90% of the goats worldwide. These goats are multipurpose animals that provide a vast range of products, including meat, milk, skin (leather) and fibre; moreover, their contribution to the livelihoods of people have not been recognized (Peacock *et al.*, 2005). In recent years, the demand for goat products has also increased in developed countries, accompanied by an increase in the number of commercial goat farmers. Therefore, it is essential for farmers to increase the reproduction efficiency (conception, fecundity and kidding rate) of goats (Greyling, 1988).

Reproductive efficiency of goats is determined by many factors, including the length of the breeding season, cyclic activity, ovulation rate, fertilization rate, the postpartum anoestrous period, and the growth and viability of the offspring. The reproductive efficiency is measured and expressed in terms of kidding rate, kidding interval, weight of kids born or weaned, length of the reproductive life of the dam, good mothering ability and extended breeding season (Greyling, 1988). The goat is an ideal species to manipulate reproduction physiology to increase reproduction efficiency (Devendra and Burns, 1983).

3.2 Reproductive Physiology

3.2.1 Male and female reproductive systems

Anatomy of the male reproductive system

The male reproductive system consists of four different organs (Fig. 3.1)

- **The testicles:** responsible for producing spermatozoa and secretes testosterone (the male hormone).
- **Reproductive excretory passage** (Epididymis, the ductus deferens, the urethra with annex glands): allows the sperm to exit and fertilize ovule.
- **The penis:** the male copulatory organ. The goat has a fibro-elastic penis.
- **The prepuce:** protects the free part of the penis.

Anatomy of the female reproductive system

The female reproductive organs consist of the following parts (Fig. 3.2.):

- **The reproductive glands:** the ovaries and tubular genital organs (uterine tubes, uterus, vagina, the vestibule and urethra).
- **The external genital organs:** vulva, clitoris and urethra.
- **The ovaries:** produce ova, as well as the hormones oestrogen and progesterone.
- **The uterus:** consists of two symmetrical horns, a body and the neck called the cervix. The function is to nourish the foetus until birth. The cervix is the communicating point between the uterus and vagina.
- **The oviducts:** also known as the fallopian tubes, they pass from the horns of the uterus to the ovaries. These tubes carry one or more ova from the ovary to the uterus.
- **The vagina:** is the female copulatory organ. It extends from the neck of the uterus (cervix) to the vulva. It is a tubular organ and extends up to the external urethral ostium.
- **The placenta:** is an organ that is responsible for the nutrition of the foetus and also exchanges nutrition with the mother. The umbilical cord extends between the embryonic umbilicus and the placenta.

3.2.2 Physiology of sexual change

Physiology of female reproduction

Puberty is the period when the reproductive organs of young animals become functional. In females the oestrus is first exhibited at about 7 months of age; however, it varies from 4 to 12 months of age. The smaller breeds attain puberty at an earlier age and

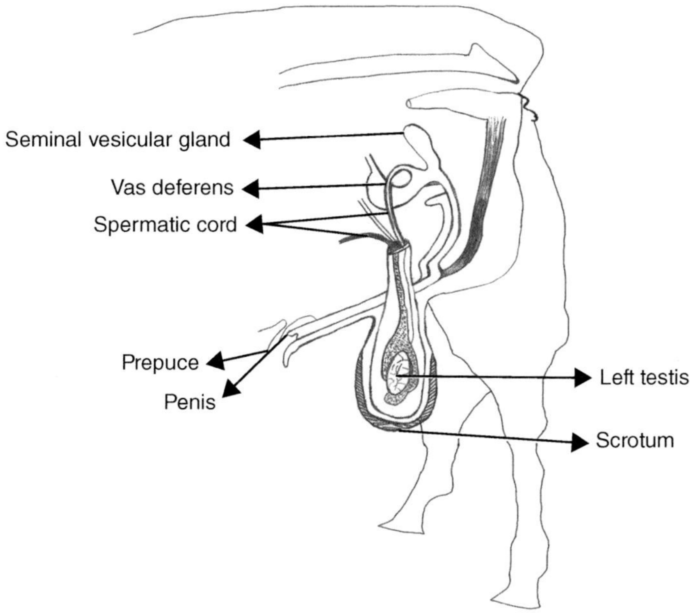

Fig. 3.1. Male reproductive organs.

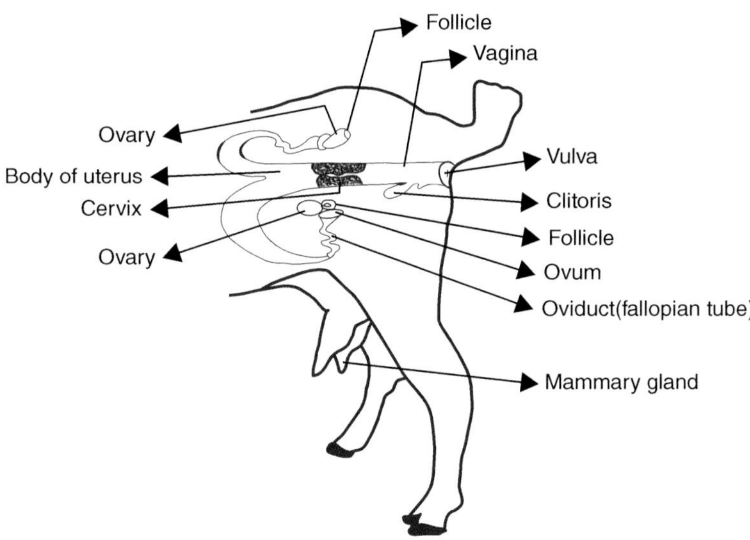

Fig. 3.2. Female reproductive organs.

larger breeds attain sexual maturity at a later age. Sexual maturity is the age when goats become sexually mature, and it depends on the breed, the type of management, level of nutrition and season of birth. The small and dwarf Indian breeds are known to attain puberty as early as 150 days in an optimum environment. The medium and large breeds attain puberty at a later age (range: 325–550 days). The weight at puberty varies considerably from 9.2–18.0 kg in small and dwarf breeds; whereas, it varies from 15.66–24.45 kg in larger breeds (Goel and Agrawal, 2004). Among Indian goats, Black Bengal was reported to attain the earliest puberty (160 ± 23.27 days, range 122–233 days) (Jalaluddin, 1992), Jakhrana goats reach sexual maturity at an age of 14 months (Goel and Agrawal, 2004) and Beetal goat attains sexual maturity at the age of 20 months (592.40 ± 20.50 days). In order to gain the maximum reproductive lifespan, breeding at the third oestrus is generally preferred. Puberty is normally attained at between 60–70% of the mature body weight, which is dependent on the level of nutrition and type of management. Age and weight at puberty in Indian goat breeds are present in Table 3.1.

The oestrus cycle is the normal sexual cycle of does under the influence of steroid sex hormones. The main signs of oestrus are bleating, wagging the tail, discharge of mucous membrane from the vagina and swelling of the vagina. Ovulation occurs towards the end of the oestrus period and is the reference point of the oestrus cycle. The oestrous is defined in four classic phases: pro-oestrus, oestrus, met-oestrus and di-oestrus. The oestrus phase is only detectable from the external signs and symptoms, but the onset and cessation of oestrus can also be precisely detected by the introduction of teaser bucks. The cyclic events from one oestrus to the next are: (i) compromising attitude of the doe towards the buck; (ii) transient harbouring of sperm in the fallopian tube; (iii) ovulation; (iv) fertilization of ovum/ova; (v) transportation of embryo to uterus; (vi) implantation of the foetus; (vii) gestation; (viii) parturition; and (ix) postpartum oestrus. Failure of fertilization or premature termination of conception leads to the next oestrus progressing through the pro-oestrus period.

Detection of the oestrus and onset of oestrus are the main concern for farmers or goat rearers as it is the time for natural mating or artificial insemination. The oestrus cycle of goats in the majority of cases is 19–21 days in length. However, the oestrus cycle varies from 3 days (short cycle) to 62 days (long cycle). In mature does, the oestrus cycle is about 20.7 ± 0.7 days with an incidence of short (13 days) and long (25 days) cycles. The oestrus cycle in does has been categorized into short, medium and long cycles with the average duration of 6.4, 19.8 and 37.5 days, respectively (Greyling, 2010). The duration of the oestrus period is also variable in length. The oestrus cycle and periods were found to be shorter during extreme cold, dry and hot humid periods. The oestrous cycle length of does has been studied in several Indian breeds, which normally ranges from 17 to 22 days (Table 3.2). Jakhrana goats in semi-arid environments showed 60% cycles of normal cycle length along with short (13.70%) and long (27.40 %) cycles (Goel and Agrawal, 2004). The meat breeds generally exhibit pronounced oestrus for longer periods as compared to the dairy breeds.

The regular duration of the oestrus period in goats is around 36 hours, varying from 22–60 hours in different breeds (Greyling, 1988). For example, the duration of the oestrus period of Angora does is 33.4 hours compared to that of the Boer goat does of 37.4 hours. (Reira, 1982; Jainudeen *et al.*, 2000). A male effect is responsible for the acceleration of the occurrence of oestrus and can have an influence in shortening the duration of the oestrus period in does (Greyling, 1988; Jainudeen *et al.*, 2000). The mature egg is released into the oviduct, known as ovulation, towards the end of the oestrus cycle. The collapsed Graafian follicle is known as the corpus luteum, which will secrete the hormone progesterone. The ovum remains viable in the oviduct for 10–12 hours. If the ovum is unfertilized, it will degenerate and pass out of the oviduct into the uterus. Ovulation rate refers to the number of eggs or ova released from the ovary during a particular oestrus period. It is very important to know the time of ovulation for

Table 3.1. Age and weight at puberty in Indian goat breeds.

Breed	Age at puberty (days)	Weight (kg) at puberty	References
Black Bengal	160.37 ± 23.27	13.0	Jalaluddin (1992)
Barbari	274.00 ± 13	12.40	Prasad and Bhattacharyya (1979)
	307.90 ± 6.02	14.50 ± 0.40	Prakash *et al.* (1988)
Jamunapari	553.00 ± 6.69	25.28 ± 0.70	Khan *et al.* (1981)
Pashmina	365.00	–	Mazumdar and Mazumdar (1982)
Malabari	437.00 ± 21.00	18.00	Mukundan *et al.* (1983)
Beetal	523.20 ± 14.60	24.00 ± 0.95	Rana *et al.* (1982)
Sirohi	498.96 ± 10.81	26.76 ± 0.29	AICRP-Goats (1985)
Marwari	311.20 ± 5.43	16.80 ± 0.89	Kaushish *et al.* (1994)
Kutchi	561.00 ± 6.52	23.50 ± 0.75	Kaushish *et al.* (1994)
Jakhrana	422.40 ± 10.85	22.18 ± 0.43	Goel and Agrawal (2004)

Table 3.2. Oestrous cycle length in some Indian goat breeds.

Breed	Oestrous length (days)[a]	References
Jamunapari	21.14 ± 0.15 (18–24)	Goel *et al.* (2003)
Black Bengal	20.00 ± 0.40	Bhattacharyya (1982)
	19.2 7± 1.94 (15–24)	Jalaluddin (1992)
Barbari	18.18 ± 3.10	Mittal (1981)
	19.00 ± 0.40	Bhattacharyya (1982)
Malabari	20.50	Mukundan (1980)
Pashmina (Changthangi)	20.39 ± 3.00 (18–23)	Mazumdar and Mazumdar (1983)
Marwari	19.33 ± 2.26 (17–21)	Mittal (1988)
Jakhrana	20.76 ± 0.20	Goel and Agrawal (2004)
Surti	19.67 ± 0.41	Suresh Kumar *et al.* (1989)

[a]Mean ± standard error; parenthesis indicates the range of observations.

successful natural mating or artificial insemination. The synchronization of ovulation and mating is critical for successful fertilization.

Kidding rate is linked to the ovulation rate and varies widely between breeds (see Table 3.3) Ovulation rate is affected by season and the level of nutrition. Ovulation rates are higher during the natural breeding season. Factors such as body weight and nutritional supplement contribute to increased ovulation rates (Greyling, 1988; Jainudeen *et al.*, 2000).

The 'male effect' is prominent in does that have been isolated from males, for example, with the introduction of bucks during the transitional period (time from the anoestrous to the breeding season). The ovarian response of anovular does is stimulated by an androgen-dependent pheromone secretion of the buck. It is generally accepted that the 'male effect' is responsible for the acceleration of the occurrence of oestrus and could also have an influence in shortening the duration of the oestrous period in the doe (Chemineau, 1983; Kassem *et al.*, 1989; Chemineau *et al.*, 1992).

GESTATION PERIOD The gestation period is defined as the period of time from conception to parturition in livestock. The gestation period in goat varies between 149 to 150.8 days, with an average of 149 days. There are several factors that affect the gestation period. The genotype of the foetus plays an important role in determining gestation length. Male kids are carried for a longer period than female kids. Similarly, single kids are carried for longer periods than twin-born kids. Nutritional status also affects the gestation period but it has been reported that the variation due to nutrition may only be around 1.5 days. Moreover, nutritional management is very important during gestation periods for maintaining the health of the dam and foetus. About 70% of the foetal growth occurs during the last trimester of pregnancy (after about 100 days) (Greyling, 1988; Jainudeen, 2000). It has been observed that the corpus luteum is essential for the maintenance of pregnancy throughout the gestation period. It does not happen in the case of ewes where the corpus luteum is essential for the first 3 months of pregnancy.

Physiology of the reproduction

The reproductive organs of male goats are shown in Fig. 3.1. The male reproductive organs consist of two testes and a spermatic cord. The main function of male reproductive organs is spermatogenesis, which generally takes about 50 to 60 days. When spermatogenesis is completed, the sperm move to the epididymis where they mature for 2–3 weeks. Therefore, it takes about 60–80 days to produce viable and mature sperm. In the case of temporary sterility (i.e. due to illness, high fever or vaccination), the time required for complete spermatogenesis should be kept in mind if optimal fertilization is required (Greyling and Grobbelar, 1988). The sperm and the male hormone testosterone are both produced in the testes, which maintain a temperature 4–7°C lower than body temperature. The testicular function is also subject to hormonal control. Generally, spermatogenesis stimulatory hormone and interstitial cell stimulating hormone are the two gonadotrophic hormones responsible for spermatogenesis in the seminiferous tubules; testosterone is produced in the Leydig cells. The sperm's quality is the major criteria for evaluating semen (see Protocol 1 at the end of this chapter). The most important aspect of the evaluation of semen quality is the fertilizing capacity of the sperm (Greyling and Grobbelar, 1983).

3.2.3 Reproductive efficiency

The profitability of a flock depends on the reproductive rate of the doe. It is necessary to produce

Table 3.3. Kidding rate of different goat breeds in tropical locations.

Breed	Location/ agroclimatic zone	Average litter	Reference
Maltese and Damascus	Athalassa, Cyprus	1.85	Maule (1949)
Damascus	Cyprus	1.71	
Kilis	Ankara, Turkey	1.27	Yarkin and Eker (1961)
Boer	South Africa	2.08	Okere *et al.* (2011)
Kambing Katjang	Malaysia	1.65	
Syrian Mountain	Syria	1.34	
Malta	Aegean	1.59	
Assam Hill	Assam	1.58	AICRP-Goats (2017)
Black Bengal	West Bengal	1.90	
Gaddi	Himachal Pradesh	1.20	
Malabari	Kerala	1.62	
Marwari	Rajasthan	1.08	
Osmanabadi	Maharashtra	1.62	
Sangamneri	Maharashtra	1.58	
Jamunapari	Uttar Pradesh	1.46	
Sirohi	Rajasthan	1.12	
Barbari	Uttar Pradesh	1.58	
Surti	Gujarat	1.68	
Ganjam	Orissa	1.09	
Changthangi	Leh-Ladakh	1.04	

kids for replacement of the flock and for the sale of excess stock for profit. The unproductive individuals should be identified and should be culled from the flock. Reproductive efficiency can be measured by fertility, prolificacy, fecundity and survival of the kids until weaning age. Subsequently, the age of first parturition (kidding) and parturition (kidding) interval should be taken into account to determine reproductive performance. A prolific animal must be highly fertile. Fertility percentage is the conception rate of does in a year out of total breedable does in a particular management system. The fertility percentage of different Indian and African breeds varies between 70.0% and 94.0%. However, it depends on various management factors. Prolificacy or litter size is defined as the number of progenies born per parturition (Haumesser, 1975; Haumesser and Gerbaldi, 1980). The average litter size can be calculated on a yearly basis to be consistent with the annual rate of fertility. The litter sizes range from 1.27 to 2.08 in native goats of sub-Saharan Africa. The highest kidding rate has been reported for Boer goats, West African Dwarf goats in the humid zone of Ghana, Sudan Desert does (Wilson, 1982) in the semi-arid zone of Mozambique and Sudan, and Rwandan does in the highlands of Rwanda (Murayi *et al.*, 1987). The kidding rate in Indian breeds varies from 1.05 to 1.9 in their native tract in farmers' management systems (Table 3.3). The highest kidding rate has been observed in Black Bengal goats in humid climate zones. The lowest kidding rate was observed in Ganjam goats in humid climate zones and Sirohi goats in arid climate zones. Jamunapari and Beetal are large-sized breeds with a kidding rate range of 1.35 to 1.49. Fecundity is defined as the number of offspring born per female of reproductive age per year. Fecundity is a product of fertility and prolificacy (i.e. fecundity = fertility × prolificacy). Certainly, it is more useful to know the level of fecundity than that of fertility and prolificacy (Haumesser, 1975).

Age at first mating for does is an important criteria because early mating may cause abortion, low birth weight and the doe may not produce enough milk for kids. The age at first mating of does is dependent on breed and nutritional status of the production system. Ideally, a doe should have one pair of permanent incisors, which is at the age of 14–17 months, before first mating. Age at first parturition is a good indicator of early sexual maturity in does. It is an economically important trait as greater population turnover and more rapid genetic progress can be obtained when goats produce their first progenies at an earlier age. Early maturing females are also known to have a relatively long and fruitful reproductive life. Age at sexual maturity is affected by genotype and various management components. The age at first parturition and parturition interval are presented in Table 3.4.

3.2.4 Reproductive disorders and infertility

Reproductive problems may be due to physiological, nutritional and management problems. If a doe is not kidding or the overall reproductive rate of the flock is poor, then it is necessary to investigate in detail to identify the probable cause. The first

Table 3.4. Age at parturition and parturition interval of different goat breeds in tropical zones.

Ecological zone	Country/location	Breed	Age at parturition (days)	Parturition interval (days)	Reference
Indian goats					
Hill Region	Assam	Assam Hill Goat	404.5	226.42	AICRP-Goats (2017)
Semi-arid	Uttar Pradesh and Rajasthan	Barbari	507	216.66	
Humid	West Bengal	Black Bengal	386.26	222.96	
Semi-arid	Uttar Pradesh	Jamunapari	814.38	338.56	
Arid	Rajasthan	Sirohi	625.4	369.38	
Arid and semi-arid	Gujarat	Surti	627.02	323.66	
Humid	Odisha	Ganjam	599.92	289.12	
Humid coastal	Kerala	Malabari	394.57	275.70	
Arid	Rajasthan	Marwari	482.72	220.53	
Semi-arid	Maharashtra	Osmanabadi Goat	488	320.0	
Semi-arid	Maharashtra	Sangamneri	410.29	255.66	
Island region, humid	Andaman and Nicobar	Andamani	403.38	97.27	
Dry cold	Leh-Ladakh	Changthangi	700.82	359.33	
African goats					
Highlands	Malawi	Small East African		314.3	Karua (1989)
Humid	Nigeria	West African Dwarf	518	283	Sumberg and Mack (1985)
	Nigeria	West African Dwarf	545	278	Mosi *et al.* (1982)
	Senegal	West African Dwarf	361	239	Faugère *et al.* (1988)
	Burkina Faso	Sahel	423	328	ILCA (1987)
	Kenya	Small East African	555.8	306.1	Wilson (1984)
	Mali	Sahel	487	298	Wilson and Traore (1988)
	Niger	Maradi		332.4	Haumesser (1975)
	Niger	Sokoto		331	Gerbaldi (1978)
	Uganda	Mubende		297	Wilson (1982)
	Sudan	Sudan desert	280	238	Wilson (1982)
	Somalia	White Somali		280.6–600.8	Bourzat *et al.* (1992)
	Zimbabwe	Zimbabwean Small East African		370	Ndlovu (1992)
Sub-humid	Chad	Kirdi	472.75		Dumas (1980)
	Tanzania	Newala	503.25		Hendy (1980)
	Swaziland	Nguni	262–340	268	Lebbie and Manzini (1989)

step is to investigate if the doe is coming to heat and showing any signs of oestrus. It is necessary to detect oestrus properly, and the oestrus detection can be problematic if there are no males in the flock. Females may be experiencing normal cycles but not exhibiting the signs of oestrus. The cause of anoestrus may be due to poor nutritional condition if there is a loss of 10–20% body weight. Does may not show signs of oestrus during the early stages of lactation (lactational oestrus).

If oestrus is not regular, it may be due to short or long oestrus cycles linked to the following factors:

- embryonic death;
- anoestrus; or
- hormonal unbalance.

A short oestrus cycle (6–10 days) occurs due to factors such as ovarian cysts, metritis, mummified kids and stress. Stress occurs due to transportation. Premature corpus luteum regression also causes a short oestrus cycle.

If oestrus is regular, then next investigation is focused on the buck fertility. Males may be infertile for different reasons:

- brucellosis infection: the bucks show swollen testicles (orchitis) and bucks are infertile;
- physical deformities such as physical damage to penis or testicles;
- mounting problems;
- overuse of bucks: bucks should serve one or two females in a day; if the buck serves more than four females, then sperm quality may not be effective for fertilization;
- congenital deformities;
- age; or
- homosexuality: some males reared in separate flocks may develop homosexual behaviour and may not respond to females.

3.2.5 Reproductive failures

Reproductive failure can be either sterility or infertility. Sterility occurs when the animal is not able to produce kids during their lifetime, whereas infertility or temporary sterility is the inability to produce live young within a specific time period. There are a number of factors associated with reproductive failure in goats.

1. Congenital genital organs: Congenital deformities in genital organs affect normal reproductive physiology and subsequently kid production. Genital abnormalities such as incomplete oviducts, an abnormal uterus or a blind cervix or vagina results in abnormal heat and pregnancy. Hermaphrodites have both male and female reproductive gonads (ovaries and testis). Cryptorchidism, where the testes do not pass through the inguinal canal into the scrotum and remain in the body cavity, is another condition that might lead to reproductive failure in males. Fertility may be maintained in unilateral cryptorchidism, but if both testicles are still inside the body (bilateral cryptorchid) then the male will be sterile. This is a recessive heritable condition that is controlled by a few pairs of genes (Warwick, 1961). Bucks with this condition should not be used for breeding and their progeny should be culled from the flock. In a slaughterhouse study in Ethiopia, 5.5% of bucks were cryptorchid and both unilateral and bilateral cases were observed.

2. Cystic ovaries, either follicular or luteal, can prevent conception. In follicular cysts, goats come into heat but ovulation fails; it is related to a problem with luteinizing hormone (LH). Even though this type of cyst is hard to treat, large doses of luteinizing hormone/human chorionic gonadotrophin (hCG) or gonadotrophin-releasing hormone (GnRH) followed by a dose of prostaglandin 10–14 days later, might be effective. In an ultrasonographic study of five goats from Japan, follicular cystic ovaries were successfully treated with GnRH followed 10 days later by prostaglandin (Medan *et al.*, 2004). Luteal cysts can occur when the corpus luteum (CL) fails to regress and becomes filled with fluid such that hormone secretion is altered. Poor uterine health resulting in low prostaglandin secretion could cause this problem. Luteal or corpus luteal cysts can lead to pregnancy loss, a dead mummified foetus, infection and even pseudopregnancy. Luteal cysts are usually treated with prostaglandin to lyse the CL.

3. Silent heat: Sometimes, females do not exhibit visible signs of oestrus but ovulation occurs. The incidence of a silent oestrus is high during transition periods in goats, such as the onset of puberty and transition into and out of the breeding season.

4. Feed toxicity and plant with phytoestrogens: Storage of feed in wet and damp stores may lead to growth of *Aspergillus*, *Fusarium* and *Penicillium* in feed material. The toxins or mycotoxins produced can lead to reproductive problems and failure. In addition, moulds and endophytes in forages can also cause problems. Fescue toxicosis can delay onset of puberty, impair luteal function, reproductive failure, decreased milk production or low foetal birth weights and cause longer than normal pregnancies (Smith *et al.*, 2004). Cotton seeds with high gossypol content can cause delayed puberty in young animals due to decreased growth and can also cause problems with sperm production in males. Certain plants such as soybeans, subterranean clovers and alfalfa produce hormonal compounds called phytoestrogens that may reduce fertility in animals ingesting them and have been associated with a weak or prolonged oestrus, lack of oestrus, irregular oestrous cycle lengths, cystic ovaries and abortion.

5. Stress: Environmental and physical stress such as heat/cold, transportation, isolation and traumatic injury may result in psychological and physiological causes of reproductive failure. It is evident that stress hormones including adrenocorticotrophic hormone (ACTH) and cortisol can reduce reproductive hormones like LH. This can delay or prevent the release of eggs in the females, which impairs reproductive performance in both females and males (reduced libido). It is, therefore, important to minimize any types of stress during the breeding season for good reproductive results.

6. Nutrition: Overfeeding, underfeeding or feeding a nutritionally imbalanced diet, or deficiency in essential minerals such as copper, selenium, iodine and zinc, will prevent goats from conceiving. Poor nutrition also includes a diet lacking in energy, protein and complex fibre. Any or all of these conditions can prevent conception. A starving doe will abort her kids or deliver weak kids (stress abortion). Stress abortions are common in undernourished goats (Smith and Sherman, 2009), associated with placental malfunction. The hormone leptin is well documented as preventing conception in improperly fed does, i.e. starving does, fat does and does that are significantly underweight.

7. Early/immature breeding: A doe that has been bred too young (less than 12 months of age) may experience dystocia (parturition difficulties), particularly if the doe is unwisely bred to a buck of a larger-frame size breed. If a doe survives kidding problems, then a torn uterus or other internal damage may prevent future conception.

3.2.6 Management of breeding bucks

- Body condition score should be at least 3 out of 5.
- Provide supplementary feeding of 300 gm/goat/day to medium-sized breeds or 400–500 gm/goat/day for a large-sized breed.
- All the necessary vaccinations should be carried out at least 6 weeks before the breeding season.
- Hooves should be trimmed before the breeding season for easy mounting and mating ability.

Foot dipping is recorded during the breeding season.

- Deworming for internal parasites should be carried out 3 weeks before mating.
- Testing for sperm fertility, i.e. semen characteristics, should be analysed.
- Screening for brucellosis and mycoplasma may be carried out 30 days before the breeding season (Smith, 1980; Gall, 1981; Devendra and Burns, 1983).

3.2.7 Basic pregnancy and lactation care

The pregnant and lactating goat needs special attention with respect to space requirements, optimum nutrition and health care. It is also necessary to vaccinate pregnant goats to increase the concentration of specific antibodies in the colostrum, thereby enhancing passive immunity in the kids. It is important to vaccinate with enterotoxaemia vaccine 4–6 weeks before the kidding (Smith and Sherman, 2009). Infestation with gastrointestinal parasites, in particular bursate worms, increases in pregnant does due to the periparturient egg rise. Therefore, pregnant does should be given a broad spectrum anthelmintic drug 2 to 3 weeks before kidding. Anthelmintic drugs at proper doses have not been shown to have abortive or teratogenic effects in goats; levamisole and albendazole can have negative effects, thus inhibiting their use in pregnant does (Smith and Sherman, 2009).

The important objectives are to provide better care to lactating goats to optimize milk production and maintain udder health. The feeding schedule needs to be modified as per milk production. The lactating doe should be kept away from the buck. Bruising of the teats and udder of goats should be avoided. Milking should start after washing the udder with lukewarm disinfected water and drying with a clean cloth. A preventive medicine procedure for milking does includes vaccinations, foot care and deworming. Only approved anthelmintic drugs should be used in lactating animals. It may be required to discard the milk at times, for example, the anthelmintic dosing may affect cheese culture and disrupt cheese making. The lactating does should have clean, well-bedded premises to avoid udder infections. Intramammary infusions of antibiotics at the beginning of the dry period can be used in herds with an active mastitis problem. Clipping the hair on the udder and hindquarters decreases sediment in the milk. The bucks should not be kept near milking parlours or milk storage areas because buck odour permeates the milk. In pasture-grazing lactating goats, some plants produce an off-flavour in the milk and need attention. Proper handling of the milk after extraction from the doe is extremely important. Excessive agitation and delays in pasteurization promote lipolysis of the fat in goat milk, adding to the so-called 'goaty' flavour of the milk.

3.2.8 Reproduction management

Efficient reproduction management is a prerequisite for any successful and profitable small stock enterprise. Generally, the reproduction performance is measured in terms of the kidding rate or weaning rate. By the implementation of an effective health and nutritional programme through breeding, gestation, lactation and the postpartum period of the doe, the production of viable offspring can be practically guaranteed. The applied aspect of reproduction management should be followed in a flock for higher productivity.

Record keeping

Record keeping is an integral part of an effective reproduction management programme. It is necessary to record mating dates, kidding dates, birth weights of the kids, sex of the kids and where natural mating is used, buck fertility, mating ability and libido of the bucks (see Protocol 2, Chapter 2) for necessary intervention to prepare the performance.

Herd health management

The management of growth at an early age is essential for the animal's future productive life. Therefore, a proactive health programme is required to manage the kids.

Nutritional management

Nutrient intake influences the body reserves and is related to the reproductive performance of the animal. It is necessary to provide a balance of energy to animals as the overfeeding of energy is related to a lower conception rate, abortions and retained placentas. Similarly, underfeeding of energy leads to delayed puberty, lower semen production, silent heats, decreased libido and irregular oestrous cycles of breeding animals. Nutritional flushing of energy 3 weeks prior to mating could have a dramatic effect on ovulation and conception rates. Similarly, nutrition during pregnancy is crucial, as 70% of foetal growth occurs during the last trimester of pregnancy.

Pregnancy diagnosis

Pregnancy diagnosis is a tool to determine the reproduction efficiency of the herd and nutritional management of does with singletons, twins or triplets. Non-pregnant animals can be isolated from the herd and not given the feed supplement supplied to pregnant does. Currently, sonography (abdominal 6–7 weeks or rectal 4–5 weeks after breeding) is the standard procedure determining pregnancy and multiple births in goats.

Mating

The availability of adequate nutrition is essential prior to the breeding season and also during the kidding season. A maiden doe exhibits breeding behaviour as the animal attains 60–75% of its mature body weight.

Dystocia

Difficult births are generally not a problem in goats. The factors contributing to this phenomenon are kid size, body condition, frame size and litter size. Unplanned mating of a higher body weight male with a smaller sized goat also predispose it to dystocia.

Buck soundness

Buck soundness management is an effective tool to maintain libido during the breeding period. Reproductive aspects such as libido, semen quality, body condition and health need to be constantly monitored. In the case of temporary sterility (irrespective of the cause), a period of 60–70 days must be allowed for the buck to once again produce mature, viable and fertile sperm. Make sure that the buck suits the breeding goal. Look at the performance records and breeding history of the animal. If the buck does not mount temporarily, then behaviour management may be necessary.

Environmental manipulation

Goats are considered to be seasonal breeders; however, some goat breeds also breed beyond the normal breeding season. The main environmental change that influences seasonal breeding in goats is the annual change in day length (goats are considered short-day breeders). Seasonal anoestrous occurs; that is, when the day length increases there is an associated absence of oestrus and ovulation and decreased secretion of the reproductive hormones. Seasonal species are responsive to a hormone called melatonin that is produced by the pineal gland in response to declining periods of light. The seasonal breeders have some distinctive advantages such as being able to provide a consistent supply of milk and meat products, and to plan for three kiddings in 2 years. Similarly, the offseason breeding may have some disadvantages such as higher costs of production, reduced fertility and pregnancy-related problems.

3.2.9 Abortion management

The magnitude of reproductive disorders in goats is relatively lower than other ruminants. Abortion is the expulsion of the immature dead foetus before completion of the gestation period. Abortion by one or more goats in a herd may signal an infectious disease or some other condition that should be controlled by proper management (Smith and Sherman, 2009).

The immediate steps to diagnose and control abortion include:

1. Analyse all the records of each aborting animal and the stage of gestation when the foetal loss occurred.
2. All aborted goats should be isolated to reduce the risk of contamination and infection.
3. The aborted foetus, placenta and uterine discharges are important sources of infection to other goats and should, therefore, be collected carefully and buried or burnt.
4. The area of abortion should be disinfected immediately with detergent/antiseptic solution.
5. Samples such as the foetus, placenta, sera, foetal tissues and fluids, impression smears of the cut surface of cotyledons, formalin-fixed tissues of the kidney, liver, spleen and heart for histology, and exudate from the vagina should be collected to aid the identification of the cause of abortion.
6. Prescribe treatment for the remaining pregnant does with suitable drugs. One possible protocol is to give three subcutaneous injections of long-lasting tetracycline at 20 mg/kg at 3-day intervals (Smith and Sherman, 2009).
7. Prepare a biosecurity programme to quarantine all new animals as long as one month before introduction to the farm, and keep postpartum and prepartum females in separate pastures.
8. Rear the flock in a stress-free, sanitary, uncrowded environment.
9. Maintain a source of supplemental energy and protein for use during emergency or high stress situations.
10. Provide a complete, good-quality trace mineral–salt mixture.
11. Keep feed and water sources free of faecal and urine contamination.
12. Reduce the number of rats, wild or free-range birds and cats.

The cause of abortion due to infection agents is listed in Table 3.5.

3.2.10 Oestrus detection and synchronization

Oestrous monitoring in the flock with an aproned or vasectomized teaser buck twice a day (morning and evening) is essential for identifying oestrous cycling does. It helps in the detection of cyclic and non-cyclic does. Oestrus symptoms are more distinct in goats compared to cattle and buffalo. Timely detection of oestrus is highly essential for successful mating or artificial insemination. It is very difficult to know the exact time point when oestrus started and therefore very difficult to predict ovulation time, which is the optimum time for mating. The ovum remains viable for 10–12 hours. Similarly, sperm remains viable for 12–24 hours. It is desirable to mate within approximately 20 hours of the onset of oestrus.

An oestrous doe can be detected by observing the following symptoms in the flock:

- restless, reduced appetite and milk production;
- acceptance of mounting of teaser buck;
- bleats frequently and wagging of tail constantly and rapidly;
- mounting over other oestrus and non-oestrous females and allowing mounting of other females;
- clear, transparent sticky mucus discharge from vagina;
- swollen and reddish vulva; and
- gathering of oestrous females around the teaser buck.

Methods of oestrus detection

There are a number of methods available for oestrous detection. The most common and practical methods are discussed below.

Table 3.5. Summary of infectious abortion in goats. (Adapted from Aiello and Moses, 2016.)

Disease	Transmission	Time of abortion	Clinical data/diagnostic sample	Vaccination/control measure
Brucellosis (*Brucella melitensis*)	Coitus, infected discharges	Late or stillbirth, weak kids	Abortion, retained placenta Organisms in foetal stomach and placenta	*B. abortus* strain 19 or *B. melitensis* Rev. 1 vaccine
Enzootic abortion (*Chlamydophila abortus*)	Ingestion	Last 2–3 weeks, Stillbirths, No sickness in goats, neonatal mortality	Organism in foetal cotyledons Degenerative changes in placenta	Killed vaccine / live attenuated vaccine; tetracyclines given orally or parentally in abortion cases
Leptospirosis (*Leptospira interrogans*)	Ingestion	Late abortion Anaemia, icterus, and hemoglobinemia	Leptospira spp. in the dam's urine, the placenta or foetal kidney	Leptospira vaccination
Listeriosis (*Listeria monocytogenes*)	Ingestion	After 3 months	Retained placenta, metritis Septicaemia in goats	In some countries killed or live attenuated vaccines are available
Toxoplasmosis	Ingestion	Late or stillbirths	Abortion, stillbirths, and neonatal mortality	Live S48 tachyzoite in some countries (e.g. UK, New Zealand); single dose 3 weeks before mating
Coxiellosis (Q-fever)	Inhalation, ingestion	Later term and weak kid	No illness in dam Neonatal mortality, necrotizing placentitis	Vaccine available in Europe but not in most other countries
Caprine herpes virus 1 (CpHV 1)	Inhalation	Late-term abortions	No illness in dam Vulvovaginitis, balanoposthitis,	Vaccines are not commercially available
Tick-borne fever	Ticks	Late, following systemic disease	Fever and abortion	None

- **Aproned teaser buck**: An adult intact, active breeding buck with a healthy body condition and good libido is covered with an apron in such a way that penile insertion in the female reproductive tract is avoided while mounting. The buck should be allowed to parade among females for 10–15 minutes once a day in open paddocks.
- **Vasectomized teaser buck**: An adult intact and active breeding buck with a healthy body condition with good libido is vasectomized and allowed to parade among females for 10–15 minutes once a day in open paddocks.
- **Androgenised does**: One or two adult, active breeding does with a healthy body condition can be injected with doses of testosterone at periodic intervals to make them masculine so that they act as bucks for heat detection. They can be housed and managed along with females without the risk of undesired mating.
- **Microscopic examination of mucus secretions for fern patterns:** The mucus secretion changes during the different periods of oestrus. Fern patterns are formed with any salt solution containing colloid. Fern patterns exhibit different forms in the mucus drawn from goats at varying stages of oestrus. This can be visualized using specially designed handheld equipment.

Heat synchronization

Goats exhibit seasonal cycles of reproduction. The oestrus cycle can be controlled by oestrus synchronization so that females express oestrus during a desirable time period (Zhao *et al.*, 2010). Hormonal oestrus synchronization is achieved either by reducing the length of the luteal phase of the oestrous cycle with prostaglandin $F_2\alpha$ or by extending the cycle artificially with exogenous progesterone or more potent progestogens (Abecia *et al.*, 2012). The efficiency of progestogens is considerably reduced when management and hygienic conditions are poor, or goats are reared under a low plane of nutrition. The response of progestogen is better during peak breeding season. Progestogens are available in various commercial forms and can be administered in different routes. See details in Protocol 2 at the end of this chapter.

3.2.11 Pregnancy detection

Monitoring of pregnancy after insemination (natural or artificial) is an essential component of successful farm management. It is essential to select pregnant does to provide better nutrition and care, and simultaneously to rebreed empty goats within the on-going breeding season. It will help to assess early embryonic losses, abortion and production of weak kids.

Rectal palpation technique

This practice is used effectively in cattle and buffalo but it is not possible in goats due to their small-sized pelvis. The conventional methods of pregnancy diagnosis under field conditions are based on non-return to oestrous and abdominal palpation. It is difficult to differentiate between pregnant and infertile goats.

Abdominal palpation

The gravid uterus or foetus can sometimes be palpated through the relaxed abdominal wall of the standing doe by placing a hand on either side of the abdomen and squeezing or lifting upward. The accuracy of pregnancy settling over 90% can be achieved 80–90 days after mating based on the experience of the veterinarian.

Specific pregnancy protein/antigen

Several pregnancy-associated glycoproteins are produced by chorionic binucleate cells from the placenta and have been detected as early as 21 days of gestation by radioimmunoassay (Gonzalez *et al.*, 2004). An ELISA test for the closely related pregnancy-specific protein B (PSPB) is available from Bio Tracking (Moscow, ID). This is a placental hormone that has been used as an indicator of pregnancy in goats (Humblot *et al.*, 1990).

Ultrasonography

Ultrasonography has been used extensively for pregnancy diagnosis as well as for ultrasonic imaging of internal organs, soft tissues, foetal number counts, sexing and foetal viability, and ovarian pathological conditions for increased goat production. Real-time B-mode ultrasonography was introduced to veterinary practice in the early 1980s and is a tool for early pregnancy diagnosis (Davey, 1986; Haibel, 1990). It also offers an accurate, rapid, safe and practical means of diagnosing the pregnancy, foetal numbers and estimation of gestational age in livestock. False positives are rare. In addition, real-time ultrasonography can also be helpful in diagnosing diseases of the reproductive tract (Buckrell, 1988). Real-time B-mode produces a two-dimensional, moving image of the uterus, foetal fluids, foetal heartbeat and placentomes (Fig. 3.3). The ideal time for transabdominal scanning is between 40 and 75 days. A positive diagnosis of pregnancy is assured by imaging the embryo/foetus or placentomes surrounded by fluid. Ultrasonography by real-time B-mode with a 5 MHz transrectal transducer was found to be a reliable, safe, accurate and practicable means of diagnosing pregnancy as early as 25 days post insemination (Singh *et al.*, 2004). However, a presumptive diagnosis of pregnancy or hydrometra can be made by imaging multiple anechoic (fluid-filled) sections of uterine lumen cranial to the bladder from days 25 to 40 of gestation using a transabdominal or transrectal approach. A false-positive pregnancy diagnosis during this period may be caused by hydrometra. This condition occurs commonly enough in goats to advise caution against making a positive diagnosis of pregnancy until the embryo/foetus can be seen (Dawson, 2015). The urinary bladder should not be confused with a fluid-filled uterus. Placentomes can be seen by 35–40 days, appearing as echogenic densities in the uterine wall. Multiple foetuses can be diagnosed with real-time ultrasonography in comparison to other ultrasound techniques. The optimal time for counting foetal numbers is between 40 and 70 days. Twins can be more accurately diagnosed than triplets and foetal numbers are frequently underestimated. Feeding management should be adjusted for

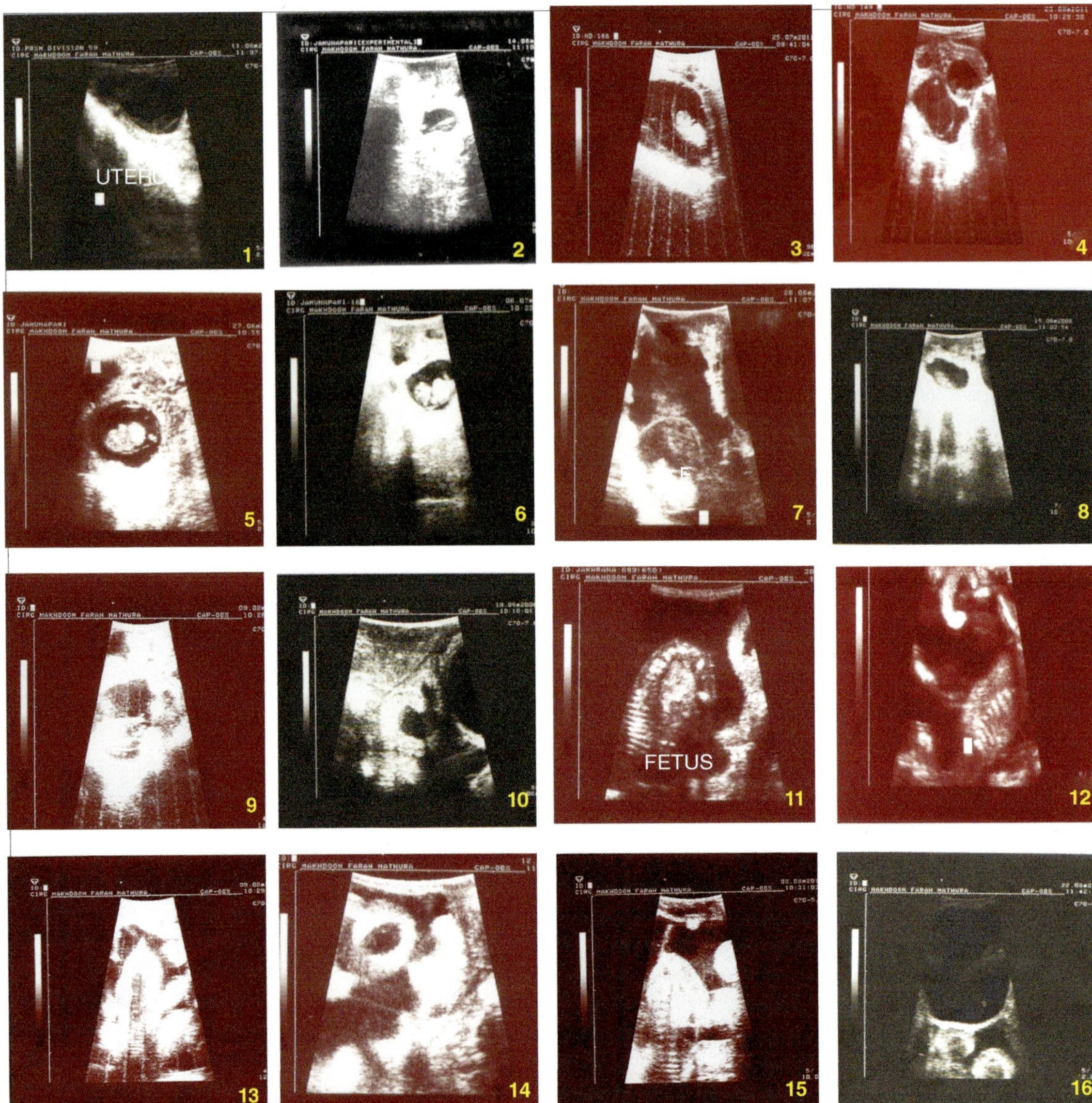

Photo plate 1: [1] Non pregnant Uterus, [2] 28 days pregnancy, [3] 31 days pregnancy, [4] 32 days pregnancy (twin foetuses), [5] 35 days Pregnancy (fetal budding), [6] 35 days pregnancy (fetal head & trunk), [7] 38 days pregnancy (hyper echoic amniotic ring), [8] 40 days pregnancy, [9] 45 days pregnancy, [10] 56 days pregnancy, [11] 60 days foetus (fetal ribs), [12] 62 days foetus (half moon shaped placentomes), [13] 69 days pregnancy (spinal column), [14] 70 days pregnancy (bowl shaped placentomes), [15] 75 days pregnancy (enlarged placentomes & spinal column), [16] 93 days pregnancy (enlarged placentomes).

Fig. 3.3. Ultrasonograms of pregnant goats at different stages. Courtesy of Goel *et al.*, 2018, personal communication.

does carrying multiple or single foetuses. This technique can distinguish a viable pregnancy from a hydrometra, pyometra and foetal mummification. Real-time ultrasonography can also be used to estimate foetal age in the goat at 40 to 100 days of gestation by measuring the width of the foetal head.

3.3 Assisted Reproduction and Reproductive Technologies

3.3.1 Artificial insemination

The artificial insemination (AI) technique has been extensively used in livestock breeding programmes to make optimum use of outstanding genetic merit males for better genetic progress. AI is difficult in goats due to the complex anatomy of the cervix.

Advantages of artificial insemination include:

- The technique helps to introduce a desirable genotype into the livestock population at a faster rate.
- High-quality semen produced by superior sires can be used to breed large numbers of does.
- The maintenance of a large number of males, which would otherwise be required, is reduced.
- AI permits reproduction when suitable males are not available for natural mating.
- The technique provides accurate breeding records for good herd management.
- Valuable sires that are incapable of mating due to age or injury may be utilized through electric ejaculation.
- AI prevents stray mating and also solves the problems of mating of dams of unequal size.
- The technique helps control disease transmission, since males used for insemination are under health control and do not circulate from one flock to another.

Successful artificial insemination depends on precise management of semen collection, its frequency and use. Semen can be collected by an artificial vagina (AV) using an anoestrus or oestrus doe as a dummy. The AV for goats is made of heavy rubber, measuring 20 cm length and 5 cm diameter. A rubber (latex) lining of 30 cm length and 5 cm diameter is inserted in the artificial vagina; the space between the rubber lining and AV is filled with 150–175 ml warm water (45°C). A graduated glass semen collection cup is held at the end nearer to the valve and the other end is lubricated with soft paraffin to avoid friction injury to the buck penis. The buck is allowed one or two false mounts before the AV is placed towards the end of the penis. The penis is deflected into AV so that a single thrust will result in ejaculation. As soon as possible after collection, the semen should be delivered to the laboratory for evaluation. Sperm concentration is measured to determine the number of spermatozoa per millilitre of undiluted semen. It can be measured by calorimetric/spectrophotometer technique, which is considered to be an efficient and convenient technique. In the sperm meter technique, semen is diluted with Tris buffer in a ratio of 1:50 and observed under 20× magnification. The total number of sperms in 100 squares will be the millions of sperm per millilitre. (Details for artificial insemination procedures is given in Protocol 3 at the end of this chapter.)

3.3.2 Superovulation

Reproductive efficiency in animals may be increased by induction of superovulation to harvest a large number of embryos followed by their transfer into synchronized recipients. The commonly used gonadotropins are pregnant mare serum gonadotropins (PMSG or eCG), porcine follicle-stimulating hormone (pFSH) and ovine FSH. Serum gonadotropin from pregnant mares is administered as a single injection during mid-luteal phase, followed by prostaglandin $F_2\alpha$ (lutalyse) 48 hours after PMSG injection. Super ovulation with pFSH requires multiple injections in the morning and evening at 12-hour intervals for 4 days and single injection of $PGF_2\alpha$ (Lutalyse) 48 hours after first pFSH injection for better superovulation and oestrous response.

3.3.3 Embryo transfer

Embryo collection and transfer form the basis of embryo transfer technology in animals and can be performed by: (i) invasive (involves surgical intervention); (ii) semi-invasive (laparoscopic); and (iii) non-invasive (intracervical) methods. Embryos for transfer can be collected non-surgically, surgically or at slaughter following removal of the reproductive tract. Slaughter collections are usually conducted by slaughtering the animal at an abattoir, removing the reproductive tract aseptically and then flushing the tract under laboratory conditions. Embryo transfers are carried out either surgically/laparoscopically or non-surgically in goats. Limitations of the surgical approach include damage to the reproductive tract in the form of adhesions and other disorders that restrict the repeated use of animals for embryo collection and transfer.

3.3.4 *In-vitro* fertilization

This is the large-scale economical embryo production method to produce high-quality animal production. It is not only possible to produce embryos on a large scale from slaughterhouse material, required for the various biotechnological experiments, but also from adult and pre-pubertal animals for faster propagation and conservation of a particular breed.

3.3.5 Cryopreservation of ova and embryos

Cryopreservation is the long-term storage of cells and tissues for future use. Cryopreservation of embryos provides a means for reducing animal space requirements, and protects valuable animal lines from potential loss due to environmental disasters, genetic drift and infectious diseases. Cryopreservation helps in the import and export of embryos and prevents vertical disease transmission. The technique allows the storage of valuable germplasm produced by in-vitro embryo production, intracytoplasmic sperm injection and cloning. There are two main methods for cryopreserving embryos/ova: conventional slow cooling/freezing and vitrification.

3.3.6 Cloning

Cloning denotes the production of an exact genotypic replica of organisms. Cloning involves collection of oocytes, culture and *in-vitro* maturation of oocytes, enucleation of oocytes, transfer of (somatic) nuclei to (or fusion of the somatic cells with) enucleated oocytes, culture of the resulting embryos and, finally, embryo transfer into recipients of the same or highly related species. In mammals, live offspring have been obtained from embryos generated from somatic cells in a number of species, i.e. sheep, cattle, mice, pigs, goats, horses, rabbits and cats.

3.3.7 Transgenesis

Transgenic animal production offers a powerful tool to introduce useful genes selectively from diverse sources into the genome of targeted organisms. Goats act as a model animal for transgenic animal production and there are successful models in different areas. Pronuclear microinjection (PMI) and nuclear transfer technique are the methods successfully used to produce transgenic livestock (see Chapter 8 for details).

3.4 Protocols

Protocol 1: Goat semen evaluation

1. Preparation of Tris buffer: Tris-based buffer can be prepared by dissolving 18.02 g Tris (hydroxymethyl aminomethane), 9.50 g citric acid and 5 g fructose in 500 ml distilled water. The pH of this buffer is adjusted to 6.75 with 10% citric acid solution. Antibiotic penicillin 1000 IU/ml and streptomycin 1000 µg/ml is added to the buffer.

2. Semen extension: On the day of collection, each ejaculated semen sample after macroscopic evaluation has to be kept in a water bath at 37°C and then diluted with suitable diluent (Tris buffer, initial dilution 1 : 8, i.e. one part semen and eight parts diluent) on the basis of sperm concentration using a slow rate of extension in three or four steps within 10 minutes and then for an equilibration period of 4 hours in a cold cabinet machine.

3. Initial motility: This is assessed at 37°C at 40× magnification using a phase-contrast microscope just after extension of semen. 5–10 µl of diluted semen is placed on a clean, grease-free, warm microscopic slide covered by a cover slip. Sperm motility estimation is performed in different microscopic fields for each semen sample.

4. Physical evaluation:

- **Volume**: It is recorded in millilitres and measured directly from calibrated collection cups with an accuracy of 0.10 ml.
- **Colour:** The colour score for each ejaculate is recorded as creamy white, milky white or whey white colour.
- **Consistency:** The consistency score for each ejaculate is recorded as thick, medium or thin.
- **Mass activity:** Mass activity is assessed at 37°C with a microscope at 4× magnification. 5–10 µl neat semen, immediately after collection, is placed directly on a microscope slide without coverslip. A mass activity or wave motion characteristic in neat semen depends upon concentration percentage of progressively motile spermatozoa and speed of progression of spermatozoa.

Table 3.6. The standard score card for sperm mass activity.

Semen activity characteristics	Mass activity score
Rapid waves and swirls	+ 5
Less rapid waves and eddies	+ 4
The swirls are slow scattered in the field	+ 3
Swirls are absent. Individual movement of spermatozoa are more evident from the field	+ 2
No wave motion are observed	+ 1
Spermatozoa are immobile	+ 0

5. Thawing and evaluation of post-thaw progressive motility: Thawing is performed at 40°C for 45 seconds and post-thaw progressive motility is assessed using a phase-contrast microscope (40×) with a warm stage maintained at 37° C. 5–10 μl of diluted semen is placed directly on a microscope slide and covered by a cover slip. Post-thaw sperm motility estimation is performed in different microscopic fields for each semen sample by at least two independent experts and the average value is considered for the final motility score.
6. Functional tests:
   - **Live and dead sperm count:** One drop of semen sample is mixed with two or three drops of eosin-nigrosine stain with a fine glass rod on a clean grease-free slide. A thin smear is prepared on a clean and dry glass slide, air dried at 37°C and observed under an oil immersion lens (100× magnification) under a phase-contrast microscope (1000× magnification). A total of 200 sperm should be counted at different fields. The dead sperm cells take on eosin colour, whereas those that are alive before staining do not take any colour.
   - **Hypo-osmotic swelling test in buck spermatozoa:** The hypo-osmotic solution (150 mosm/L, 50 mosm/L) should be prepared fresh and stored at 4°C till use. A 0.5 ml aliquot of hypo-osmotic solution is mixed with 0.05 ml of undiluted semen and incubated at 37°C for 1 hour. A drop of diluted semen is then placed on a clean sterilized dry glass slide and covered with a cover slip. A total of 100 spermatozoa should be examined under phase-contrast microscopy for tail curling.
   - **Acrosomal integrity test in buck spermatozoa:** A thin smear of diluted semen is made on a clean, grease-free slide. The smear is dried and kept in Hancock fixative solution for 30 minutes at 37°C. The slide is then washed in double-distilled water for 5–7 minutes. Then, 3 ml of Giemsa stain solution is added drop by drop; 2 ml of Sorensen buffer solution and 35 ml of double-distilled water is added and the solution is mixed thoroughly. The slide is then kept in the staining jar solution for 3 hours at 37°C in an incubator. The stained slide is washed with distilled water and air dried. Two hundred spermatozoa per slide is examined under a high power phase-contrast microscope for acrosome morphology (classed as intact, altered, completely lost and abnormality).

$$\text{Acrosome integrity (\%)} = \frac{\text{number of acrosome stained spermatozoa}}{\text{total number of spermatozoa}} \times 100$$

Protocol 2: Standard procedures for oestrus synchronization

Progesterone-based protocol

1. A sponge or CIDR (controlled internal drug-release dispenser) impregnated with the hormone progesterone or with synthetic derivatives of progesterone known as progestogens (melengestrol acetate (MGA), altrenogest or norgestomet) is introduced into the vagina of the doe.
2. These agents mimic the action of the CL, a progesterone- secreting structure located on the ovary. Elevated levels of progesterone released by the CIDR prevent the doe from coming into oestrus until the device is removed.
3. After CIDR removal, the doe will come into heat about 36–72 hours later and can be bred by AI. CIDRs can be administered for varying periods, typically ranging from 9 to 21 days.

Prostaglandin-based protocols

1. Prostaglandin F_2 alpha ($PGF_2\alpha$), a naturally occurring hormone, is effective for oestrous synchronization only if females are normally cycling and have a CL present in their ovaries.
2. This method typically involves administration of two $PGF_2\alpha$ injections, the first given on day 1 and the second on day 10. Does will come into heat

after either the first or second injection and can be bred by AI at either time. Most of the treated does will show oestrus 36–96 hours after the second $PGF_2\alpha$ injection

Combined methods utilizing both progesterone and prostaglandin-based treatments

1. A CIDR-$PGF_2\alpha$ protocol involves a short, 6-day CIDR treatment. A CIDR is inserted on day 1 in the morning and 6 days later is removed. An injection of $PGF_2\alpha$ must be given after CIDR removal. If PGF is given at the time of CIDR removal, does will come into oestrus 36–72 hours later and can be bred by AI.

Combined methods utilizing both progesterone and PMSG

1. A sponge or CIDR impregnated with the hormone progesterone shall be introduced into the vagina of the doe and shall be removed after 17 days.
2. An intramuscular injection of PMSG shall be given to the doe at the time of sponge/CIDR removal.
3. All the does will exhibit oestrus simultaneously around 30 hours after removal of the sponge/CIDR.

Protocol 3: Standard operating procedures for artificial insemination services in goats

Artificial insemination always requires a high standard of technical efficiency. It should be ensured that AI technicians follow the breeding policy of the state/country for achieving breeding goals.

Storage and distribution of frozen semen straw (FSS)

- Store frozen semen doses in a well-ventilated and safe storage area.
- Ensure a proper identification system for each semen container, canister and goblet.
- While transferring semen doses, frozen semen should not be exposed above liquid nitrogen as it may cause irreversible damage to sperm viability.
- Liquid nitrogen should be replenished in both storage and distribution containers at regular intervals to ensure the proper level of liquid nitrogen.
- Details of semen doses supplied to centres at the time of dispatch should be recorded.

Storage and distribution of liquid semen

- Store liquid semen containers/vials in a refrigerator at 5°C and ensure a proper identification system for each liquid semen container.
- Liquid semen containers vials should not be exposed to outside temperature, as it may cause irreversible damage to sperm viability.
- Details of semen doses supplied to centres at the time of dispatch should be recorded.

Liquid nitrogen procurement, storage and delivery

- Prepare a schedule of liquid nitrogen replenishment on a fortnightly/monthly/quarterly basis and maintain records.
- Centres should have a dip stick with critical level marks and a ready reckoner for assessing the liquid nitrogen levels and quantity of liquid nitrogen in litres.

Procedures for artificial insemination technicians

SEMEN HANDLING

1. Keep the liquid nitrogen container with semen straws in a location that allows seeing into the neck tube of the container, and ease in withdrawing and replenishment of semen and liquid nitrogen. The surrounding should be well ventilated, dry and dust free.
2. Clean the AI gun, scissors and other accessories whenever they get soiled or at least once a week with hot water and air dry them. Sanitize the AI gun and the scissor with isopropyl alcohol after drying. The AI equipment should be kept clean after each insemination. Surgical spirit and soaps are lethal to semen, hence should not be used to clean equipment.
3. Always carry the required semen doses in the liquid nitrogen container to the farmer's/farm supervisor's doorstep. Never carry semen straws in pockets/thermos-flasks/polythene bags filled with water/ice, and so on.

4. Always attach the paper tag provided with each goblet to the requisite canister of the container to identify the type of semen in each canister.

PROCEDURE FOR THAWING

1. Thaw straws in a water bath at 37°C for 30 seconds. Set the temperature of water to 37°C using a thermometer. Lift the required canister and remove the required straw with the help of forceps. Plunge the straw for 30 seconds in the water in the vessel. Wipe the straw with tissue paper or a clean cloth.
2. It may sometimes be appropriate to assess post-thaw motility using a microscope.
3. Load the straw in an AI gun, cut the sealed end of the straw and prepare the AI gun for insemination by placing a sheath over it and proceed to inseminate the does.

ARTIFICIAL INSEMINATION PROCEDURE In goats, cervical artificial insemination or trans-cervical artificial insemination procedures may be performed as per the process described below.

1. The doe in oestrus is separated from the flock. Does detected to be in oestrus in the morning are inseminated in the evening and those detected in the evening are inseminated the next morning.
2. Lift the hind legs of the doe in a 45° angle towards the direction of sunlight.
3. The inseminator should gently introduce a lubricated speculum into the vagina to determine the stage of oestrus. If a large quantity of vaginal mucus is preventing the correct location of the cervix, the assistant should tilt the animal down to allow the mucus to run out of the vagina with the aid of the speculum.
4. Locate the cervix using the speculum with the help of sunlight or AI light.
5. The inseminator passes the insemination gun as deep as possible into the cervix (generally 0.5–1 cm) and deposits the semen. The assistant should hold the doe firmly. Ensure that no urine or faecal matter contaminates the tip of the AI gun.
6. The doe should then be gently lowered and returned to the flock while keeping it calm.
7. The speculum should be properly disinfected/sterilized between inseminations.
8. Ask the farmer/farm supervisor to keep the animal under observation for the next 12–24 hours.
9. Ask the farmer/farm supervisor to prevent the animal from coming into contact with any buck during the remaining part of the present heat.
10. If signs of heat persist even after 18–24 hours, repeat AI, otherwise observe for heat symptoms after 18–21 days and also after 36–42 days.
11. Follow each and every animal inseminated after around 21 days to find out whether it has repeat heat.
12. Follow each and every animal inseminated for pregnancy diagnosis after 1 month and record the date and result of pregnancy diagnosis in the register.
13. Follow each and every pregnant animal and record kidding/lambing details of the animals inseminated in the register.
14. Maintain all records related to artificial insemination, pregnancy diagnosis, and kidding/lambing.
15. Advise farmer/farm supervisor on proper heat detection, feeding, management and healthcare of animals.

3.5 References

Abecia, J.A., Forcada, F. and Gonzalez-Bulnes, A. (2012) Hormonal control of reproduction in small ruminants. *Animal Reproduction Science* 130, 173–179.

Aiello, S.E. and Moses, M.A. (eds) (2016) *Merck Veterinary Manual*, 11th edn. Elsevier Health Sciences, London, UK.

AICRP-Goats (1985) Terminal Report-1985, All India Coordinated Research Project on Goats. CIRG, Makhdoom, Farah (Mathura) India.

AICRP-Goats (2017) Annual Report. All India Coordinated Research Project on Goats. CIRG, Makhdoom, Farah (Mathura) India.

Bhattacharyya, N.K. (1982) Physiology of reproduction in Indian and small dwarf goats. Paper presented in Physiology section, LXIX session of Indian Science congress Association, Mysore (India), pp. 1–8.

Bourzat, D., Gautsch, K.D., Baumann, M.P.O. and Zessin, K.H. (1992) Major demographic parameters for White Somali goat and Black-headed Somali sheep in the Somalia central rangeland. In: Rey, B., Lebbie, S.H.B. and Reynolds, L. (eds) *Small Ruminant Research and Development in Africa. Proceedings of the First Biennial Conference of the African Small Ruminant Research Network*. ILRAD, Nairobi, Kenya pp. 159–175.

Buckrell, B.C. (1988) Application of ultrasonography in reproduction in sheep and goats. *Theriogeniology* 29, 71–84.

Chemineau, P. (1983) Effect on oestrus and ovulation of exposing Creole goats to the male three times of the year. *Journal of Reproduction & Fertility* 67, 200–208.

Chemineau, P., Daveau, A., Maurice, F. and Delgadillo, J.A. (1992) Seasonality of estrus and ovulation is not modified by subjecting female Alpine goats to a tropical photoperiod. *Small Ruminant Research* 8(4), 299–312.

Davey, C.G. (1986) An evaluation of pregnancy testing in sheep using a real time ultrasound scanner. *Australian Veterinary Journal* 63, 347–348.

Dawson, L. (2015) Pregnancy diagnosis in goats. *In: Proceedings of the 17 Annual Goat Field Day,* Langston University, Langston, Oklahoma, USA, pp. 41–44.

Devendra, C. and Burns, M. (1983) *Goat Production in the Tropics*. Commonwealth Agricultural Bureaux, Farnham Royal, UK.

Dumas, R. (1980) Contribution à l'étude des petits ruminants du Tchad. *Revue d'élevage et de médecine vétérinaire des pays tropicaux* 33(2), 215–233.

Faugère, O., Leforban, Y., Landais, E., Faugère, B., and Merlin, P. (1988). L'élevage traditionnel des petits ruminants dans la zone de Kolda (Haute Casamance). *Document de travail No. 1/Kolda*. ISRA, Dakar, Senegal.

Gall, C.L. (1981) *Goat Production*. Academic Press, London.

Gerbaldi, P. (1978) Divers rapports sur les petits ruminants au Niger. IEMVT, Maisons-Alfort, France.

Goel, A.K. and Agrawal, K.P. (2004) Studies on the onset of puberty and oestrus behaviour in Jakhrana goats. *Indian Journal of Animal Sciences* 74(11), 1130–1131.

Goel, A.K., Kharche, S.D., Sinha, N.K. and Jindal, S.K. (2003) Oestrus behaviour of Jamunapari goats under tropical environment. In: Souvenir, Annual Convention and National Symposium on Current Reproduction Technology for Improvement of Livestock Production in India. Salt Lake City, Kolkata, India, 22–24 August 2003 (Ab. No. BPL-01), p.86.

Gonzalez, F., Cabrera, F., Batista, M., Rodriguez, N., Alamo, D. *et al.* (2004) A comparison of diagnosis of pregnancy in the goat via transrectal ultrasound scanning, progesterone and pregnancy associated glycoprotein assays. *Theriogenology* 62, 1108–1115.

Greyling, J.P.C. (1988) *Certain Aspects of Reproductive Physiology in the Boer Goat Doe*. PhD thesis, University of Stellenbosch, South Africa.

Greyling, J.P.C. (2010) *Applied Reproductive Physiology*, PhD Thesis, University of Stellenbosch, South Africa, pp. 139–155.

Greyling, J.P.C and Grobbelaar, J.A.N. (1983) Seasonal variation in semen quality of Boer and Angora goat rams using different collection techniques. *South African Journal of Animal Science* 13(4), 250–252.

Greyling, J.P.C. and Grobbelaar, J.A.N. (1988) Artificial insemination in sheep: farming in South Africa. *Mutton Sheep* B7/1988.

Haibel, G.K. (1990) Use of ultrasonography in the reproductive management of sheep and goat herds. *Veterinary Clinics of North America: Food Animal Practice* 6(3), 597–613.

Haumesser, J.B. (1975) Quelques aspects de la reproduction chez la chèvre rousse de Maradi. *Comparison avec d'autres races tropicales ou subtropicales. Revue d'élevage et de médecine vétérinaire des pays tropicaux* 28(2), 225–234.

Haumesser, J.B. and Gerbaldi, P. (1980) Observations sur la reproduction et l'élevage du mouton Oudah Nigérien. *Revue d'élevage et de médecine vétérinaire des pays tropicaux* 33(2), 205–213.

Hendy, C.R.C. (1980) A single visit survey of animal performance in goat flocks of Mtwara region. In: *Tanzanian Society of Animal production Morogoro (Tanzania).* Proceedings of the 7th Scientific Conference of the Tanzanian Society of Animal production held at Moshi 7, 96–108.

Humblot, P., De Montigny, G., Jeanguyot, N., Tetedoie, F., Payen, B. *et al.* (1990) Pregnancy-specific protein B and progesterone concentrations in French Alpine goats throughout gestation. *Journal of Reproduction and Fertility* 89, 205–212.

ILCA (1987) *ILCA's Strategy and Long-term Plan*. Addis Ababa, Ethiopia.

Jainudeen, M.R., Wahid, H. and Hafez, E.S.E. (2000) Sheep and goats. In: Hafez, B. and Hafez, E.S.E. (eds) *Reproduction in Farm Animals*, 7th edn. Lippincott Williams & Wilkins, Philadelphia, Pennsylvania, USA.

Jalaluddin, A. (1992) Reproductive pattern of black Bengal goats. In: *Proceedings, Pre Conf. Proc. (Vol. I) V Indian Conference on Goats*, New Delhi, March 2–8, 324.

Karua, A.K. (1989) Some performance indicators of Malawi indigenous goats under village and ranch conditions. In: Wilson R.T. and Azeb M. (eds) *African Small Ruminant Research and Development*. Proceedings of a Conference held at Bamenda, Cameroon, pp. 23–28.

Kassem, R., Owen, J.B. and Fadel. I. (1989) The effect of pre-mating nutrition and exposure to the presence of rams on the onset of puberty in Awassi ewe lambs under semi-arid conditions. *Animal Science* 48, 393–397.

Kaushish, A.K., Singh, D. and Pal, S. (1994) Productive and reproductive performance of different breeds of goats under semi-arid conditions. *Indian Veterinary Journal* 71(6), 565–568.

Khan, B.U., Sinha, N.K., Wani, G.M. and Sahni, K.L. (1981) A note on breeding performance in Jamunapari, goats. *Indian Veterinary Journal* 50(3), 251–252.

Lebbie, S.H.B. and Manzini, A.T. (1989) The productivity of indigenous goats under traditional management in Swaziland. In: Wilson, R.T. and Azeb, M. (eds) *African Small Ruminant Research and Development*. Proceedings of a Conference held at Bamenda, Cameroon, pp. 39–50.

Maule, J.P. (1949) Breeding for milk in a Mediterranean environment. Part II: Goats. *Empirical Journal of Experimental Agriculture* 17, 37.

Mazumdar, N.K. and Mazumdar, A. (1982) Breeding and production performances of the Indian Pashmina goats. *Proceedings of the III International Conference of Goat Production and Disease*, Tucson, Arizona, USA, 10–15 January 1982 p. 529.

Mazumdar, N.K. and Mazumdar, A. (1983) Breed characteristics of some Indian Pashmina goats. *Indian Journal of Animal Sciences* 53(7), 779–782.

Medan, M.S., Watanabe, G., Sasaki, K. and Taya, K. (2004) Transrectal ultrasonic diagnosis of ovarian follicular cysts in goats and treatment with GnRH. *Domestic Animal Endocrinology* 27, 115–124.

Mittal, J.P. (1981) A comparative Study on the Reproductive Performance of Barbari and Jamunapari Goats. *Indian Veterinary Journal* 58(10), 794–799.

Mittal, J.P. (1988) Breed characterization of Barware Goats of Arid Western Rajasthan. *Indian Journal of Animal Sciences* 58(8), 357–361.

Mosi, A.K., Opasina, B.A., Hayward, B., Carew, B.A.R. and Valez, M. (1982) Productivity of the West African Dwarf goat at village level in south west Nigeria ILCA, Addis Ababa, Ethiopia.

Mukandan, G. (1980) *Genetic studies on some economically important production characteristics in Malabari breed of goats and its- exotic crosses*. PhD thesis, University of Rohilkhand, Bareilly (India).

Mukundan, G., Bhat, P.N. and Khan, B.U. (1983) Reproduction traits in Malabari goats and their half breds with Sannen. *Indian Journal of Animal Sciences* 53(1), 90–93.

Murayi, Th., Sayers, A.R. and Wilson, R.T. (1987) *La productivité des petits ruminants dans les stations de recherche de l'Institut des sciences agronomique du Rwanda*. CIPEA. Rapport de recherche No. 15. CIPEA, Addis Ababa, Ethiopie.

Ndlovu, L.R. (1992) Reproductive performance of indigenous goats in traditionally managed flocks in north-east Zimbabwe. In: Rey, B., Lebbie, S.H.B. and Reynolds, L. (eds) *Small Ruminant Research and Development in Africa. Proceedings of the First Biennial Conference of the African Small Ruminant Research Network*. ILRAD, Nairobi, Kenya, pp. 177–184.

Okere, C., Streter, J., Solaiman, S and Gurg, N. (2011) Critical analysis of reproductive performance in Boar and Kiko meat goats used under semi-intensive market system. *International Journal of Applied Research* 2, 145–151.

Peacock, C., Devendra, C., Ahuya, C., Roets, M., Hossain, M. *et al.* (2005) Goats. In: Owen, E., Kitalyi, A., Jayasuriya, N. and Smith, T. (eds) *Livestock and Wealth Creation*, Nottingham University Press, Nottingham, UK, pp. 361–385.

Prakash, B., Singh, S.K. and Khan, B.U. (1988) *Effect of system of feeding management on some early reproductive parameters in Barbari goats*. National Symposium and IV Annual Conference Society of Animal Physiology India, CIRG, Makhdoom, Mathura, Sept. 24–26, pp. 65–66.

Prasad, S.P. and Bhattacharyya, N.K. (1979) Oestrus cycle behaviour in different season in Barbari nannies. *Indian Journal of Animal Sciences* 49(12), 1058–1062.

Rana, Z.S., Chopra, S.C. and Balain, D.S. (1982) Comparative performance of Beetal goats and its Crosses with French Alpine and Aug 10 Nubian. *Proceedings of III International Conference on Goat Production and Disease*, Tucson, Arizona, USA. 10–15 January 1982, 350pp. (Abstract).

Reira, S. (1982) Reproductive efficiency and management in goats. *Proceedings of the 3rd International Conference on Goat Production and Disease*, Tucson, Arizona, USA.

Singh, N.S., Gawande, P.G., Mishra, O.P., Nema, R.K., Mishra, U.K. *et al.* (2004) Accuracy of ultrasonography in early pregnancy diagnosis in Doe. *Asian Australian Journal Animal Sciences* 17(6), 760–768.

Smith, M.C. (1980) Caprine reproduction. In: Marrow, D.A. (ed.) *Current Therapy in Theriogenology Diagnosis, Treatment and Prevention of Reproductive Disease in Animals*, W.B. Saunders Company, Philadelphia.

Smith, M.C. and Sherman, D.M. (2009) *Goat Medicine*, 2nd edn. Wiley-Blackwell, Ames, Iowa, USA.

Smith, G.W., Rotstein, D.S. and Brownie, C.F. (2004) Abdominal fat necrosis in a pygmy goat associated with fescue toxicosis. *Journal of Veterinary Diagnostics Investigation* 16(4), 356–359.

Sumberg, J.E. and Mack, S.D. (1985) Village production of West African Dwarf goat and sheep in Nigeria. *Tropical Animal Health Production* 17, 135–140.

Suresh Kumar, P.K., Patel, A.V. and Mehta, V.M. (1989) A note on oestrous cycle characteristics in Surti and Marwari goats. *Indian Journal of Animal Reproduction* 10(2), 173–174.

Warwick, B.L. (1961) Selection against cryptorchidism in Angora goats. *Journal of Animal Science* 20, 10–14.

Wilson, R.T. (1982) Husbandry, nutrition and productivity of goats and sheep in tropical Africa. In: Gatenby, R.M. and Trail, J.C.M. (eds) *Small Ruminant Breed Productivity in Africa*. Proceedings of a seminar held at ILCA, Addis Ababa, Ethiopia, pp. 61–74.

Wilson, R.T. (1984) Indigenous Goats: Productivity in traditional livestock systems of semi-arid Africa. *International Goat and Sheep Research* 2(3), 243–251.

Wilson, R.T. and Traore, A. (1988) Livestock production in Central Mali: Reproductive performance and reproductive wastage in ruminants in the agro-pastoral system. *Theriogenology* 29(4), 931–944.

Yarkin, I. and Eker, M. (1961) The study on physiological breed characteristics of Kilis dairy goats. *Ankara University Agriculture Faculty Annual* Bulletin 11, 143.

Zhao, Y., Zhang, J., Wei, H., Sun, X., Mu, B. *et al.* (2010) Efficiency of methods applied for goat estrous synchronization in subtropical monsoonal climate zone of Southwest China. *Tropical Animal Health Production* 42, 1257–1262.

4 Nutritional Management

4.1 Basics of Goat Nutrition

The major input in livestock production systems is feed, and it is essential to feed the goat for promoting health, immune function and for efficient economical production of meat, milk and fibre. The feeding behaviour of goats is different from other ruminants. The goat is a natural browser, browsing preferably on tree leaves, bark and pods. Goats naturally prefer to graze at a height of 20–120 cm above ground level. They can stand on their hind legs for long periods and choose to pick at the desired parts of trees (Fig. 4.1). Goats have very mobile lips and an agile tongue with which they are able to select specific parts of plants and leaves between thorns. They are very selective in choosing the best portions from available feed and fodder resources. Goats have a wide range of taste and can thrive on bushes, tree leaves and unusual food in extreme conditions. They are highly selective, do not consume soiled feed and fodder, and do not consume the same feed every day. Feeding behaviour of goats in the tropics is quite different from European countries and the US. Farmers use natural grazing and available crop by-products to feed them. Therefore, the nutritional requirement of goats can be formulated based on the available feed and fodder, and farmers need to make efficient use of local resources by formulating complete and concentrated feed.

Ruminants show a preference for certain plants or specific parts of plants during grazing or stall-fed conditions. Cattle and goats appear to recognize specific parts of a plant and have a greater preference for sweet and salty tastes; pygmy goats and local-breed goats have a higher preference for salty tastes (Goatcher and Church, 1970). Goats can distinguish between bitter, sweet, salty and sour tastes and are more tolerant than sheep and cattle to a bitter taste (Bell, 1959). Goats prefer a wide variety of plants that are distasteful to other ruminants. Grazing goats prefer grasses to legumes, and clover over alfalfa (De Rosa *et al.*, 1997, 2002). Goats also show a preference for forage with higher available carbohydrate content (Burns *et al.*, 2001). Goats have the capacity to recognize nutrients and toxic plants (euphagia) and also consume certain plants because of their pleasant smell and taste (hedyphagia). In most cases, body morphology and size may be determinant factors for food preference. Goats have a small body size, and thus prefer more digestible nutritious feeds. Goats as grazing animals with a longer snout or with eyes atop their head would have a selective advantage.

Saliva is essential as a lubricant to facilitate swallowing. Goats have a pointed tongue and jaws, mobile thin lips and deep mouth. Grazed forage is gathered between the labial surface of the lower incisor teeth and the upper dental pad (ruminants lack upper incisors), and forward movement of the muzzle cuts through the forage. In contrast, during rumination, the cud is chewed more slowly and regularly. Premolars and molar teeth aid in mastication. Mastication results in particle size reduction which increases ports of entry for fermenting microorganisms. In addition, movements of teeth excite mechanoreceptors in the mouth that provide stimuli for production of saliva by the salivary glands. Mouth movement also increases the rate and amplitude of primary and secondary cycle contractions of the reticulo-rumen. Feedstuffs that fail to promote chewing (low particle size) result in reduced rumination, saliva secretion and fore stomach motility

Although saliva production is relatively continuous, volumes are greater when ruminants are eating and ruminating. Daily saliva production averages 6–16 litres by sheep and may be more by goats. Saliva produced by ruminants contains no enzymes, but it is particularly rich in buffers (HCO and HPO), which account for its alkalinity (pH 8.1). Being swallowed, saliva aids in preventing ruminal pH from becoming too low for microbial growth. In the rumen, pH is maintained within a range

Fig. 4.1. Browsing of goats with bipedal stance.

(5.5–7.0) ideal for microbial growth. The HPO in saliva allows phosphate to be recycled for rumen microorganisms to synthesize nucleoproteins, phospholipids and nucleotides; recycled urea, being up to 77% of total salivary nitrogen, also provides ammonia for the formation of microbial protein. Urea, recycled in saliva, together with efficient kidney renal tubular urea resorption appears critical for survival of ruminants consuming forage or feeds with very low protein content (Leek, 1993; NRC, 2007). In addition, recycling of urea reduces water excretion, combined with very efficient resorption of water by the large intestine, which helps ruminants survive when the quantity or quality of water is limited.

As browsing ruminants, goats have relatively large salivary glands and higher rates of secretion than sheep. The primary functions of saliva are to provide a copious and continuous supply of alkaline buffers to counterbalance the volatile fatty acids (VFA; primarily acetate, propionate, and butyrate) produced during fermentation, and to provide an aqueous suspension for rumen solids. Secondary functions of saliva include urea recycling as a source of non- protein nitrogen (NPN) for microbial protein synthesis and for synthesis of microbial nucleic acid and membrane phospholipids. Saliva also acts as a wetting agent for ingesta, provides an anti-foaming agent to prevent frothy bloat and supplies proline-rich proteins that bind and deactivate dietary tannins. Saliva provides a medium to short-term adaptation to changes in diet composition, namely, the presence of plant secondary metabolites such as tannins. Salivary proteins influence taste and digestive function. Proline-rich proteins in saliva that bind tannins are present in saliva of browsers but not grazers. The protein profile of goat saliva is in the 25–35 kDa range and differs from sheep saliva (Lamy *et al.*, 2008). Tannin-binding proteins are present in the saliva of deer but not of sheep and cattle (Austin *et al.*, 1989). The mass of salivary glands is correlated positively with body mass but negatively correlated with the ratio of grass to browse in the diet, perhaps reflecting the need for complex salivary compounds, such as tannin-binding proteins, by browsing ruminants (Hofmann *et al.*, 2008). The ratio of salivary gland mass to body size was high for goats, indicating their capacity to browse plants in the foregut (Silanikove, 2000).

4.1.1 Digestive system of the goat

Functional anatomy

In ruminants, the digestive system works first as a storage for forage which is then digested in a safe location to avoid any type of predator. The digestive system starts from the oral cavity and ends at the anus. The system has two functions: the first part of the digestive system is responsible for grazing, mastication and salivation, and the second part is responsible for the digestion of food and elimination of wastes. The digestive system is divided into the following parts to understand the functional anatomy.

- The **oral cavity** is the mouth of animal. The first part is the lips, which are highly mobile. The lips and incisive teeth are responsible for prehension of food. The tongue and teeth are the major structures of the mouth cavity. The lateral walls of the oral cavity are the masseter muscle which helps with mastication. The salivary glands help with salivation.
- The **tongue** is a highly mobile part of the mouth but is not used for prehension of food.
- **Teeth:** Ruminants have four incisor teeth. The milk teeth are 20 in number and permanent teeth are 32 in number.
- The **salivary gland** produces saliva, which mixes with the forage during mastication and helps with the formation of alimentary bolus. The major salivary glands are the parotid, mandibular and sublingual salivary glands.
- The **oesophagus** consists of three segments: the cervical, thoracic and a very short abdominal segment.
- The **stomach** consists of four parts: the rumen, reticulum, omasum and abomasum. The entire stomach is surrounded by visceral peritoneum (Fig. 4.2). The goat stomach has a mechanical and chemical role in the digestion process.
    - The **rumen** lies in the left-hand side of the abdominal cavity along with the reticulum and part of the abomasum (Fig. 4.2). The major function of the rumen is to mix the ingesta with water, breaking it into small particles and starting microbial digestion.
    - The **reticulum** is observed on both sides of the abdomen. It looks like an elongated ball with honeycomb-like mucosa.
    - The **omasum** is mainly located on the right-hand side of the abdominal cavity. It looks like a flattened ball and has 35 laminae on its inner side.
    - The **abomasum** lies in the floor of the right-hand side of the abdominal cavity. It is pear shaped and the inner mucosa has characteristic spiralled folds.

  The stomach starts from the omasum and ends with the small intestine called pylorus. Its major function is to perform chemical digestion.
- The **intestine** is comprised of the small and large intestine in goats. The average length of the large intestine is 7 m and the small intestine is 20–30 m. The small intestine has three parts: the duodenum, jejunum and ileum. These parts mostly facilitate digestion. The large intestine consists of the cecum, colon and rectum. Its main function is reabsorption of ingesta water. It is also responsible for digestion of ingesta and eliminates waste material from body.
- The **liver** is the longest gland of the body. The liver is connected to the stomach and duodenum by the lesser omentum.
- The **pancreas** makes contact with the rumen, omasum, spleen, liver, duodenum, right kidney and jejunum. The pancreatic duct opens into the major duodenal papilla with the bile duct.

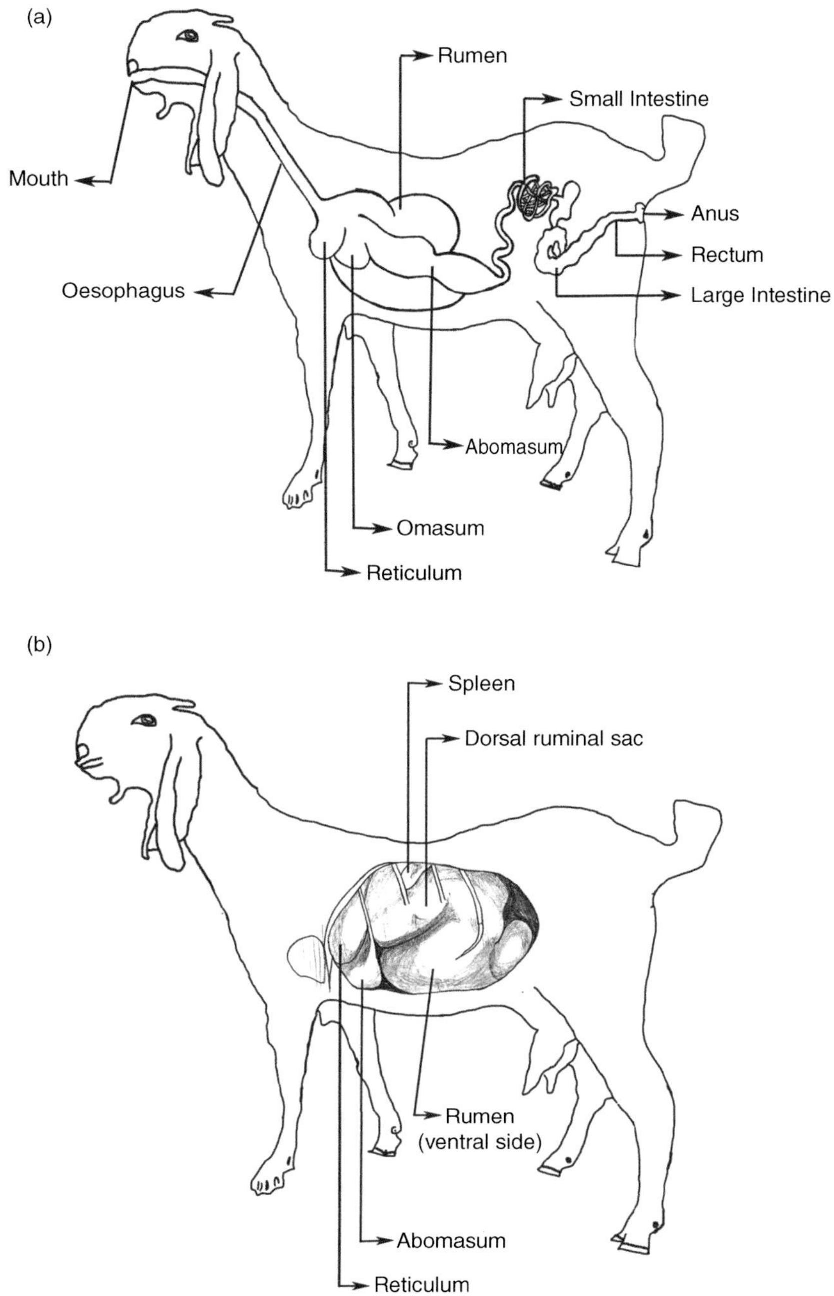

Fig. 4.2. Anatomy of the digestive system of a goat. (a) The details of digestive tract from mouth to anus. (b) The structure of different parts of the stomach.

- The **spleen** belongs to the cardiovascular system but is connected with the stomach. The spleen is a flat organ and lies in the dorsal aspect of the rumen.

The function of the stomach

The rumen has a capacity of 13–30 litres and occupies the left-hand side of the abdominal cavity along with the reticulum and parts of the abomasum. The rumen contents include bacteria, protozoa and fungus, which help in microbial digestion that continues in the reticulum and omasum. The primary function of the rumen is the mincing of ingesta, mixing it with water and starting microbial digestion. The reticulum has a capacity of 1–2 litres and lies on the floor of the abdominal cavity. The typical structure of the reticulum is a honeycomb-like mucosa. The main function is the squeezing of ingesta by contraction to move the ingesta through the rumen. The omasum is located on the right-hand side of the abdominal cavity and has a capacity of less than 1 litre. Its main function is squeezing ingesta and water absorption. The gastric groove running from the reticulum and omasum ends at the abomasum. It directly passes the bolus into the omasum. In kids, during the suckling stage, the gastric groove is transformed into the gastric duct, which allows milk to move from the oesophagus to the abomasum where milk is digested. The gastric groove is functional during the kid and adult stages. The abomasum is located on the right-hand side of the abdominal cavity and has a capacity of 2–3 litres. It starts from the omasum and connects to the small intestine via the pylorus. The major function of the abomasum is to carry out chemical digestion (HCl hydrolysis). Most of the VFAs are absorbed in the abomasum. VFAs are produced during microbial digestion and fulfil the major protein component of the ruminant's energy needs.

The rumen and reticulum are the parts of the digestive tract responsible for microbial fermentation. After the intake of food, rumination occurs. Rumination is the process that includes regurgitation, salivation, re-mastication and re-swallowing. The major part of the fermentation occurs in the rumen mat. The host animal and the microbes are in a symbiotic relationship. The anaerobic fermentation of feed by microbes is desired and has several functions.

- The microbes ferment structural (cellular and hemicellulose) and non-structural (sugar and starch) carbohydrates into metabolized energy sources such as VFAs.
- Conversion of non-protein nitrogen from plants and urea to a high value microbial protein.
- Synthesis of vitamin K and B complex.
- Conversion of metabolically recycled phosphate to microbial nucleic acid.

The omasum helps in fermentation and the absorption and flow of digestion from the reticulo-rumen to the abomasum. VFAs are absorbed in the omasum along with electrolytes. Chlorides (instead of bicarbonate) play a major role in VFA absorption in the omasum. The abomasum transfers partially digested feed material and breaks down ingesta by enzymatic action. HCl, pepsin and lysozymes act on ingesta and digest microbial protein and allow the feed material to move to the duodenum for further digestion

Microbial fermentation

Feed type affects the rate of digestion, rate of passage and retention time of digesta in the rumen. Diets rich in protein stimulate growth of protein-digesting microorganisms but can lead to production of ammonia toxicity or other toxins in the feed. High-fibre diets will increase the proportion of cellulose digesting microorganisms within the rumen and high starch diets increase starch-digesting microorganisms. Diets that contain less than 15% roughage have a negative impact on fermentation and digestion (Vieira *et al.*, 2008). Diets having concentrate feed up to 80% have a better fermentation rate and better digestibility of dry matter. Goats have a higher tolerance to a downwards shift in rumen pH (i.e. more acidic) compared to an upwards shift in pH (Silanikove, 2000). Nitrogen, sulphur and minerals must be obtained from microbial fermentation in the rumen for microbial growth and fibre digestion. Saliva increases the buffering capacity of the rumen via bicarbonate of sodium and potassium. Similarly, grinding of feed material increases passage rates of particles and negatively affects fibre digestion.

Intestine digestion

The intestine consists of the small and large intestine. The small intestine is 20–30 m long and the large intestine is 7 m long. The small intestine has three parts: duodenum, jejunum and ileum. The duodenum is the part where the digestion process is facilitated by the opening of the bile duct and the pancreatic duct. The jejunum is the longest segment of the small intestine and the ileum is the shortest section. The ileum joins to the caecum of the large intestine. The large intestine has three segments (caecum, colon and rectum). The main function is reabsorption of water and some microbial fermentation

The intestine is the primary site for digestion and absorption. The bile duct and pancreatic duct open in the small intestine. In new-born goats, food material passes from the oesophagus to the abomasum via the reticular groove. Colostrum passes directly into the abomasum and then into the duodenum where immunoglobulins are absorbed by the intestinal mucosa. The process is activated by phagocytosis. Colostrum contains an antitrypsin factor that prevents the digestion of antibiotics in the duodenum. This absorption occurs until 48 hours after birth and provides immunity to the kid. Colostrum also provides lactose, which provides energy during the first 48 hours. If the kids are not able to use lactose as a source of energy, they will die. New-born kids are in the nursing phase until they reach 3 weeks of age. Milk flows directly to the omasum and abomasum through the reticular groove of the reticulo-rumen. Suckling stimulates the saliva, which contains esterase. The abomasum secretes renin (chymosin) and hydrochloric acid during the suckling period. Butter fat is hydrolyzed by lipase from esterase in saliva. Curd proteins are hydrolyzed by renin and acid. From 3–4 weeks of age, the kids start nibbling green or dry fodder. This activity stimulates development of the salivary glands and the reticulo-rumen. Ruminal microbes start growing due to ingestion of food and water. At this juncture, the metabolism shifts from a glucose-driven to a VFA-driven system. Fermentation of feed starts and rumination also begins. After 8 weeks of age, the reliance on milk reduces and the animal takes different feed and fodder. The salivary esterase diminishes and salivary urea is present. The abomasal renin is replaced with pepsinogen.

4.2 Characterization of Feed

4.2.1 Feed and feed composition

The basic aspect of nutrition is to assess the nutrient requirement of an animal, to analyse the composition of available feed and to formulate the feed to meet the nutritional requirements of that animal. Nutrient requirements of goats for growth, lactation, pregnancy and production stages have been published by the NRC (2007). For practical use for farmers, some nutritional requirements have been provided in Protocol 3 at the end of this chapter. Goats consume a wide variety of shrubs, bark, leaves, woody plants and weeds. The grazing behaviour of goats allows them to survive in harsher climates as compared to sheep and cows. Feeds for goats can be divided into two groups: forage/roughage and concentrates. The proportion of forage to concentrate in the diet depends on the different stages of growth, physiological stages and production requirements.

In tropical countries, it is necessary to manage milk production for kids up to 2 months of age for meat goats and 4–5 months for dairy goats. The concentrate feed requirement will vary depending on the objective and type of production system.

Feed composition

Feed analysis should be carried out individually by a process of proximate analysis. The proximate analysis parameters include dry matter, nitrogen, ether extract and ash. Proximate analysis partitions feed compound into different categories and the details of measurements are given in Table 4.1.

The nutrient content and chemical composition of feed stuff are described in Fig. 4.3.

4.2.2 Forage/roughage

The major livestock production system in the tropics depends on grazing and mostly mixed farming systems. Goats consume large amounts of forage during grazing. They select long hay over chopped hay and leaves over stem portions of forage. Fresh forage are preferred first; similarly, hay is preferred over silage. When goats graze in pastures or browse, they typically choose better quality forage. Goats consume more legumes than grasses; they also choose the parts within a plant with the highest nutritional value if they have a selection opportunity.

Goats select tree foliage, shrubs, bark, seeds and nuts during browsing. They prefer foliage from trees and shrubs that is rich in different nutrients. The chemical composition of browsing material differs widely between different grasses. Grass is rich in unlignified cell wall neutral detergent fibre

Table 4.1. Proximate analysis parameters of feed.

Parameter	Components
Moisture	Water
Ash	Minerals
Crude protein	Protein, amino acids, vitamin B
Ether Extract	Fat, oils, vitamins A, D, E, K
Crude fibre	Cellulose, hemicellulose, lignin
Nitrogen-free extract	Cellulose, hemicellulose, lignin sugars, tannins, water soluble vitamin

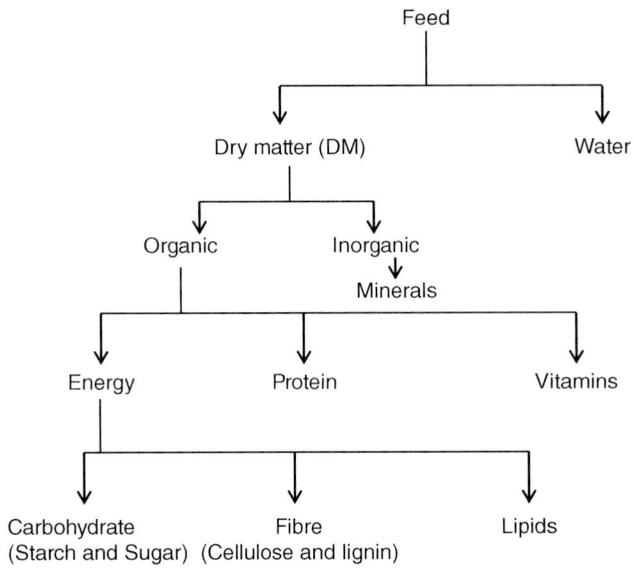

Fig. 4.3. The general outline of the chemical composition of feed.

(NDF = 50–70%) that is readily available for cellulite fermentation in the rumen. Browsing materials have relatively lower NDF (NDF = 30–50%) and are highly lignified and rich in readily fermented cell contents. They also contain secondary plant metabolites that are not present in grasses. Goats are resistant to many plant toxins and anti-nutritive factors. Therefore, they can eat more browsing materials compared to sheep and cattle. Goats prefer browsing (65–70%), with fewer amounts of grasses (20–25%) and forbs (5%). They have a preference for woody plants, bushes and use their upper mobile lips to select the tender leaves that are more digestible. Due to the high tannin content in their diets, goats are less prone to parasitic infection (Min and Hart, 2003). Browsing goats have high blood urea nitrogen concentrations, which reflect the reduced consumption of carbohydrates or the fact that carbohydrates and proteins are bound with tannin.

Forage or roughage includes dry fodder (straw, husks, hay and other by-products) and wet forage (green fodder, tree leaves, pasture). Forage is rich in fibre (18% crude fibre) and low in energy. As the goats usually graze tree leaves and vegetation from other sources, it increases the butter fat content of goat milk. Dry fodder includes straw and husks from leguminosae fodder. Feeding of leguminosae fodder like channa (*Cicer arietum*) and arhar (*Cajanus cajan*) straw will also reduce the protein content of concentrate (grain mix) feed. The hay should be good quality as it needs to be harvested during the leafy stages. Legume hay has a better feeding value. Hay intake is dependent on the fibre content, which increases as plants mature. Similarly, the feed intake increases when the hay is chopped or pelleted. Groundnut (*Arachis hypogaea*), channa (*Cicer arietum*), arhar (*Cajanus cajan*) and alfalfa (*Medicago sativa*) are good for goats. In African countries, *Sericea lespedeza* was effective in reducing parasitic loads in goats and it has better digestibility (Terril *et al.*, 2007). Straw is considered to be low-quality feed but the value of straw increases when fed in a mixture with other feed.

Silage should be given to goats gradually. A higher intake of silage has been associated with metabolic problems such as acidosis, demineralization and other problems such as enterotoxaemia and listeriosis. Silage produced from legumes has higher dry matter (DM) protein content and lower cell wall content than silage produced from grass. Silage feeding produces less milk in goats as compared to green fodder. A combination of hay and silage reduces digestive and metabolic problems and increases feed intake and production performance.

4.2.3 Grains and concentrate

Goats prefer pelleted or cubed and coarse feed compared to finely ground feed and will refuse to consume mouldy or fermented feed. The nutrient requirements of goats should be met from forage consumption. If the forage requirement is not sufficient to support performance, then it is necessary to provide supplementary feeding to fulfil the dietary shortages. Feeding grain generally decreases forage consumption and digestibility; however, the combination of low-quality forage with concentrate feeding has a positive effect on forage intake (Huston, 1994).

Energy concentrate

This feed includes cereal grains, root crops, bakery by-products and other by-products such as cauliflower waste, pomegranate pulp, tomato pulp, orange leaves, olive leaves and soya bean hulls. While formulating the energy concentrate, it is important to be cautious so that the constituents can be provided judiciously. For example, feeding whole cotton seed in high amounts has a negative effect on male growth and reproduction. Cotton seed contains gossypol, which has deleterious effects on male reproduction. Cotton seed supplementation up to 15% of DM diet had no adverse effect on growth or sperm quality in growing male goats (Solaiman *et al.*, 2009). Again, wheat proportion in feed formulation should be less than 50% of grain mix due to its rapid fermentation activity. Some carbohydrate sources promote renal calculi and should be used in a balanced feed with different minerals. Similarly oil, vegetable fat or animal fats are also used to promote growth activity and increase the energy density of diet. However, feeding more than 7–8% of fat intake may negatively impact fermentation and depress fibre digestion.

Protein supplements

Commercial protein supplements are available in different forms and it is the most costly component of feed. Fishmeal, bone meal and blood meal must not be fed to goats, as there is a chance of transferring prions and this can cause human health problems. Similarly, urea is used as a source of NPN and a source of nitrogen. As a rule of thumb, urea should not be more than 25% of the required protein in feed for milking does. Urea does not provide energy; it is useful when the diet contains fermentable carbohydrate in adequate quantities. Urea should not be fed to young ruminants before the rumen is fully functional. Urea is hydrolyzed to ammonia by rumen microorganisms. In turn, the ammonia is used by rumen microorganisms during fermentation for synthesis of microbial protein. Grain-based commercial supplements may not be economical for growing and finishing meat goats. Some starch-based feed of cattle when fed to goats may reduce ruminal pH and fibre digestibility.

Mineral and vitamin supplements

It is necessary to provide minerals and vitamins for optimum performance. Mineral supplements are provided as salt or salt mixed with macro and trace elements. Generally, calcium, phosphorus and sodium are supplemented in feed formulations. Salt should be supplemented in feed. Calcium is present in different fodder and forages. Similarly, phosphorus deficiency is observed in forages during some periods of the season. Magnesium deficiency is also observed in certain areas. In areas where soil is deficient in some minerals, it is necessary to supplement these minerals in the diet. Copper and zinc deficiency are common in soil. If molybdenum or sulphur is present in soil, it affects the availability of copper in the diet. Molybdenum (5 or 10 mg) present in the diet can deplete copper in the liver and negatively impact the immune response in goats (Solaiman *et al.*, 2008). Generally, soil or forage should be examined for copper, molybdenum, sulphur, cobalt, iodine and sodium deficiency.

Feed additives

Feed additives are products that are used for improving the feed conversion efficiency, growth and health of animals. Commercially, two additive classes are available (ionophores and probiotics/prebiotics).

- Ionophores generally improve the health of animals and may increase milk production and help to prevent ketosis and bloat. Ionophores reduce amino acid catabolism, fat mobilization, non-esterified fatty acids and ketone body formation. Lasalocid and monensin are the ionophores used for coccidiosis control during the growth period and in meat goats. Ionophores are antibiotics produced by actinomycete bacteria in the rumen. They reduce the number of Gram-positive bacteria by altering the flux of ion in rumen bacteria. The fermentation effect due to ionophore increases the proportional production of propionate and decreases acetate and butyrate production (Russel and Strobel, 1989).
- Probiotics and prebiotics: probiotics are viable cultures of microorganisms fed directly to animals that prevent the establishment of pathogenic

bacteria in the digestive tract. Probiotics after rumen fermentation optimize the rumen pH and improve digestibility. They also reduce stress and improve immune function. The probiotics are derived from fungal cultures (*Aspergillus oryzae* and *Saccharomyces crevisiae*) and lactic acid bacteria (*Lactobacillus* and *Streptococcus*). Prebiotics such as mannan oligosaccharides and beta glycan have also been added to feed to improve gut immunity in kids and adult goats.

4.3 Feeding Practices

Feeding practices describe the process by which the nutritional requirements of an animal are fulfilled by formulating feed ingredients of varied nutrient content. This is required for optimal animal performance, minimizing feed cost and the sustainable use of feed resources. Feeding practices in goats are highly variable in different parts of the world, for example, extensive systems, intensive systems, browsing in range land, high-altitude grazing and simple tethering near a household (Fig. 4.4) (see also Chapter 7).

Carbohydrates (fibre, starch and sugars) are converted into VFAs and methane. VFA is absorbed in the omasum or rumen. VFAs are the main source of energy for growth. Rumen produces these types of VFA, that is, acetic acid, propionic acid and butyric acid. Starch/sugars relates to milk production, while increasing fibre content increases fat content of milk. Feeding whole grains increases milk production and also produces more methane. Fibre is broken down slowly and helps to maintain an efficient rumen condition. Proteins are mainly digested by microorganisms and produce peptides, ammonia and amino acids. Microbial proteins are again absorbed in the small intestine (duodenum). A non-protein nitrogen source should be added to the feed if required. Protein that is digested in the rumen is called rumen-digested protein, which is a source of nitrogen to microbes. Protein that is digested in the small intestine is called by-pass protein and is utilized by goats. The DM requirement should be fulfilled by supplying green/dry feed and concentrate. In the tropics, goats eat about 2–3% of their body weight as dry matter. Small-sized goats such as West Africa Dwarf, Small East Africa

Fig. 4.4. Goats managed near a household by tethering to coconut trees.

and Black Bengal may need 2.5% of their body weight, but large-sized goat breeds such as Jamunapari and Beetal will require 3% of their body weight. Dairy goats from temperate countries require about 4.5% their body weight. It is therefore necessary to have an idea about the dry matter content of each feed. The DM content of feed can be obtained from published reports or by estimation of the dry matter by standard laboratory methods. The dry matter requirement can be calculated as follows:

Suppose, an adult goat weighs 30 kg.

Then, it requires 3% of body weight as dry matter requirement, i.e. $30 \times \frac{3}{100} = 0.9$ kg dry matter per day.

If the feed sample or grass contains 30% dry matter, then $0.9 \times \frac{100}{30} = \frac{9}{3} = 3$ kg of grass requirement per day.

Therefore a 30 kg goat will require 3 kg grass feed per day.

Formulation of different feed compositions should be carried out as per Protocol 1 at the end of this chapter.

4.3.1 Feeding for meat, milk and fibre production

Feeding for meat production

Creep feed is generally provided to kids for getting a better market price. Creep feed is a commercial grower feed mix containing 14–18% protein and about 10–12 % fibre. Creep feeding should be provided to animals in an individual feeder and the feeder should be kept in the area where animals spend most of their time. More fibre in creep feed will promote rumen development, and a mineral mix and salt should also be provided. Different ethnic groups or locations require good quality animals before festivals or holidays. Therefore a planned body growth attainment programme should be followed and a specific diet formulation should be provided. The composition of a grower mixture is provided in Table 4.2.

Generally, male kids in the tropics are castrated early in life and it is suggested that castrated kids will grow faster. However, castration will reduce their average daily gain, reduce the growth rate and increase the time period for kids to reach marketable weight. Premature castration also increases the incidence of urinary calculi by reducing the development of the urinary tract. It has been observed that buck kids gain 20–30% more weight than castrated kids (Solaiman, 2010). If the market age is 6–12 months of age for meat, then castration can be omitted so that weight gain will be more rapid. However succeeding with better meat production requires the skill and management practices of individual farmers. Goats fed a concentrated diet will have a higher dressing percentage at slaughter due to more fat and better muscle mass with a smaller digestive tract (Solaiman, 2010).

Table 4.2. Composition of grower mixtures.

Particulars	Percent in feeds 1	2	3	4
Maize	50	–	30	12
Dal chunies	–	–	30	35
Oats, barley	–	50	30	–
Wheat bran	30	20	–	30
Ground nut cake	10	10	–	5
Molasses	7	17	7	5
Mineral mixture	2	2	2	2
Salt	1	1	1	1

Feeding for milk production

The most challenging task is to maintain the health of animals and maximize the milk yield after the 5-month gestation period. Therefore, it is necessary to effectively manage the doe from the start of the pregnancy period. Pregnancy, parturition and lactation are the phases when animals need high energy for maintenance of body weight. It is important to manage the advanced pregnancy period for maintaining the body reserve of does and better health of the kids. During early lactation, the animal is in negative energy balance. Milk production in goats generally peaks at 45–60 days (6–8 weeks) of lactation. The energy intake peaks during 3 months of lactation. The animal manages to produce milk by relying on body reserves and loses 0.5 kg of body weight per week for the first 2 months of lactation. The animal restores body weight after 3 months of lactation and body weight gain is about 1–2 kg per month. The feed should contain 3–4% of fat of DM content of feed to promote optimum milk fat synthesis. The higher fat percentage has a negative effect on fibre digestion and calcium absorption. The higher protein percentage during early lactation is required for microbial protein synthesis amino acid milk. A complete feed formulation for lactating goats is presented in Table 4.3.

Table 4.3. Recommended nutrient contents for complete mixed rations for dairy goats.

Item	Amounts
Total digestible nutrients	75% for early lactation and reduced to 60% during late lactation
Crude protein	18% for early lactation and reduced to 14% during late lactation
Acid detergent fibre	18% for early lactation and increased to 20–22% during late lactation
Fat	3–4% added fat particularly for early lactation
Concentrate:forage ratio	60:40 for early lactation and changing to 40:60 during late lactation
Calcium:phosphorus ratio	At least 1.5:1
Phosphorus	0.4%
Added salt	0.5%
Sulphur	0.25–0.35%
Nitrogen:sulphur ratio	10–12:1, with added sulphur usually needed when diets contain non-protein nitrogen
Magnesium	Higher magnesium may be needed with N and K fertilized pastures to prevent hypomagnesemia
Copper	15 ppm
Selenium	3 ppm

Source: Solaiman (2010).

Dairy goats can consume between 4% and 6% of dry matter per 100 kg body weight as compared to 3–4% dry matter consumption for dairy cows. This high level of intake allows the dairy goat to have an abundance of nutrients readily available for the synthesis of milk. Good quality dry fodder and a balanced grain mix appear to be the best approach for maintaining high levels of milk production. However, too much poor quality fibre will lead to lowered levels of milk production. Legumes make excellent hay and are usually superior to most other hay crops because of their higher protein content. Good quality leguminous hay should have a green colour, a small stem, be adequate in leaves and have 17–20% of protein. The protein concentrate mixture (17–20%) is needed as a supplement as average quality grass hays are low in digestible protein. With good quality hay (15–20% protein) or the use of small grain pastures such as oat, rye and ryegrass, a 14–16% protein concentrate mixture is generally adequate. High concentrate, low roughage diets are associated with decreases in rumen fluid pH and in the ratio of acetate to propionate produced in the rumen (Lu *et al.*, 2005). These changes, in turn, are associated with decreased butterfat production (Davis *et al.*, 1964). It may therefore be necessary to add a buffer, such as sodium bicarbonate at 4%, to the concentrate portion of the ration to maintain butterfat production (Hadjipanayiotou, 1982).

Feeding for fibre production

Nutrition for fibre goats will be similar to that used for meat goats. Fibre goats are small to medium in size and their growth rate is comparative to meat goats. Forage should be the major component of fibre-producing goats. Fibre protein contains higher levels of sulphur and it is necessary to maintain a nitrogen:sulphur ratio of between 10:1 and 12:1. A diet containing higher protein for goats will increase fibre diameter and produce a lower quantity of coarse fibre. Fibre yield may increase but quality will decrease. Therefore, it is necessary to be careful when designing feed formulation for fibre goats. The fineness in fibre quality will be at a maximum when providing forage and in adverse nutritional conditions. The basic nutritional requirement should be optimized for better body condition score, growth and less kid mortality.

4.3.2 Feeding of kids

Feeding of kids is different compared to adult goats as the rumen is not well developed. The feeding of kids is similar to simple stomach (i.e. monogastric) animals, but it is designed to help in the development of the rumen microbial system along with proper growth of the kids.

Colostrum is the first milk from the doe; it is high in nutrient value comprised of vitamins A and B, proteins (immunoglobulins) and minerals. The protein content of colostrum is about 20% as compared to 3.5% for normal milk. The antibodies (immunoglobulins) in colostrum provide protection from diseases during the neonatal stage. Colostrum should be ingested or bottle-fed as soon as kids have a suckling reflex. Ideally all new-born kids should get colostrum within the first hour after birth and no later than within the first 6 hours because the percentage of antibodies found in colostrum decreases rapidly after parturition. The ability of the new-born kids to absorb antibodies also decreases rapidly 24 hours after birth. New-born kids should ingest 10% of their body weight in colostrum during the first 24 hours of life for optimum immunity. The extra colostrum should be stored in the refrigerator and fed at about body temperature at a later time.

If the mother of the kids is not available, colostrum from another dam may be given to them. If natural colostrum is not available then artificial colostrum can be given. The ingredients should be mixed well and fed at a temperature of 40°C. Milk feeding should be continued after colostrum feeding until 90 days of age. The milk should be fed in regular intervals and concentrate or green fodder can be introduced at 2–3 weeks of age. The amount of milk should be reduced as the animal starts consuming concentrate mixture.

Milk replacer (Table 4.4) can be introduced to kids in some breeds that have high multiple births and low milk yields.

Some precautions should be taken while feeding the milk replacer to the kids. Milk replacer should be thoroughly mixed in lukewarm water (40°C). It should be prepared fresh in clean/sterile feeding bottles. Feeding milk replacer twice a day is sufficient.

The kids start nibbling at grass from 15 days of age. Good quality green fodder should be given to kids. This will help in the proper development of microbial digestion in the rumen. Green fodder like berseem, cowpea, maize or lucerne can be given at this time. Tree leaves can also be fed to kids after lopping. Creep feed is also introduced at the same time. This will help with the rapid growth of the kids and hasten the development of the rumen. Creep mixture is a concentrated mixture containing high energy and easily digestible protein. This concentrate mixture can be formulated at home using locally available ingredients. Two types of creep mixture can be offered depending on the type of roughage available. If grasses and cereal fodders are available, a creep feed with 18% digestible crude protein (DCP) and 75% total digestible nutrient(TDN) can be offered. If leguminous fodders are available, a creep feed with 12% DCP and 70% TDN should be offered. The concentrate mixture can be formulated by mixing maize (67%), ground nut cake (20%), wheat bran (10%), mineral mixture (2%) and salt (1%). Feeding of kids in this way reduces the mortality of kids and results in higher weight gain.

Table 4.4. Composition of milk replacer.

Ingredients	%	Crude protein (%)	Energy (kcal/100 g)
Skimmed milk powder	47	16.44	168
Soya bean flour	09	03.92	39
Wheat flour	35	03.85	121
Coconut oil	07	0.00	63
Mineral mixture	02	0.00	0
Add butyric acid 0.2 ml, citric acid 2 g, probiotics in 1 kg of milk replacer powder			

Source: Dutta (2009).

4.3.3 Feeding growing goats (3–12 months)

Normally growing goats of 3–12 months of age consume dry matter equal to 4% of their body weight. The feed should have 12–14% of crude protein and 60–65% total digestible nutrients. Feeding kids in these proportions will reduce mortality and result in a better growth rate.

4.3.4 Feeding of bucks

Bucks need to be in good body condition (i.e. a body condition score [BCS] greater than 3) before the breeding season because feed intake may be relatively low during that time, resulting in weight loss. Thus body condition should be evaluated 3 months before the breeding season. The necessary supplemental nutrition should be provided for the buck to achieve the desired BCS. Under most conditions, supplementation with concentrate mixture at the rate of 0.5–1.0% of body weight will be adequate when pasture is scarce. A mineral mixture should always be given to the breeding bucks. This will improve the semen quality of the bucks and also maintain their reproductive efficiency.

4.3.5 Feeding pregnant goats

There is a very slow growth of the foetus during the first 100 days of gestation, with more than 80% growth taking place during the last 8 weeks. Low levels of energy during late pregnancy can lead to pregnancy toxaemia, particularly with goats. However, high levels of feeding through pregnancy can also lead to pregnancy toxaemia that can result in kidding difficulties.

4.4 Water Requirements

Goats are efficient water users and are the best when compared to other ruminants. The amount of water a goat needs depends on the breed, size, the climatic conditions, the type of feed and fodder available, and body function. The water requirement is different during lactation, growth and fibre production. Tropical breeds in arid and semi-arid environments may drink once every 2 days in a season (Peacock, 1996). The goats have the ability to use their rumen for water storage and reduce water intake. It is generally recommended that goats consume four times the amount of water compared to dry matter (i.e. 4 kg of water for 1 kg DM feed) and the requirement will increase if the environmental temperature is high. Similarly, it is recommended to provide 1.3 litres of water per litre of milk produced by lactating goats.

The normal body water content of a goat exceeds 60% of its live weight. In certain desert-adapted breeds, water content may reach 76% of body weight (Shkolnik *et al.*, 1980). Goats in confinement generally drink three to five times daily. Lactating goats typically ingest 3.5–4 kg of water for each kg of dry matter consumed at environmental temperatures below 15°C (Jarrige *et al.*, 1978). Water ingested per kg DM consumed increases by 30%, 50% and 100% at 20°C, 25°C and 30°C, respectively (Jarrige, 1988). Kids raised in confinement systems should have water available to drink from a pan or bucket from 1 week of age, to assist with rumen development. Milk bypasses the rumen because of the closure of the oesophageal groove, so plain water must be provided to mix with grain and hay in the rumen for the best digestion. Supplying water from a nipple bottle can lead to haemolysis because too much is consumed by the kid at any one time (Middleton *et al.*, 1997).

4.5 Digestive Disorders

The major digestive disorders in goats will be discussed briefly in this section to get an idea of normal rumen function.

4.5.1 Bloat

Ruminal tympany, or bloat, is frequently observed in goats in both the frothy (nutritional) and free gas forms. Frothy bloat is usually caused by overeating lush green feeds such as clover, alfalfa or legume pastures and excessive concentrate feeding. The free gas distension may also be associated with consumption of green fodder and formation of gases in the rumen. The tympany or bloat may be easily detected by observing acute distension of the left paralumbar fossa and a drum-like sound on percussion. The rumen expands with foam and the goat can die of respiratory or circulatory failure due to excessive pressure on the diaphragm. In acute cases, immediate release of gases is needed by insertion of trocar and canula in the left paralumbar fossa and the administration of anti-zymotic and anti-flatulent drugs either orally or through a trocar and canula inserted in the rumen.

4.5.2 Rumen acidosis

Rumen acidosis develops in growing kids and adult goats due to excess feeding on carbohydrates (i.e. grain overload). This condition may be moderate to severe in nature. Goats are highly susceptible to acute ruminal acidosis induced by sorghum at a dose rate of 50 g/kg body weight. (Mohamed Nour *et al.*, 1998). Wheat at a dose rate of 100 g/kg body weight is required to produce the classical signs in goats (Tanwar and Mathur, 1983). The preventive measure can be taken up by providing magnesium hydroxide as a feed mix at the rate of 3–4% to control lactic acid formation. In preventive strategies, good feeding management is the key to reduce the mortality rate due to rumen acidosis. Prophylactic approaches have demonstrated that lactic acidosis can be reduced by immunization against *Streptococcus bovis*, and that live Sb-5 vaccine is effective in invoking mucosal as well as systemic antibody responses in sheep (Gill *et al.*, 2000).

4.5.3 Displaced abomasum

This condition may also occur after ingestion of excess carbohydrates and causes abdominal distension due to gas accumulation and fluid pooling in the abomasum, which results in left abomasal displacement. The presence of fluid can be established by ballottement of the ventral rumen and by reflux through a stomach tube. The single case of left displaced abomasum reported in a goat was associated with intermittent bloat (West *et al.*, 1983).

4.5.4 Urea/ammonia toxicity

Urea is mixed in feed as a source of non-protein nitrogen but toxicity may occur in over or improper mixing of urea in feed. Toxicity occurs if 30–50 g urea/100 kg body weight is consumed in a single meal by an unadapted or hungry ruminant (Smith and Sherman, 2009). Non-protein nitrogen is particularly dangerous when animals are fed high-fibre diets lacking readily digestible carbohydrates. Excess urea makes the rumen pH alkaline. Urea toxicity can be best managed by administration of acetic acid (vinegar) at the rate of 200–500 ml by stomach tube, which reduces the rumen pH and retards absorption of urea.

4.5.5 Cyanide and nitrate/nitrate toxicity

Hydrocyanic acid (HCN) is present as a glycoside in certain plants, and after digestion within the rumen releases free HCN. Plants heavily fertilized with nitrogen and plants affected by drought, wilting or frosting contain higher concentrations of glycosides. HCN toxicity is caused by ingesting plants with cyanogenic glycosides such as *Sorghum* spp., *Prunus* spp., *Zea mays* (corn) and *Acaia leucophoea* (Remza).

4.5.6 Mimosine toxicity

Leucaena leucocephala (Lam.) is a rapidly growing, drought tolerant, palatable and high-yielding tropical or sub-tropical fodder tree which provides 25–35% crude protein and other nutritional components. In spite of being an excellent source of nutrients, *L. leucocephala* forage and seeds contain a number of toxic constituents that may severely limit its utilization in livestock. Mimosine is a non-protein amino acid found in *L. leococephala*. The level of mimosine in the leaf is about 2–6% and varies with seasons and maturity (Kumar, 2003). *Leucaena* feeding causes poor growth, loss of hair (Chakraborty and Ghosh, 1988) and depressed serum thyroxin level leading to goitre. Hiremath (1981) suggested that use of *Leucaena* fodder may be restricted to 30% of the green forage in the case of cattle and buffalo, and 50% for goats. The effect of *Leucaena* and mimosine can be reduced by heat treatment and by supplementation with amino acids. Several strains of 3,4-DHP-degrading ruminal bacteria have been characterized and have been given the genus and species designation, *Synergistes jonesii.* Ruminal inoculation with ruminal contents from adapted animals, enriched cultures of 3,4-DHP-degrading ruminal bacteria, and pure cultures of *S. jonesii* have all been used successfully to establish ruminal populations that are capable of degrading 3,4-DHP and preventing *Leucaena* toxicosis (Hammond, 1995).

4.5.7 *Jatropha curcas* toxicity

Toxicity to *Jatropha curcus* has been reported with sequential development of the clinical signs and lesions in the organs of desert sheep and Nubian goats dosed with *Jatropha curcas* seeds at 0.05, 0.5 and 1 g/kg/day. Diarrhoea, reduced water consumption, dehydration, sunken eyes, inappetence and loss in condition were the important signs of *J. curcas* poisoning in sheep and goats. The main pathological changes are haemorrhage in the rumen, reticulum, lungs, kidneys and heart, catarrhal and/or haemorrhagic enteritis, hepatic fatty change, pulmonary congestion and oedema, and straw-coloured fluid in serous cavities (Ahmed and Adam, 1979).

4.5.8 Oak toxicity

Oak toxicity occurs after excess or sole feeding of oak leaves. Oak contains a compound called phenol, which includes tannins, that is toxic in large amounts. Animals show depression, abdominal cramping and constipation followed by bloody, mucoid faeces, and they often succumb to uraemia. Goats appear highly resistant to oak toxicity, possibly because of comparatively increased levels of tannase enzymes in the rumen mucosa (Smith and Sherman, 2009).

4.6 Metabolic Disorders

4.6.1 Grass tetany or grass staggers or hypomagnesemia

This is a condition of low magnesium concentration in the blood, caused when animals are heavily milked and graze in a field that is highly fertilized. The pasture, which is rich in nitrogen and potassium, causes low magnesium concentrations in the blood. This disease usually occurs in springtime in lush pastures, but it can also appear following rainy weather when grasses and green cereal grains are beginning to grow. Acute death often occurs within 2 to 3 hours of onset of the disease. The signs include wide-eyed staring, muscle twitching in the ears and flanks, hypersensitivity to sound and touch, staggering, foaming at the mouth and convulsions. Goats most at risk are lactating does and older goats. An affected goat responds to slow intravenous (IV) administration of calcium and magnesium.

4.6.2 Milk fever

Milk fever is a metabolic disorder due to low calcium concentration in the blood of does after kidding. This disorder is also usually related to low blood phosphorus and low blood magnesium. The initial symptoms are characterized by nervous signs such as hyperactivity, twitching of the lips, eyelids and ears, and being cold to touch. The doe quickly becomes sternally recumbent and then laterally recumbent in the final stages. The head may be turned back to the flank (opisthotonos posture). The animal becomes lethargic, anorexic with poor milk production. If milk fever is suspected, calcium borogluconate (60–100 ml of a 20–25% solution) can be given very slowly intravenously or subcutaneously (Smith and Sherman, 2009)

4.6.3 Ketosis

The primary cause of pregnancy toxaemia is inadequate nutrition during late gestation, usually because of insufficient energy density of the ration and decreased rumen capacity as a result of foetal growth. The kids require an increasing amount of carbohydrates in the last trimester. Does bearing twins have a 180% higher energy requirement than those with just a single foetus. Does carrying triplets have a 240% greater energy requirement. Toxaemia and ketosis are also seen in obese pregnant does. Underweight animals that are fed a poor quality feed are also prone to toxaemia. The affected pregnant does showed inappetence, aimless walking, muscle twitching, opisthotonos, grinding of the teeth, partial to complete blindness, ataxia and, finally, sternal recumbency and coma, which may lead to death. The timely intravenous therapy of 25% glucose saves the life of the animal.

4.7 Feed Resources Development and Conservation

Goats are recognized as a machine that can convert bushy shrubs and plants into nutritious products like milk and meat. Feeding of some browse materials like trees, leaves, bushes, twigs, and so on, appears to be advantageous and enjoyed by goats.

First the farmer should be convinced about the value of forage development for their goat enterprise. Long-term forage development strategies for a goat enterprise should be initiated by selecting appropriate forage in the specific agro-climatic zones. There should be a well-planned strategy to plant trees and cultivate legumes and grasses in different seasons. Herbaceous legumes should be cultivated and a community common-property resource should be developed.

In the tropics, the period when the grass grows coincides with the rainy season, making it difficult for farmers to dry the grass for hay. For better quality hay, grass should be cut young. The grass should not be harvested after the flowering stage. The quantity of hay may increase but it will be of low quality. Similarly, silage can be prepared but it is usually used in temperate climates; it is not a good option for tropical countries. The foliage is fermented by bacteria in anaerobic conditions. The tropical forage will need additional energy for better fermentation. The quality of fibre fodder can be improved by chopping and treating with NaOH or urea. Moreover, mineral supplementation should be provided with feed. Grazing management should be practiced to manage the natural parasites in the area.

Continuous supply of feed and fodder is one of the most important activities in goat production. When there is a lack of cropping area for fodder cultivation, then the shortage of fodder supply can be achieved by best utilization of highly nutritious top fodder. It is usually rich in minerals and vitamins and is important for development of the animals. The preferred tree fodders are presented in Table 4.5, and local names along with botanical names are presented in Table 4.6. These tree fodders are commonly

Table 4.5. The preferred tree fodder of goats.

Preference	Leaves	Pods
High	*Acacia nilotica, Ailanthus excelsa, Dichrostachyes nutan, Leucaena leucocephala, Sesbania grandiflora, Ficus glomerata, Prosopis cineraria, Zizyphus numularia*	*Pithecellobium dulcee, Zizyphus numularia*
Medium	*Acacia tortilis , Acacia Senegal, Albizia amara, Albizia lebbeck, Hardwickia binata, Dichrostachys cinerea, Sesbania sesban, bauhinia variegata, Parkinsonia aculeate, Pithecellobium dulcee*	*Sesbania grandiflora, Ficus glomerata, Parkisonia aculeate, Prosopis cinerea*
Low	*Albizia procera*	

Table 4.6. Local and botanical names of tree fodder available in the tropics.

Local name	Botanical name
Desi babool, Kikar	*Acacia Nilotica*
Israeli babool	*Acacia tortilis*
Ardu, Arua, Mahrukh	*Ailanthus excelsa*
Siris	*Albizia lebbek*
Kadam	*Anthocephalus*
Kathal	*Artocarpus*
Neem	*Azadirachta indica*
Kachnar	*Bauhinia variegata*
Lasura, Bhokar, Borla	*Cordia dichotoma*
Bar, Bargad	*Ficus bengalensis*
Gular, Umar	*Ficus glomerata*
Pakar	*Ficus lacor*
Pipal	*Ficus religiosa*
Anjan	*Hardwickia binata*
Papri, Kanju, karanji, Chilla, Banchilla, Kunj, Bisenda	*Holoptelea integrifolia*
Subabool, Koo babool	*Leucaena leucocephala*
Aam	*Magnifera indica*
Bakain	*Melia azedarach*
Shahtoot(Mulberry)	*Morus alba*
Jungle Jalebi	*Pithecelobium dulce*
Jhak, Kharjal, Pilu	*Salvadora persica*
Agasti, Bak, Basama, Basna, Chogache, Hatiya	*Sesbania grandiflora*
Jainti, Jait, Rawasan	*Sesbania sesban*
Jamun	*Syzygium cumini*
Imli	*Tamarindus indica*
Ber	*Zizypus mauritiana*
Bordi	*Zizyphus nummularia*
Khejri	*Prosopis cineraria*
Khumta	*Azadirachta indica*
Subabool	*Leucaena leucocephala*
Orinja	*Acacia leucophloea*
Bajra	*Pennisetum americanum*
M.P.Chari	*Sorghum bicolor (fodder variety)*
Cowpea	*Vigna sinensis*
Guar	*Cyanopsis tetragonoloba*
Jowar	*Sorghum bicolor (grain variety)*
Israeli Babool	*Acacia Tortilis*
Sahjan	*Moringa olifera*

available in grazing areas. Although top feeds have optimum nutrients, their utilization is restricted due to the presence of anti-nutritional factors. The effects of these anti-nutritional factors differ from species to species and from place to place. Therefore, tree foliage should be considered to be supplemental feed instead of sole feeding.

Table 4.7. Major fodder trees and their top feed yield.

Tree species	Lopping period	Leaves/tree/year (kg)
Ailanthus excelsa	January-February	100
Albizia lebbek	October-November	70
Albizia procera	February-June	11.30
Albizia amara	February-June	13.54
Acacia nilotica (leaf) (Pod)	January-February March-May	20 70
Azadirachta indica	December-January	50
Dichrostachys cinerea	December-January	0.75
Hardwickia binata	November-December	15
Haloptelia integrifolia	November-December	10–15
Leucaena leucocephala	Ever green	7.5 t/ha or 1.5 kg/tree
Prosospis cineraria (Pod)	November-May	30–40 15–30
Ziziphus numularia	November-December	–
Acacia tortilis	February-June	10–15
Sesbania grandiflora	February-June	0.90
Sesbania sesban	February-June	0.30

Source: Singh (1987) and Rai *et al.* (2007).

The leaf fodder yield per tree varies considerably and it depends on species, initial age, lopping intensity and interval as well as agro-climatic conditions. The average quantity of leaves available from a tree in a year is presented in Table 4.7.

4.8 Strategies for Feed Formulation in Agro-Climatic Zones

Goats consume many different types of feed in the tropics and browse and graze freely throughout the day. It is also important to know the different types of feed and fodder available in the local area. In arid and semi-arid tropic regions, the tree grows better and the tree fodder is fed to goats in large amounts. In humid tropics, leaves from shrubs and trees can have high moisture content which depresses the intake in goats. Goats consume pods in different areas. Goats eat grass when there is nothing else available. In a pastoral area, the grass will quickly become dry and goats do not like dry grass. Crop residues (e.g. maize, sorghum, millet, rice, wheat, barley and oats) are less digestible and unsuitable feed for goats. Legumes are quality feed and rich in protein and highly digestible. Crop by-products include rice or wheat bran, mustard cake, ground nut cake, linseed cake, sunflower and safflower cake, sugar cane tops, and so on. These are a very good source to fulfil the energy and protein requirement of goats. Crops such as maize, barley, oats, millets, bajra and sweet potatoes can be fed directly or processed for formulation of concentrate feed or complete feed.

Balanced feed formulation is necessary for better growth and milk yield as well as maintaining a better immune status in goats. It is also necessary to know the availability of fodder and forage resources in different agro-climatic conditions as this will serve purposes such as easy availability, low cost and better nutritive value. Local nutrient availability will be better with respect to nutrient absorption by the goat. Goats have an affinity for tree leaves, bushes and grasses; moreover, the feed formulation can be assembled from available cereal crop, legume crop and forest resources. Deciduous forests provide a good source of tree leaves and can be used for complete feed preparation. Tree legumes have high digestibility as well as a high protein content. Similarly, herbaceous legumes also have high digestibility and high protein content; and grasses have medium digestibility and low protein content. Legumes are able to trap atmospheric nitrogen in the soil and convert it into vegetative material with a high level of protein content. Grasses need to be supplemented with legumes to make a balanced diet. The best-adapted forages in specific climatic conditions such as frost, drought and water logging conditions are presented in Table 4.8.

Tables 4.9 and 4.10 list the cereal and legume crops available in different countries for sustainable use of natural resources for formulation of concentrate and complete feed. Farmers need to plan and process these crop residues and legumes for preparation of concentrate feed and complete feed.

Specific crop information, including cereals, legumes, fruit trees and forest information for the Indian sub-continent with respect to different agro-climatic conditions are presented in Table 4.11. Farmers should have strategies for crop procurement in different seasons for formulating different types of feed for their livestock.

4.9 Protocols

Protocol 1: How to formulate feed for your goat

Feed formulation for goats

A goat requires energy (food) for different activities such as growth, reproduction, milk production, fibre production, maintenance of normal activities and grazing.

How do you calculate the quantity of feed required to feed a goat to meet the different requirements?

Dry matter requirement

A goat needs a certain quality of dry matter to fulfil its basic requirements.

It needs energy for maintenance activity to stay alive; that is, to fulfil the basic requirements of body functions such as breathing, digestion, blood circulation and the excretory system.

In tropical climates, goats consume about 3–4% of their body weight as dry matter requirement per day. The smallest types of meat goat (West African Dwarf, Small East African, Katjang, Black Bengal) consume 2.5–3.5% of their body weight as dry matter requirement. Large-sized breeds such as Jamunapari, Beetal and Sirohi consume 3–4% of their body weight. Exotic goat breeds in tropic zones consume about 4.5–5% of their body weight as dry matter requirement. For tropical goats, it is necessary to include 3–4% dry matter requirement as a guide for practical calculation.

A goat at 12 months of age weighs about 26 kg and will consume 4% of its weight as dry matter consumption.

Dry matter requirement at 4% = $\frac{26 \times 4}{100}$ = 0.104 kg dry matter per day

As a general guide, tropical grasses contain about 25–30% dry matter depending on the growth stage.

1 kg of dry matter will be available for 3.33 kg grass/green fodder

To obtain 0.104 kg of dry matter a goat will need to eat 0.346 kg grass/green fodder per day (0.104 × 3.33=0.346 kg grass/green fodder per day).

How much energy and protein?

The energy value of feed is expressed in terms of metabolizable energy (ME), which is the

Table 4.8. Herbaceous legume, trees and grasses for frost and drought tolerance and water logging conditions.

Common name	Botanical name	Nutritive value
FROST TOLERANCE		
Tree legumes		
Tree lucerne	*Chamaecytisus prolifer*	Good
Herbaceous legumes		
Alfalfa	*Medicago sativa*	Very good
White clover	*Trifolium repens*	Very good
Grasses		
Cocksfoot	*Dactylis glomerata*	Good
Tall fescue	*Festuca aruninacea*	Good
Perennial ryegrass	*Lolium perenne*	Good
Phalaris	*Phalaris aquatica*	Good
Fodder crops		
Oats	*Avena sativa*	Good
Fodder beet	*Beta vulgaris*	Good
DROUGHT TOLERANCE		
Tree legumes		
Pigeon pea	*Cajanus cajan*	Very good
Leucaena	*Leucaena leucocephala*	Good
Herbaceous legumes		
Cassia	*Cassia rotundifolia*	Good
Alfalfa	*Medicago sativa*	Very good
Tepary bean	*Phaseolus acutofolius*	Good
Verano stylo	*Stylosanthes hamata*	Good
Seca stylo	*Stylosanthes scabra*	Fair
Grasses		
Buffel grass	*Cenchrus ciliaris*	Good
Bambatsi panic	*Panicum coloratum*	Good
Phalaris	*Phalaris aquatica*	Good
Fodder crops		
Fodder beet	*Beta vulgaris*	Good
WATERLOGGING TOLERANCE		
Tree legumes		
Calliandra	*Calliandra calothyrsus*	Good
Sesbania	*Sesbania grandifolia*	Good
Herbaceous legumes		
Greenleaf desmodiu	*Desmodium intortum*	Very good
Lotononis	*Lotononis bainesi*	Good
Puero	*Pueraria phaseoloides*	Good
Grasses		
Signal grass	*Brachiaria decumbens*	Fair
Para grass	*Brachiaria mutica*	Fair
Tall fescue	*Festuca aruninacea*	Good
Bambatsi panic	*Panicum coloratum*	Good
Hamil grass	*Panicum maximunm*	Fair
Paspalum	*Paspalum plicatulum*	Fair
Phalaris	*Phalaris aquatica*	Good
Setaria	*Setaria sphacelata*	Good
Fodder crops		
Oats	*Avena sativa*	Good

Source: Peacock (1996)

Table 4.9. Sustainable crops available in different tropical countries.

Countries	Agro-ecological zones	Sustainable crops
Zambia	Semi-arid	Maize, millets, sorghum, rice, soybean, cotton, sugarcane, sunflower, wheat
Sudan	Desert Semi-desert Low rain fall savannah	Maize, millets, sorghum, wheat, barley and pulses
Ethiopia	Semi-arid	Wheat, beans, sugarcane, maize
Somalia	Semi-arid	Maize, sorghum, cowpea
Kenya	Afro-alpine High potential Medium potential Semi-arid	Corn, beans, peas
Nigeria	Semi-arid Dry sub-humid Sub-humid Humid Very humid Ultra humid Mountainous Plateau	Wheat, groundnut, rice, cocoa bean, peanut, cashew nuts, maize
Ghana	Sudan savannah Guinea savannah Costal savannah Forest/savannah transitional zones Deciduous forest Pain forest zone	Maize, yams, cassava
Mexico	Dry desert Semi-arid	Corn, sugarcane, sorghum, wheat, tomatoes, beans, barley
Uganda	Dry desert Semi-arid	Crop residues feed: sweet potato vines, maize stover, cassava peels and leaves, banana, peels, banana leaves

Source: Trivedi (2002) and Kumar *et al.* (2016).

energy available in the feed to be used for metabolism by the goat. ME energy is expressed in mega joules (MJ). ME is expressed as MJ per kg dry matter.

The amount of energy required by goats for different body functions is shown in Table 4.12

Similarly, the requirements of protein are normally presented as digestible crude protein (DCP) and expressed in grams per day. The DCP requirement for maintenance, growth and pregnancy are presented in Table 4.13.

As the protein requirement will increase during lactation, it is necessary to increase water, calcium and phosphorous consumption. It is particularly important to provide water, calcium and phosphorous to lactating goats as the animal uses body resources and subsequently may develop metabolic diseases.

How to calculate energy and protein requirement

Let us use a growing kid weighing 18 kg and with an average growth rate of 90 g/day as an example.

1. Dry matter is 3% of body weight, the requirement is 0.54 kg. Suppose the animal is provided green fodder with 20% dry matter, then we have to provide 0.54 kg × 5 = 2.70 kg.
2. Energy requirement for an 18 kg goat with 90 gm/per day growth rate.
Energy requirement, i.e. ME = 7.3 ≅ 7.0 MJ/day.
3. The goat, having a body weight of 18 kg, needs digestible crude protein of 46 g/day.

Therefore, it is necessary to formulate the best balanced feed to fulfil the requirement of the goats. The feed formulation will include grass, tree fodder, cereals, oilcake and mineral mixture.

Table 4.10. Important legumes and grasses in the tropics.

Region	Legumes
Legumes	
Brazil, Argentina, United States, India, Thailand, Malaysia and Indonesia.	Chara mungfali (*Arachis glabrata* Benth.) (crude protein: 10–18%)
Brazil, Australia, the United States, India, Thailand, Malaysia and Indonesia	Wild moongphali (*Arachis hagenbackii*)
India	Wild kulthi, van kulthi (*Atylosia scarabaeoides* (L.) (CP: >11%)
Tropical South America, South-East Asia, Indonesia, Africa, Pacific Islands and India	Centro, butterfly pea (*Centrosema pubescens* Benth.) (CP: 12%)
Tropical America and warmer parts of the world	Butterfly pea (*Clitoria ternatea* (L.)
Islands of the West Indies	Caribbean stylo
Coastal regions of North and South America, Australia, Burma and India	(*Stylosanthes hamata* (L.) Taub.) (CP: 12–16%)
Tropical Africa and Asia in sub-humid regions, semi-arid regions and periodic waterlogging conditions	Sesbania, sanai (*Sesbania aegyptica* Poiret)
Sudan, East Africa, Syria, Arab countries, Mexico, Texas, Argentina and India	Ryncho (*Rhynchosia minima* L. DC.)
Tropics and sub-tropics of the new world from Florida to Argentina	Hedge lucerne, dwarf koa, desmanthus, ground tamarind and dashrath ghas (*Desmanthus virgatus* (L.) Wild.)(CP: 10.5–15.5%)
The West Indies, Florida, Central and South America and also tropical Africa	Green leaf desmodium, beggarlice (*Desmodium tortuosum* (Sw.) DC) (CP: 20.9%)
India, sub-tropical areas of Africa, Central America, the West Indies and many part of South-East Asia and Indonesia	Hyacinth bean, lablab bean, field bean, dolichos, poor man's bean and Sem (*Lablab purpureus* (L.) Sweet) (CP: 21–38%)
Southern coastal Texas, United States to Baja California and Sinaloa in Mexico	Sticky stylo (*Stylosanthes viscosa* Sw.) (CP: 6–12%)
Through Central America and some Caribbean islands into tropical South America and as far south as Paraguay	
Grasses	
India, Australia, Africa	Forest blue grass
Pacific Islands, Nepal, Tropical Asia, China and Tropical Africa	(*Bothriochloa intermedia* (R.Br.) A. Camus) (CP: 7%)
Kenya, Uganda, southern and north-eastern tropical Africa, Southeast Asia, Caribbean archipelago, Pakistan, Sri Lanka, South-East Asia, Indo China, Thailand, Indonesia and India	Seymour grass, (*Bothriochloa pertusa* (L.) A. Camus) (CP: 10.5–15.5%)
The West Indies; widespread in Florida, Central and South America and also in tropical Africa	Green leaf desmodium, beggarlice (*Desmodium tortuosum* (Sw.) DC) (CP: 20.9%)
Australia, South-East Asia, Pacific Islands and India	Atro, siratro, purple bean, purple bush bean (*Macroptilium atropurpureum* (DC.) Urb)
Myanmar, Africa and Australia	Marvel grass, (*Dichanthium annulatum* (Forssk.) Stapf) (CP: 6–15%)
Ceylon, Mauritius and India	Musiyal or machuri (*Iseilema laxum* Hack.)
Australia, India, semi-arid regions of Afghanistan and Iran	Blue panic, giant panic, bansi (*Panicum antidotale* Retz.) (CP: 4–15.6%)

Continued

Table 4.10. Continued.

Region	Legumes
Brazil, Mexico, Texas, South America and India	Bahia grass (*Paspalum notatum* Flugge.) (CP: 14%)
Tropical Africa and India	Setaria, Golden bristle grass, Golden millet (Schumach.) (CP: 6–20%)
Uganda	Rhodes grass ; guinea grass, elephant grass, green leaf desmodium, *Acacia tortillas*

Source: Trivedi (2002) and Kumar *et al.* (2016).

Table 4.11. Feed resources for preparation of concentrate, dry fodders and complete feed pellets in India.

Agro-climatic zones	Climatic conditions	Crop and other information for formulating feed for goats
Western Himalayan division	Average temperature 5°C and 30°C in winter and summer, respectively; winter season may dip to −5°C	Concentrate feed: rice, wheat, maize, millets, smaller millets, barley can be used for preparing concentrate feed Dry fodder: legume crop lentil, arhar, moong, mesh, sesamum, peas, groundnut, sesamum Tree leaves from deciduous forest and dry deciduous shrubs. Leaves from fruit trees and other waste fruit can be used for complete feed development
Eastern Himalayan division	Hilly area; rugged topography; temperature ranges between 10°C and 30°C	Concentrate feed: rice, wheat, maize, millets, jowar, barley Dry fodder: pea, lentil, gram, green gram, black gram, summer moong, summer black gram, soybean, urd, moong, moth, arhar, til, groundnut, kulthi
Lower Gangetic plain region	Temperature ranges between 26°C and 41°C in summer, it dips to 9°C in winter	Concentrate feed: boro rice, rabi maize, Tree leaves: litchi, mango, guava
Middle Gangetic plain region	Average temperature in summer/rainy season varies from 26°C to 41°C and in winter from 9°C to 24°C; average annual rainfall is between 100 cm and 200 cm	Concentrate feed: rice, wheat, rabi maize, barley Dry fodder: gram, arhar Tree leaves: litchi, papaya, mango, guava and Jamun
Upper Gangetic plain region	Sub-humid continental; temperature in summer between 26°C and 41°C and in winter between 7°C and 23°C	Concentrate feed: wheat, rice, sugarcane, bajra, maize, jowar, barley, minor millets, millets, urd, moong, arhar, gram, peas, lentil, groundnut, rapeseed-mustard, til, cotton, Dry fodder: gram Tree leaves: mango, papaya, guava, peach, jamun, aonla
Trans-Gangetic plain region	Semi-arid, intensive agriculture; temperature in summer ranges from 5°C to 40°C and in winter from 10°C to 20°C	Concentrate feed: rice, wheat, maize, bajra, jowar, barley, gram, arhar, cow pea, pigonpea, moong, mothbean, rapeseed-mustard, groundnut Dry fodder: gram, arhar, cow pea, pigonpea, moong, moth bean, rapseed-mustard, groundnut Tree leaves: kinnu, guava, aonla, mango

Continued

Table 4.11. Continued.

Agro-climatic zones	Climatic conditions	Crop and other information for formulating feed for goats
Eastern plateau and hill region	Mainly rainfed agriculture; temperature in summer between 26°C and 34°C, winter between 10°C and 27°C; average annual rainfall of 80–150 cm	Concentrate feed: rice, wheat, maize, ragi, barley, finger millets, jowar, biri, sorghum, arhar, gram, bengal gram, lentil, sesamum, groundnut, linseed, urd, moong, tur, pea, marwa, til, cotton, rapeseed-mustard Dry fodder: gram Tree leaves: mandarin, mango, guava, Aonla, lac, lemon, pomegranate, custard, apple, papaya, jatropha and *Moringa olifera*
Central plateau and hill region	Semi-arid climatic conditions; temperature in summer ranges from 26°C to 40°C, winter from 7°C to 24°C; low rainfall, 50–100 cm	Concentrate feed: rice, wheat, maize, barley, jowar, pearl millet, sorghum, kodo-kutki, gram, black gram, sorghum, tur, sesame, soybean, urd, moong, moth, arhar, til, lentil, linseed, groundnut, castor, rapeseed, mustard, sunflower, safflower, dry fodder: arhar, gram, pigeon pea Dry fodder: arhar, gram Tree leaves: aonla, ber, mango, mandarin & jatropha
Western plateau and hill region	Region of the regur (black) soil; temperature in summer 26°C to 41°C, winter 6°C to 23°C; low rainfall 25–75 cm	Concentrate feed: rice, wheat, maize, ragi, jowar, kodra, sorghum, pearl millet, soybean, gram, groundnut, sugarcane, niger, tur, urd, cotton, safflower Dry fodder: gram Tree Leaves: mandarin, lemon, malta, pomegranate, papaya, banana, grapes, mango, jatropha
Southern plateau and hill region	Dry zone agriculture; temperature in summer 25°C to 40°C, winter 10°C to 20°C; average annual rainfall 50–100 cm	Concentrate feed: rice, wheat, maize, jowar, pearl millets, finger millet, kodo millet, small millets, sorghum, groundnut, tur, arhar, red gram, green gram, black gram, horse gram, soybean, sugarcane, cotton, castor, sunflower, safflower Dry fodder: Gram, arhar Tree leaves: mango, banana, grapes, guava, sapota, citrus
East coast plains and hill region	Temperature in summer/rainy season ranges from 25°C to 35°C, winter from 20°C to 30°C; annual rainfall 75–150 cm	Concentrate feed: rice, sorghum, pearl millet, finger millet, maize, cholam, cumbu, ragi, red gram, black gram, green gram, horse gram, moong, urd arhar, groundnut Dry fodder: moong, arhar, gram Tree leaves: cashew nut, mango, sapota, banana, custard apple, pineapple
West coast plains and hill region	Humid region; temperature in summer 25°C to 30°C, in winter 18°C to 30°C; annual rainfall more than 200 cm	Concentrate feed: rice, foxtail millet, sunflower, cowpea, horse, gram, green gram, cotton, groundnut, sorghum, urd, coconut, seasamum, Dry fodder: gram, arhar Tree leaves: cashew nut, mango, citrus

Continued

Table 4.11. Continued.

Agro-climatic zones	Climatic conditions	Crop and other information for formulating feed for goats
Gujarat plains and hill region	Arid and semi-arid region; summer temperature 30°C and winter about 25°C; mean annual rainfall is between 50 cm and 100 cm	Concentrate feed: rice, wheat, barley, gram, pearl millets, sorghum, maize, kodra, ragi, chickpea, pigeon pea, groundnut, sesamum, castor, rapeseed, mustard Dry fodder: gram Tree leaves: sapota, banana, guava, dates, mango, grapes
Western plains and hill region	Desert climate; summer temperatures 28°C to 45°C and 5°C to 22°C in winter; average annual rainfall less than 25 cm	Concentrate feed: wheat, barley, maize, pearl millets, rice, rabi, sorghum, gram, chickpea, soybean, rapeseed, mustard, sunflower, linseed, safflower, sesame, guar Dry fodder: gram, guar Tree leaves: ber, pomegranate, kinnu, mausambi
Island region	Typically equatorial climate; annual rainfall 300 cm, the mean summer and winter temperature being 30°C and 25°C, respectively	Concentrate feed: rice, maize and cow pea Dry fodder: Guar Fruit crops: papaya, mango, sapota

Source: Final report of the working group on agro-climatic zonal planning including agriculture development in north-eastern India for XI five year plan (2007–12) and http://farmech.dac.gov.in/farm.html.

Table 4.12. Total energy requirements (MJ ME per day) for goats.

Live weight	Maintenance	Maintenance + some activity	Maintenance + a lot of activity	Maintenance + growth (50 g/day)	Maintenance + growth (100 g/day)	Maintenance + growth (150 g/day)	Maintenance + pregnancy
10	2.3	2.8	3.2	4.0	5.8	7.5	5.1
15	3.2	3.8	4.4	–	–	–	6.9
20	3.9	4.7	5.5	5.5	7.3	9.0	8.5
25	4.6	5.5	6.5	–	–	–	10.0
30	5.3	6.4	7.4	6.8	8.6	10.3	11.5
35	5.9	7.1	8.3	–	–	–	13.0
40	6.6	7.9	9.2	8.0	9.8	11.6	14.3
45	7.2	8.6	10.1	–	–	–	15.6
50	7.8	9.3	10.9	9.0	10.8	12.6	16.9
55	8.3	10.0	11.7	–	–	–	18.2
60	8.9	10.7	12.5	10.3	12.0	13.8	19.4

Source: Peacock (1996).

Table 4.13. Digestible crude protein requirements for maintenance and growth (g per day).

Live weight	Maintenance	Maintenance + 50 g/day	Maintenance + 100 g/day	Maintenance + 150 g/day	Pregnancy
10	15	25	35	45	30
20	26	36	46	56	50
30	35	45	55	65	67
40	43	53	63	73	83
50	51	61	71	81	99
60	59	69	79	89	113

Source: Peacock (1996).

HOW TO CALCULATE THE RIGHT MIXTURE OF FEEDS

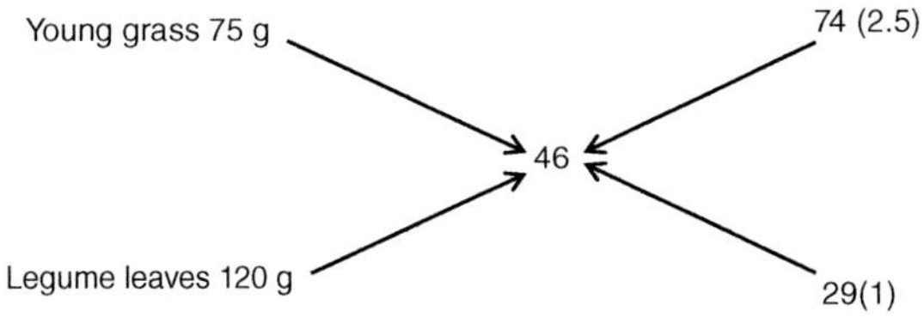

A balance of young grass and legume leaves is required to provide 46 g of protein/day to a growing kid. The left-hand side is the protein content of two types of fodder, i.e. young grasses and legume leaves. The difference in protein content is written as the target figure. In this case 1 part legume with 2.5 parts of grass should be fed to provide 46 g/protein/day. For 2.7 kg fodder, 0.8 kg of legume and 1.9 kg of grass is required.

DIET (FEED) FORMULATION There are two main methods (algebraic and Pearson's square) for calculating a balanced diet of one to two nutrients using two or more ingredients at fixed levels.

The algebraic method

Example: balancing for one nutrient with two ingredients.

Balance a 100 kg diet so that it contains 16% crude protein (CP) using grain barley (11.6% CP) and soybean meal (45.8% CP).

Let X kg of soybean meal in 100 kg of feed.

Let Y kg of barley grain in 100 kg of feed.

$X + Y = 100$ kg, $Y = 100 - X$

$X(0.458) + Y(0.116) = 16$

$X(0.458) + (100 - X)(0.116) = 16$

$0.458X + 11.6 - 0.116X = 16$

$0.342X = 16 - 11.6$

$X = \dfrac{4.4}{0.342}$

$X = 12.8$

$12.8 + Y = 100$ kg

$Y = 100 - 12.8 = 87.2$ kg

Soybean meal = 12.8 kg

Barley = 87.2 kg

Protocol 2: Complete feed

Protocol for complete feed preparation

The complete feed formulations can be transformed into:

- mash;
- pellets; and
- blocks.

The chief components of complete feed formulations as per Raghuvansi *et al.* (2007a, b, c):

- **Energy sources:** crop residues are the major energy sources for ruminant animal production systems. For goats, these includes straws (wheat, barley, oat, rice, gram, arhar etc.) and grasses (rangeland and monsoon).
- **Protein sources:** green forages and top feed resources are the major protein sources for production on grass based production systems. These include green forages (sorghum, maize, oat, barseem etc.), top feeds (locally available treetops) and *Moringa olifera* (Fig. 4.5).

Table 4.14. Constituents of complete feed formulation for supplementary feeding.

	Category of animals			
	Summer		Monsoon	
Constituents	Adult animals	Growing	Adult animals	Growing
Dry roughage	60	–	70	60
Barley	8	40	8	10
Deoiled rice bran	9	16	9	5
Wheat bran	9	16	9	5
Groundnut cake	2	3	–	2
Mustard cake	2	8	2	5
Cottonseed cake	2	6	2	5
Soya bean meal	2	3	–	1
Mineral mixture	0.5	1	0.5	1
Sodium chloride	0.5	1	0.5	1
Molasses	5	6	4	5

Fig. 4.5. *Moringa olifera.*

- **Vitamin sources and mineral sources:** although goats are capable of meeting their vitamin and mineral requirements from energy and protein sources available in the form of straws, grasses, green forages and top feed resources, soil conditions and availability of feed resources limit their accessibility. Additional mineral mixtures and vitamin supplements may be added.

Methods for preparation of pellet or block:

1. Collect all ingredients as per requirement.
2. Dry roughages chopped up to 6–8 mm in size.
3. Add the ground ingredients such as cereals and oil cakes.
4. Add the salt, mineral and vitamin mix as per the ratio.
5. Add molasses.
6. All the above should be added in serial order and mixed properly and put into the pellet-/block-making machine (Fig. 4.6).
7. The complete feed block/pellets prepared using the technology may be packed in gunny bags and stored at room temperature for up to 12 months in semi-arid regions in a dry environment (Raghuvansi *et al.*, 2007a). High moisture levels cause the blocks/pellets to deteriorate.

Preparation of licking blocks for supplementary feeding

Licking blocks provide a unique opportunity as a supplement to the basal diet in harsh environmental conditions that prevail during winter/summer or during drought/flood situations. They have greater benefits for animals being raised in crop residue-based production systems. The benefit of using a licking block comes through enhancing the efficiency of rumen fermentation, which increases the digestibility and intake of forage, leading to greater supply of microbial protein for production purposes.

The process for manually manufacturing the blocks is as follows:

1. Mix bran and molasses (see Table 4.15 for ratios), add salt to the mixture – this is mixture A.
2. Dissolve urea in water and add cement – this is mixture B.
3. Quickly mix mixtures A and B together thoroughly – this is mixture C.
4. Place 3–4 kg of mixture C in a wooden/iron block frame.
5. Apply pressure to the frame.
6. Leave the block frame for 6–8 hours to settle the material.
7. Take out the licking block from the frame and dry in a shed or oven at 40°C.
8. Pack these blocks individually in aluminium foil or with thin sheets.
9. Properly dried and packed licking blocks may be stored for several months at room temperature. However, exposure of blocks to humidity may attract fungus infestation. In general licking blocks are very safe for animals.

Fig. 4.6. Complete feed pellet for goats.

Table 4.15. Components (kg) of a licking block.

Constituents	Types of licking blocks I	II	III	IV	V
Molasses	45	38	25	35	35
Cement	10	10	5	–	–
Wheat bran	30	40	35	33	36
Urea	10	10	10	10	10
Salt	5	1	5	5	5
Water	4	5	7	4	6
Lime	–	–	10	10	–
Clay	–	–	5	5	10
Mineral mixture	–	–	2	2	2

Source: FAO (2007).

Protocol 3: Nutrient requirements (NRC, 2007)

Table 4.16. Nutrient requirements for goats (maintenance – breeding).

	Energy			Minerals		Vitamins	
Weight (kg)	Total digestible nutrients (kg/day)	Metabolizable energy (Mcal/day)	Crude protein (g/day)	Calcium (g/day)	Phosphorus (g/day)	A (RE/d)	E (IU/day)
Mature does dairy maintenance[a]							
40	0.53	1.91	67	1.9	1.5	1256	212
50	0.62	2.25	79	2.1	1.7	1570	265
60	0.72	2.58	90	2.4	2.0	1884	318
70	0.80	2.90	101	2.6	2.2	2198	371
80	0.89	3.20	112	2.8	2.4	2512	424
90	0.97	3.50	122	3.0	2.6	2826	477
Mature bucks dairy maintenance[a]							
50	0.72	2.59	86	2.4	2.0	1570	265
75	0.97	3.51	116	3.0	2.6	2355	398
100	1.21	4.36	144	3.7	3.2	3140	530
125	1.43	5.15	170	4.2	3.8	3925	663
150	1.64	5.91	195	4.8	4.3	4710	795
Mature does (dairy) breeding							
40	0.58	2.10	73	2.0	1.6	1256	212
50	0.69	2.48	87	2.3	1.9	1570	265
60	0.79	2.84	99	2.6	2.1	1884	318
70	0.88	3.19	111	2.8	2.4	2198	371
80	0.98	3.53	123	3.1	2.6	2512	424
90	1.07	3.85	134	3.3	2.9	2826	477
Mature bucks (dairy) breeding							
50	0.79	2.85	94	2.6	2.1	2275	280
75	1.07	3.86	128	3.3	2.9	3413	420
100	1.33	4.79	158	4.0	3.5	4550	560
125	1.57	5.67	187	4.6	4.1	5688	700
150	1.80	6.50	214	5.2	4.7	6825	840

[a]Feed all non-dairy breeds 90% of requirement of dairy breed.

Table 4.17. Nutrient requirements of growing kids (doelings and wethers).[a]

		Energy			Minerals		Vitamins	
Weight (kg)	Daily gain (g/day)	TDN (kg/day)	ME (Mcal/day)	Crude protein (g/day)	Calcium (g/day)	Phosphorus (g/day)	A (RE/day)	E (IU/day)
10	0	0.20	0.72	26	0.9	0.5	1000	100
10	25	0.24	0.86	36	1.6	0.8	1000	100
10	100	0.35	1.27	69	3.5	1.6	1000	100
10	150	0.43	1.55	90	4.8	2.2	1000	100
10	200	0.51	1.82	112	6.2	2.7	1000	100
15	0	0.27	0.98	35	1.3	0.9	1500	150
15	25	0.31	1.11	46	1.8	1.0	1500	150
15	100	0.42	1.53	78	3.6	1.7	1500	150
15	150	0.50	1.80	100	5.0	2.3	1500	150
15	200	0.58	2.08	121	6.3	2.8	1500	150
20	0	0.34	1.21	43	1.4	1.0	2000	200
20	25	0.37	1.35	54	2.2	1.4	2000	200
20	100	0.49	1.76	86	4.0	2.0	2000	200
20	150	0.57	2.04	108	5.1	2.4	2000	200
20	200	0.64	2.31	130	6.4	3.0	2000	200
20	250	0.72	2.59	151	7.7	3.5	2000	200
25	0	0.40	1.43	51	1.6	1.2	2500	250
25	25	0.44	1.57	62	2.3	1.5	2500	250
25	100	0.55	1.95	94	4.1	2.2	2500	250
25	150	0.63	2.26	116	5.5	2.8	2500	250
25	200	0.70	2.54	137	6.5	3.1	2500	250
25	250	0.78	2.81	159	7.8	3.6	2500	250
30	0	0.46	1.64	59	1.8	1.4	3000	300
30	25	0.49	1.78	69	2.5	1.7	3000	300
30	100	0.61	2.19	102	4.2	2.3	3000	300
30	150	0.68	2.47	123	5.6	2.9	3000	300
30	200	0.76	2.74	145	7.0	3.5	3000	300
30	250	0.84	3.02	166	7.9	3.7	3000	300
30	300	0.91	3.30	188	9.3	4.4	3000	300
35	0	0.51	1.84	66	1.9	1.5	3500	350
35	25	0.55	1.98	77	2.7	1.9	3500	350
35	100	0.66	2.39	109	4.8	2.9	3500	350
35	150	0.74	2.67	130	5.7	3.0	3500	350
35	200	0.82	2.95	152	7.1	2.6	3500	350
35	250	0.89	3.22	174	8.5	4.2	3500	350
35	300	0.97	3.50	195	9.4	4.4	3500	350
40	0	0.56	2.04	73	2.1	1.7	4000	400
40	25	0.60	2.17	83	2.8	2.0	4000	400
40	100	0.72	2.59	116	5.0	3.0	4000	400
40	150	0.79	2.86	137	5.8	3.1	4000	400
40	200	0.87	3.14	159	7.2	3.7	4000	400
40	250	0.95	3.42	181	8.6	4.4	4000	400
40	300	1.02	3.69	202	10.0	5.0	4000	400

[a]Dairy breeds require less protein (15–20%) and more energy (5–10%) than Boer goats. Indigenous breeds require less protein (15–20%) and energy (5–10%) than Boer breeds. Intact require 5–10 % more energy for growth than wethers.
Source: NRC (2007).

Table 4.18. Nutrient requirements for goats (gestation).

Weight (kg)	Energy: TDN (kg /day)	Energy: ME (Mcal/day)	Crude protein (g/day)	Minerals: Calcium (g/day)	Minerals: Phosphorous (g/day)	Vitamins: A (RE/d)	Vitamins: E (IU/d)
Mature does (dairy) early gestation – single kid							
40	0.64	2.31	101	4.2	2.5	1256	212
50	0.75	2.71	117	4.4	2.8	1570	265
60	0.86	3.09	132	4.7	3.1	1884	318
70	0.96	3.46	147	5.0	3.3	2198	371
80	1.06	3.81	161	5.2	3.6	2512	424
90	1.15	4.14	174	5.5	3.8	2826	477
Mature does (dairy) early gestation – twin kids							
40	0.70	2.52	117	5.8	3.2	1256	212
50	0.82	2.96	135	6.2	3.6	1570	265
60	0.94	3.38	153	6.5	3.9	1884	318
70	1.04	3.77	169	6.7	4.1	2198	371
80	1.15	4.16	187	7.0	4.4	2512	424
90	1.26	4.53	202	7.3	4.7	2826	477
Mature does (dairy) early gestation- three or more kids							
40	0.74	2.68	127	7.3	3.9	1256	212
50	0.87	3.13	147	7.6	4.2	1570	265
60	0.99	3.56	165	8.0	4.5	1884	318
70	1.10	3.97	183	8.2	4.8	2198	371
80	1.21	4.38	201	8.5	5.1	2512	424
90	1.32	4.77	218	8.8	5.4	2826	477
Mature does (dairy) late gestation – single kid							
40	0.88	3.18	141	4.3	2.7	1820	224
50	1.03	3.70	177	5.2	3.5	2275	280
60	1.17	4.20	199	5.5	3.9	2730	336
70	1.30	4.69	220	5.9	4.2	3185	392
80	1.42	5.14	239	6.2	4.5	3640	448
90	1.55	5.57	257	6.5	4.8	4095	504
Mature does (dairy) late gestation – twin kids							
40	1.05	3.79	177	6.2	3.6	1820	224
50	1.23	4.42	203	6.6	4.0	2275	280
60	1.39	5.02	229	6.9	4.3	2730	336
70	1.55	5.57	251	7.2	4.6	3185	392
80	1.70	6.15	276	7.6	4.9	3640	448
90	1.85	6.66	325	8.8	6.2	4095	504
Mature does (dairy) late gestation – three or more kids							
40	1.17	4.21	189	7.4	4.0	1820	224
50	1.36	4.89	215	7.7	4.3	2275	280
60	1.53	5.53	256	8.6	5.1	2730	336
70	1.71	6.16	2.82	8.9	5.5	3185	392
80	1.88	6.77	308	9.3	5.8	3640	448
90	2.04	7.37	334	9.6	6.2	4095	504

Source: NRC (2007).

Table 4.19. Nutrient requirements for goats (lactation).

	Energy			Minerals		Vitamins	
Weight (kg)	TDN (kg/day)	ME (Mcal/day)	Crude protein (g/day)	Calcium (g/day)	Phosphorous (g/day)	A (RE/day)	E (IU/day)
Mature does (dairy) early lactation – single kid							
40	0.89	3.19	93	5.9	3.9	2140	224
50	1.03	3.70	109	6.3	4.3	2675	280
60	1.16	4.20	124	6.7	4.6	3210	336
70	1.29	1.65	138	7.0	5.0	3745	392
80	1.41	5.09	151	7.3	5.3	4280	448
90	1.53	5.51	164	7.6	5.6	4815	504
Mature does (dairy) early lactation – twin kids							
40	1.24	4.48	120	10.0	6.4	2140	224
50	1.43	5.17	139	10.5	6.9	2675	280
60	1.61	5.80	157	10.9	7.3	3210	336
70	1.77	6.40	174	11.4	7.7	3745	392
80	1.93	6.97	190	11.8	8.1	4280	448
90	2.09	7.53	206	12.2	8.5	4815	504
Mature does (dairy) early lactation – three or more kids							
50	1.83	6.61	142	13.7	8.5	2675	280
60	2.05	7.41	160	14.1	8.9	3210	336
70	2.26	8.15	177	14.6	9.4	3745	392
80	2.46	8.85	229	16.3	11.0	4280	448
90	2.64	9.53	247	16.7	11.5	4815	504
Mature does (dairy) late lactation – single kid							
40	0.68	2.46	78	5.4	3.4	2140	224
50	0.80	2.88	92	5.7	3.7	2675	280
60	0.91	3.27	105	6.0	4.0	3210	336
70	1.01	3.65	117	6.3	4.2	3745	392
80	1.11	4.02	129	6.5	4.5	4280	448
90	1.21	4.36	140	6.8	4.8	4815	504
Mature does (dairy) late lactation – twin kids							
40	0.83	3.01	90	8.9	5.3	2140	224
50	0.97	3.50	105	9.3	5.7	2675	280
60	1.10	3.96	119	9.6	6.0	3210	336
70	1.22	4.40	132	9.9	6.3	3745	392
80	1.34	4.82	145	10.2	6.6	4280	448
90	1.45	5.23	158	10.5	6.9	4815	504
Mature does (dairy) late lactation – three or more kids							
50	1.14	4.13	118	12.8	7.7	2675	280
60	1.29	4.65	133	13.2	8.0	3210	336
70	1.43	5.15	148	13.6	8.4	3745	392
80	1.56	5.63	162	13.9	8.7	4280	448
90	1.69	6.09	176	14.2	9.1	4815	504

Source: NRC (2007).

4.10 References

Ahmed, O.M.M. and Adam, S.E.I. (1979) Toxicity of Jatropha curcas in sheep and goats. *Research in Veterinary Science* 27(1), 89–96.

Austin, P.J., Suchar, L.A., Robbins, C.T. and Hagerman, A.E. (1989) Tannin-binding proteins in saliva of deer and their absence in saliva of sheep and cattle. *Journal of Chemical Ecology* 15, 1335–1347.

Bell, F.R. (1959) Preference thresholds for taste determination in goats. *The Journal of Agricultural Science* 52, 125–128.

Burns, J.C., Fisher, D.S. and Mayland, H.F. (2001) Preference by sheep and goats among hay of eight tall fescue cultivars. *Journal of Animal Science* 79, 213–224.

Chakraborty, T. and Ghosh, T.K. (1988) Chemical composition and nutritive value of Subabul (Leucaena leucocephala) foliage in Black Bengal goats. *Indian Journal of Animal Nutrition* 5, 237–239.

Davis, C.L., Brown, R.E. and Beitz, D.C. (1964) Effect of feeding high-grain restricted-roughage rations with and without bicarbonates on the fat content of milk produced and proportions of volatile fatty acids in the rumen. *Journal of Dairy Science* 47, 1217–1219.

De Rosa, G., Fedele, V., Napolitano, F., Gubitosi, L., Bordi, A. *et al.* (1997) Dietary preferences in adult and juvenile goats. *Animal Science* 65(3), 457–463.

De Rosa, G., Moio, L., Napolitano, F. Grasso, F., Gubitosi, L. *et al.* (2002) Influence of flavour on goat feeding preferences. *Journal of Chemical Ecology* 28, 269–281.

Dutta, T.K. (2009) Nutritional strategies for improved production and reproduction in goats. In: Kharche *et al.* (eds) *Manual on Advanced in Production and Reproduction in Goats*. CIRG Publications, India, pp. 66–91.

FAO (2007) Urea-molasses multi-nutrient blocks: simple and effective feed supplement technology for ruminant agriculture. In: Makkar, H.P.S., Sánchez, M. and Speedy, A.W. (eds) *Animal Production and Health Division*. Food and Agriculture Organization of the United Nations, Rome, Italy, pp. 257.

Gill, H.S., Shu, Q. and Leng, R.A. (2000) Immunization with Streptococcus bovis protects against lactic acidosis in sheep. *Vaccine* 18, 2541–2548.

Goatcher, W.D. and Church, C.D. (1970) Taste response in ruminants. III. Reactions of pygmy goats, normal goats, sheep and cattle to sucrose and sodium chloride. *Journal of Animal Science* 31, 364–372.

Hadjipanayiotou, M. (1982) Effect of sodium bicarbonate and of roughage on milk yield and milk composition of goats and on rumen fermentation of sheep. *Journal of Dairy Science* 65, 59–64.

Hammond, A.C. (1995) *Leucaena* toxicosis and its control in ruminants. *Journal of Animal Science* 73, 1487–1492.

Hiremath, N.B. (1981) Subabool (*Leucaena leucocephala*) – A wonder plant. *Indian Dairyman* 33, 351–356.

Hofmann, R.R., Streich, W.J., Fickel, J., Hummel, J. and Clauss, M. (2008) Convergent evolution in feeding types: Salivary gland mass differences in wild ruminant species. *Journal of Morphology* 269(2), 240–257.

Huston, J.E. (1994) Effects of supplemental feeding on intake by kid, yearling, and adult angora goats on rangeland. *Journal of Animal and Feed Science* 72, 768–773.

Jarrige, R. (1988) Ingestion et digestion des aliments. In: Jarrige, R. (ed.) *Alimentation des Bovins, Ovins et Caprins*. Institut National de la Recherche Agronomique, Paris, pp. 29–56.

Jarrige, R., Morand-Fehr, P. and Hoden, A (1978) Consommation d'aliments et d'eau. In: Jarrige, R. (ed.) *Alimentation des Ruminants* INRA Publications, Versailles, France, pp. 177–206.

Kumar, R. (2003) Anti-nutritive factors, the potential risks of toxicity and methods to alleviate them. Available at: http://www.fao.org/docrep/003/t0632e/T0632E10.htm (accessed 23 May 2019).

Kumar, S., Singh, H.V., Singh, S., Singh, K.K., Kumar, R.V., *et al.* (2016) *Tropical Range Grasses and Legumes: Production Packages and Nutritional Profile*. Published by ICAR-Indian Grassland and Fodder Research Institute, Jhansi, Uttar Pradesh, India.

Lamy, E., da costa, G., e Silva, F.C., Potes, J. Coelho, A.V and Baptista, E.S. (2008) Comparison of electrophoretic protein profiles from sheep and goat parotid saliva. *Journal of Chemical Ecology* 34, 388–397.

Leek, B.F. (1993) Clinical diseases of the rumen: a Physiologist's view. *The Veterinary Record* 113, 10–14.

Lu, C.D., Kawas, J.R. and Mahgoub, O.G. (2005) Fibre digestion and utilization in goats. *Small Ruminant Research* 60, 45–52.

Middleton, J.R., *et al.* (1997) Hemolysis associated with water administration using a nipple bottle for human infants in juvenile pygmy goats. *Journal of Veterinary Internal Medicine* 11, 382–384.

Min, B.R. and Hart, S.P. (2003) Tannins for Suppression of internal parasites. *Journal of Animal Science* 81, 102–109.

Mohamed Nour, M.S., Abusamra, M.T., Hago, B.E.D. (1998) Experimentally induced lactic acidosis in Nubian goats: clinical, biochemical and pathological investigations. *Small Ruminant Research* 31, 7–17.

NRC (2007) *Nutrient requirements of Small Ruminants, Sheep, Goats, Cervids, and New World Camelids*, The National Academies Press. National Academies Press, Washington DC, USA.

Peacock, C (1996). *Improving Goat Production in the Tropics*. Farm-Africa and Oxfam, Oxford, UK

Raghuvansi, S.K.S., Prasad, R., Mishra, A.S., Chaturevedi, O.H., Tripathi, M.K. *et al.* (2007a) Effect of tree leaves inclusion on nutrient utilization, rumen fermentation pattern, protozoal population and blood bio-chemical constituents of sheep fed straw based complete feed blocks. *Bioresource Technology* 98, 511–517.

Raghuvansi, S.K.S., Prasad, R., Tripathi, M.K., Mishra, A.S., Chaturvedi, O.H. *et al.* (2007b) Intake, performance, nutrient utilisation, rumen fermentation and rumen microbial enzymes of lambs maintained on complete feed blocks or grazing and supplementation. *Animal (Cambridge)* 1, 221–226.

Raghuvansi, S.K.S., Tripathi, M.K., Mishra, A.S., Chaturevedi, O.H., Prasad, R. *et al.* (2007c) Feed digestion, rumen fermentation and blood-biochemical constituents of Malpura rams on feeding of tree leaves included pearl millet stover based complete feed block diets. *Small Ruminant Research* 71, 21–30.

Rai, P., Gupta, A. and Samanta, A.K. (2007) Tree leaves, their production and nutritive value for ruminants: A review. *Animal Nutritive and Feed Technology* 7, 135–159.

Russell, J.B. and Strobel, H.J. (1989) Effect of ionophores on rumen fermentation. *Applied Environmental Microbiology* 55, 1–6.

Shkolnik, A., Maltz, E. and Choshniak, I. (1980) The role of the ruminant's digestive tract as a water reservoir. In: Ruckebusch, Y. and Thivend, P. (eds) *Digestive Physiology and Metabolism in Ruminants*. MTP Press, Lancaster, UK.

Silanikove, N. (2000) The Physiological basis of adaptation in goats to harsh environments. *Small Ruminant Research* 35, 181–193.

Singh, P. (1987) Rangeland reconstruction and management for optimising biomass production. Presidential address in National Rangeland Symposium, 9–12th November 1987, IGFRI, Jhansi, Uttar Pradesh, India.

Smith, M.C. and Sherman, D.M. (2009) *Goat Medicine*, 2nd edn. Wiley and Blackwell Publication, Ames, Iowa, USA.

Solaiman, S.G. (2010) *Feeds and Feeding Management in Goat Science and Production 2010*. Wiley Blackwell Publication, Hoboken, New Jersey, USA.

Solaiman, S., Roper, S., Beguesse, K., Reddy, G., Gurung, N. *et al.* (2008) Effects of induced copper deficiency with added molybdenum on health and immune response of male goat kids. *Journal of Animal Science* 86 (suppl.3).

Solaiman, S.G., Gurung, N.K., Mccrary, Q., Goyal, H. and McElhenney, W.H. (2009) Feeding performance and blood parameters of male goat kids fed EasiFlo Cottonseed. *Small Ruminant Research* 81, 137–145.

Tanwar R.K. and Mathur P.D. (1983) Biochemical and microbial changes in experimentally induced rumen acidosis in goats. *Indian Journal of Animal Science* 53, 271–274.

Terrill, T.H., Mosjidis, J.A., Moore, D.A., Shaik, S.A., Miller, J.E. *et al.* (2007) Effect of pelleting on efficacy of sericea lespedeza hay as a natural dewormer in goats. *Veterinary Parasitology* 146, 117–122.

Trivedi, B.K. (2002) Grasses and legumes for tropical pastures. Published by ICAR-Indian Grassland and Fodder Research Institute, Jhansi, Uttar Pradesh, India.

Vieira, E.L., Batista, A.M.V., Guim, A., Carvalho, F.F., Nascimento, A.C. *et al.* (2008) Effects of hay inclusion on intake, in vivo nutrient utilization and ruminal fermentation of goats fed spineless cactus (*Opuntia ficus-indica* Mill) based diets. *Animal Feed Science and Technology* 141, 199–208.

West, G.A., Dale, T. and Mayhew, R.F. (1983) Left displacement of the abomasum in a goat. *Veterinary Medicine & Small Animal Clinician* 78(12), 1919–1921.

5 Health Management

Diseases affect the productivity of an animal and cause losses due to morbidity, mortality, veterinary cost and production factors. A wide range of bacterial, viral and parasitic infections are associated with morbidity and mortality, resulting in economic losses to the goat-farming community. Disease problems are inevitable in any livestock production programme, hence it is mandatory to frame a suitable herd health management and preventive medicine programme to enhance production performance. Disease is defined as the state that prevents optimal productivity, while health is a state of harmonious well-being that allows optimum productivity. This chapter will discuss the range of goat diseases in the tropics and the possible approaches to managing disease in farmers' fields and in intensive production systems. The emphasis in this chapter has been placed on diseases of economic importance in intensive production and subsistence farming systems.

5.1 How to Screen Healthy Goats

It is necessary to screen the flock to identify ailing animals. It is general practice to observe the flock or an individual animal from a distance, followed by a detailed physical examination. A physical examination includes observation of the animal, vital clinical signs, medical history and sample collection.

5.1.1 Physical examinations

Respiration rate (RR)

Respiration rate is the chest movement per minute when an animal is at rest. A stethoscope is placed on the chest to record the cycle of inhalation and exhalation (Fig. 5.1).

Heart rate (HR)

Heart rate is the number of heart beats per minute, measured by placing the stethoscope on the 3rd–4th intercostal space (Fig. 5.2).

Temperature

This is the rectal temperature (RT), recorded by inserting the mercury/digital thermometer into the rectum for 1 minute. The recorded temperature should be in degrees centigrade (°C) or in Fahrenheit (°F) (Fig. 5.3).

Mucous membrane examination

The eye mucous membranes are observed for any discolouration in comparison to a normal pinkish mucous membrane. An abnormal discolouration of the mucous membrane is indicative of a clinical condition in animals.

Rumen contractions

The ruminal contraction or movement are recorded to assess ruminal activity. It is best recorded by placing the fisted hand on the left paralumbar fossa (Fig. 5.4). The normal rumen contraction is 1–2 times per minute.

History taking (anamnesis)

Identification of a sick goat is of paramount importance in good rearing practice in order to ensure an early diagnosis and consideration when adding a new animal to the herd. History taking is an important task for the clinician to aid in predicting the animal's clinical condition. It should be exhaustive, involving all information such as age/sex; general symptoms and changes in behaviour; change in feeding and habits; pregnancy status; symptoms in other animals; history of vaccination and recent preventive medication, and so on.

Sample collection

It is important to collect samples (e.g. faeces, urine, skin scrapings, blood) for laboratory examination

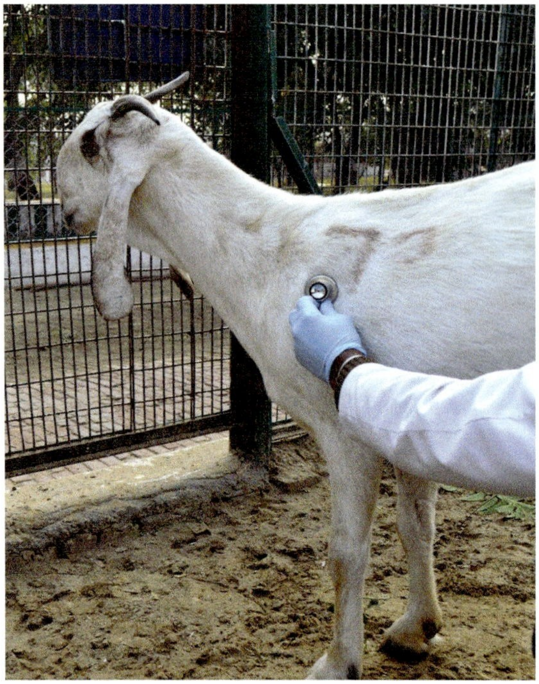

Fig. 5.1. Recording of respiration rate by stethoscope.

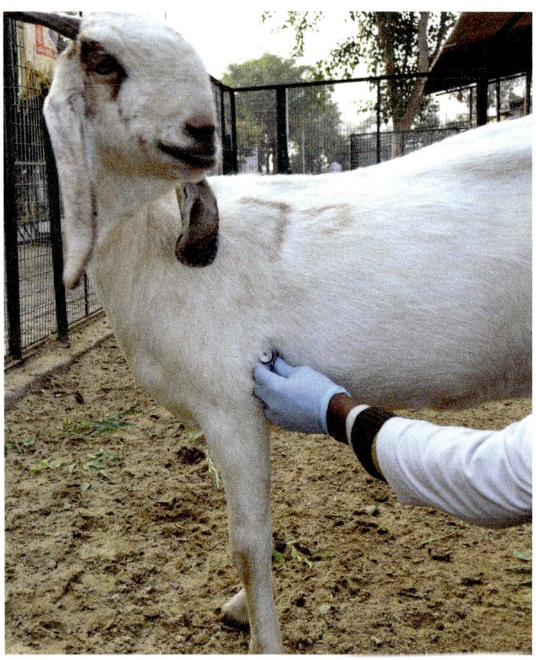

Fig. 5.2. Recording of heart rate by stethoscope.

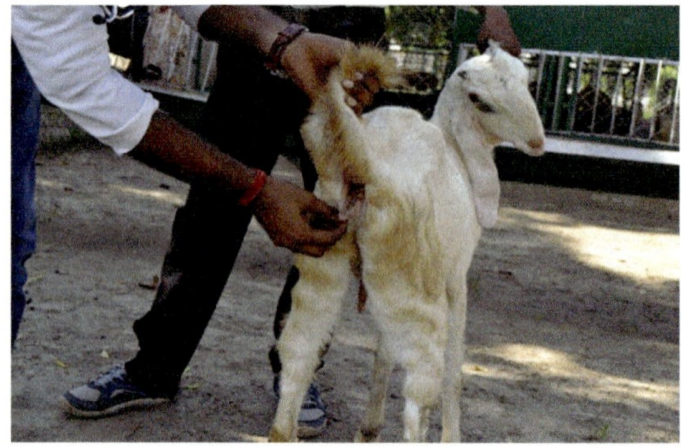

Fig. 5.3. Recording of rectal temperature using a thermometer.

in order to make an accurate diagnosis. Samples can be taken from living or dead goats to identify the cause of disease. This will help determine the appropriate preventive measures for the other animals in the flock or village. (See Protocol 2 at the end of this chapter.)

5.1.2 Signs of a healthy goat

- A healthy goat should be bright, alert, inquisitive and interested in its surroundings.
- Normal goats show some distinctive behaviours in both extensive and intensive rearing systems. Animals tend to flock together and establish a hierarchical structure and social dominance.
- Healthy goats are energetic, with a tendency to climb and browse in the pasture.
- A depressed attitude is exhibited by listlessness, dullness, separation from the herd and unresponsiveness to the environment.
- The normal temperature, heart rate and respiration rate should be noted for healthy physiological condition (Table 5.1).
- A clinical illness is likely if a goat is not drinking or eating normally, pressing its head against a wall or fence, or not urinating or having difficulty urinating.
- Goats usually have pelleted faeces. If the goat has loose faeces, this indicates diarrhoea.
- Limping or staggering, nasal/eye discharge, coughing, abnormal breathing or unusual crying suggest an unhealthy goat.
- Examination of superficial lymph nodes can reveal regional infections if they are enlarged. The intermandibular, parotid, retropharyngeal,

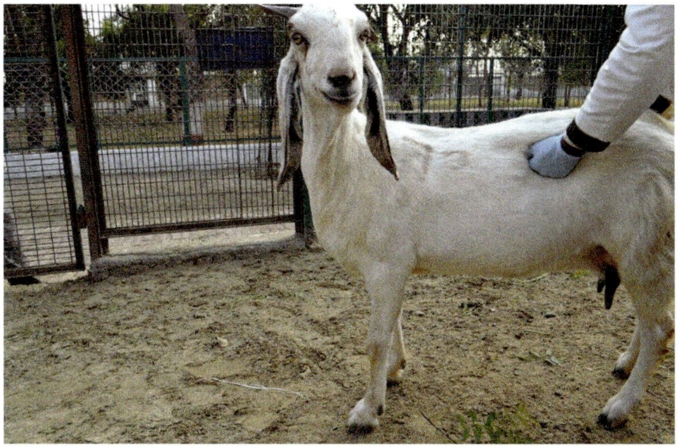

Fig. 5.4. Recording ruminal contractions.

Table 5.1. Normal values for vital signs in goats in normal environmental conditions.

Parameter	Values
Rectal temperature	39–40°C
Heart rate	70–90 beats per minute
Respiratory rate	20–30 breaths per minute
Rumen motility	1–2 motility per minute

prescapular, prefemoral and supramammary lymph nodes must be digitally palpated for enlargement.

- A rough, dry, leathery coat, flaky skin, patchy hair loss or pruritic skin is indicative of poor nutritional status, parasitic infestation or a chronic clinical condition.
- Conjunctiva and mucous membranes are normally pinkish. If the conjunctiva are pale the animal may have anaemia caused by *Haemonchus* infection (mainly); if they are yellow/jaundiced this may indicate a hepatic dysfunction. Congestion may indicate an acute infection, fever or toxaemia.
- A hot udder is indicative of mastitis; a cold udder may be gangrenous mastitis; pitting on digital pressure is an udder oedema; and nodular formation occurs in chronic mastitis. The milk should be observed for colour, consistency and flake formation even if the udders are healthy.
- The scrota should have a bilateral symmetry. A bipartite or split scrotum is a common congenital condition that is considered to be a breeding fault.
- The blood profile should be within the physiological limits (Table 5.2).

Table 5.2. Normal values of blood, urine and cerebrospinal fluid parameters in goats.

Parameter	Range (units)
Red blood cells	
Red blood cell count	10–18 × 10^{12}/l
Haemoglobin	8–15 g/dl
Packed cell volume	17.0–34.0%
Mean corpuscular haemoglobin	7–9 pg
Mean corpuscular haemoglobin concentration	31–42 g/dl
Mean corpuscular volume	16–34 fl
Thrombocytes	300–600 × 10^{9}/ul
White blood cells	
White blood cell count	6–14 × 10^{9}/l
Neutrophils	1.2–7.2 × 10^{9}/l
Lymphocytes	2–9 × 10^{9}/l
Monocytes	0.0.5 × 10^{9}/l
Eosinophils	0.05–0.5 × 10^{9}/l
Basophils	0–0.1 × 10^{9}/l
Urinalysis	
Specific gravity	1.015
pH	7–8
Colour	Yellow
Turbidity	Clear
Volume	10–40 ml/kg bodyweight/day
Cerebrospinal fluid	
Specific gravity	1.005
Colour	Colourless
Total protein	0–0.39 g/l
Glucose	3.0–4.0 mmol/l
White blood cells	0–4 cells/µl

Source: Matthews (1999) and Constable *et al.* (2017).

5.1.3 Biosecurity and goat diseases

Biosecurity is a system of practices adopted in the goat farm that prevents infectious diseases from being carried into herds or premises. It is necessary to develop a management programme to prevent the transmission of disease-causing agents to animals by direct or indirect means. Effective biosecurity management practices prevent the spread of disease by minimizing movement of biological organisms and their vectors into and within the premises. These approaches help in cost-effective health management, consumers' confidence in farm products, and improved animal welfare issues. (See Protocol 12 at the end of this chapter for more details.)

The protection of animals from infection is the main goal of biosecurity measures. The immunity of animals depends upon host factors (immune status), environmental factors (temperature, stocking rate, pasture condition, etc.), and the disease agent itself. The risk of infection increases due to:

- contact with other livestock;
- management practices;
- incoming/returning animals;
- vehicle/equipment/traffic flow around the farm;
- feed and water sources;
- manure management; or
- movement of other animals (e.g. dogs, cats, wildlife, rodents) and farm visitors.

Therefore, it is essential to follow region-specific practices for effective biosecurity measures. The following points need to be kept in mind when implementing biosecurity measures.

- Implement an individual animal identification system.
- Effective quarantine practices should be implemented. The introduction of diseased goats to healthy flock should be done after meticulous clinical examination, vaccination and deworming.
- Keep accurate records of every animal bought and sold, medication and vaccination status, breeding dates, birth dates, death dates and means of disposal of dead animals.
- Newly introduced animals should be acclimatized to environmental conditions and feed/fodders.
- Follow the region-specific vaccination programme.
- Provide a clean area for restraint, treatment and isolation of sick animals.
- Create a regular practical schedule to observe the animals and identify any sick animals in a flock.
- Do not re-use needles for injections and ensure used syringes and needles are correctly disposed of.
- Ensure prompt removal of placenta and aborted foetuses as they can harbour thousands of infectious organisms that can spread to other goats. A specific measure to adopt to incinerate and dispose of placenta is given in Protocol 2 at the end of this chapter.
- Consult with a veterinarian or animal health personnel when goats are ill or die unexpectedly.
- Ensure farm workers practice proper personal hygiene.
- Carry out proper manure management by adopting the practice of compost and vermin-compost preparation.
- Routinely clean and disinfect feeding equipment, which can be done with chlorine, iodine or quaternary ammonia products.
- Routinely clean and disinfect equipment used to medicate animals, especially equipment used on multiple animals.
- Check all possibilities of contamination by vehicles, equipment, clothing and other contaminated inanimate objects.
- Ensure safe and timely disposal of carcasses of dead animals as per regional guidelines of the state government.
- Proper storage of feed and feeder, and prevention of water contamination.
- Devise a programme to control the non-livestock vector threat for disease transmission (birds, rodents, insects, cats, etc.).
- Transport properly vaccinated animals in clean vehicles.
- Visitors should be permitted through an antiseptic foot pathway.
- In most countries, there is a list of the diseases (reportable infectious diseases or notifiable diseases) that need to be reported to the state government or national veterinary authority as soon as they are confirmed. Help should be immediately sought from an advanced or legal diagnostic laboratory.

The following is a list of notifiable diseases determined by the World Organization of Animal Health (2019):

- caprine arthritis/encephalitis;
- contagious agalactia;
- contagious caprine pleuropneumonia;
- infection with *Chlamydophila abortus* (enzootic abortion of ewes, ovine chlamydiosis);

- infection with *peste des petits ruminants* virus;
- maedi-visna;
- Nairobi sheep disease;
- ovine epididymitis (*Brucella ovis*);
- salmonellosis (*Salmonella abortusovis*);
- scrapie; and
- sheep pox and goat pox.

5.2 Common Diseases in Goats

The common diseases affecting different organs and systems are presented in Table 5.3.

5.2.1 Viral and bacterial diseases

Enterotoxaemia

Enterotoxaemia is frequently the cause of mortality in all age groups of goats. The principal cause of caprine enterotoxaemia worldwide is *Clostridium perfringens* type D. There are five main types of *C. perfringens* based on the lethal exotoxins (alpha, beta, beta-2, epsilon, iota and enterotoxin). This organism is a common resident of the intestine and rapidly proliferates under favourable conditions, resulting in the production of toxins that cause disease in the host. It is usually precipitated by consumption of a carbohydrate-rich diet. The outbreaks occur in goats raised under intensive or semi-intensive management conditions. Well-fed kids and feedlot kids on heavy grain rations are found to be more susceptible. A sudden change in the weather or a change to a carbohydrate-rich feed or grazing on fodder crops (e.g. green wheat, green lucerne, etc.) may predispose animals to enterotoxaemia. Fibre-producing goats (e.g. South African or Texas Angoras) have been affected with the disease during periods of corralling or restricted grazing in grain fields (van Tonder, 1975; Ross, 1981). The disease may also affect goats turned out to lush pastures, those fed bread or other bakery products, after excessive grain consumption following accidental

Table 5.3. Summary of system-wise diseases in goats.

Organ/system	Diseases
Eye	Cyclopia, conjunctivitis , infectious keratoconjunctivitis, blindness, anterior uveitis
Skin	Contagious ecthyma, capripox, peste des petits ruminants, bluetongue, staphylococcal folliculitis, dermatophilosis, dermatophytosis, sarcoptic mange, zinc deficiency, pemphigus foliaceus and protothecosis
Head	Big head disease, osteodystrophia fibrosa, hydrocephalus
Digestive tract	Anorexia, tympany (bloat), abomasal impaction, peste des petits ruminants, bluetongue, enterotoxaemia, salmonellosis, yersiniosis, paratuberculosis, coccidiosis, cryptosporidiosis, nematode gastroenteritis, paramphistomiasis, intestinal cestodiasis or tapeworms, neonatal diarrhoea complex
Respiratory system	Nasal bots, nasal schistosomiasis, peste des petits ruminants, goat pox, progressive interstitial retroviral pneumonia, contagious caprine pleuropneumonia, chlamydiosis, q fever, pleuropneumonia, *Pasteurella* and *Mannheimia* pneumonia, caseous lymphadenitis abscesses, cryptococcosis, parasitic pneumonia (dictyocaulus pneumonia), cyanide poisoning, nitrate poisoning
Reproductive Tract	Abortion, pregnancy toxaemia, prolapsed vagina, prolapsed uterus, retained placenta, metritis, vulvovaginitis, orchitis and epididymitis, trichomoniasis
Musculoskeletal system	Arthritis, bowed limbs, caprine arthritis encephalitis, foot-and-mouth disease, foot scald, foot rot, foot abscesses, nutritional muscular dystrophy, rickets, osteodystrophia fibrosa, fluorosis, laminitis
Nervous system	Rabies, pseudorabies, scrapie, west Nile encephalomyelitis, listeriosis, tetanus, botulism, setariasis, strongyloidiasis, coenurosis, ataxia and swayback, lead poisoning, organophosphate poisoning
Urinary system	Obstructive urolithiasis, dysuria, urinary infections
Circulatory system	Anaemia, anaplasmosis, eperythrozoonosis, tick-borne fever, anthrax, babesiosis, theileriosis, trypanosomiasis, copper poisoning
Mammary gland	Mastitis, skin wart, bloody milk, udder oedema, agalactia
Nutritional deficiency and Metabolic Diseases	Floppy kid syndrome, pregnancy toxaemia, rickets, parturient paresis (milk fever), ketosis, urolithiasis, milk fat depression

access to feed storage sheds, and feeding of garden greens to goats unaccustomed to green feed (King, 1980). Feed changes, however, are not a prerequisite for enterotoxaemia to occur (Smith and Sherman, 2009). Death is attributable to damage of vital neurons, generalized toxaemia and shock (Kimberling, 1988).

The toxins (epsilon) produced during the pathogenesis of the disease increase the permeability and absorption of the intestinal mucosa, facilitating the entry of the toxins into the circulatory system. This leads to rapid fatal toxaemia affecting the vital organs, including the nervous symptoms. In type D infections, pathognomonic signs are hyperglycaemia and glycosuria. Infection spreads from animal to animal through contaminated faeces and soil.

Mortality in vaccinated kids/lambs is 0.5% and high in unvaccinated kids/lambs (5–10%). Peracute, acute and chronic forms occur more frequently in young goats compared to adults (Smith and Sherman, 2009). Sudden death may occur without any previous symptoms – a goat may be normal at night, but is found dead in the morning. The main signs are sudden abdominal pain, distress, loss of appetite, depression, lack of coordination, continuous changing of sitting posture, champing of jaws, hyperesthesia, rapid breathing, excitement and convulsions (Fig. 5.5). In the last stage, the faeces may be mixed with blood due to intestinal haemorrhage, followed by sudden death. High fever (105°F/40.5°C) is reported. In the chronic form, intermittent, recurring bouts of illness are observed over several weeks. Mature animals are usually affected. There is progressive weight loss with intermittent episodes of pasty or loose faeces. The chronic form is extremely difficult to recognize (Shanks, 1949).

History of overfeeding of grains or concentrate feed and change in diet provide diagnostic clues to the owners and veterinarians. Post-mortem lesions consist of serous effusions in the peritoneum, epicardium, endocardium, intestinal serosa, abdominal muscles, diaphragm and thymus gland. The kidney may appear swollen, reddish, soft and pulpy (pulpy kidney). Congestion of the lungs, trachea, abdominal lymph nodes, haemorrhagic abomasum and enteritis (Fig. 5.6) are noticed. The liver is friable, congested and spotted. The stomach is packed with food. The diagnosis can be made by impression smears of affected mucosa, which reveal a predominant population of large Gram-positive rods typical of *C. perfringens*. Multiplex polymerase chain reaction (PCR) toxin typing is conducted to identify the *C. perfringens* strains present in the gastrointestinal contents or faeces of goats (Uzal *et al.*, 1996; Singh *et al.*, 2018) or from paraffin-embedded goat tissues (Warren *et al.*, 1999).

Prognosis is mostly grave in acute conditions. It requires intensive management with intravenous (IV) fluid therapy (electrolytes with bicarbonate) and steroid or non-steroidal anti-inflammatory drugs particularly flunixin meglumine (1 mg/kg IV hourly) to combat shock and alleviate pain. Parenteral antibiotics including penicillin, tetracycline or trimethoprim-sulpha combinations may also be useful. *C. perfringens* strains are often resistant to aminoglycosides (Songer, 1996). Commercially available type C and D antitoxins should be administered parenterally, preferably intravenously, in severe cases; 15–20 ml has been effective in Australian practices (King, 1980).

When an outbreak occurs, all previously vaccinated goats should receive a booster vaccine and unvaccinated goats should receive a prophylactic dose of antitoxin in conjunction with an initial vaccination, followed by a booster shot 2–3 weeks

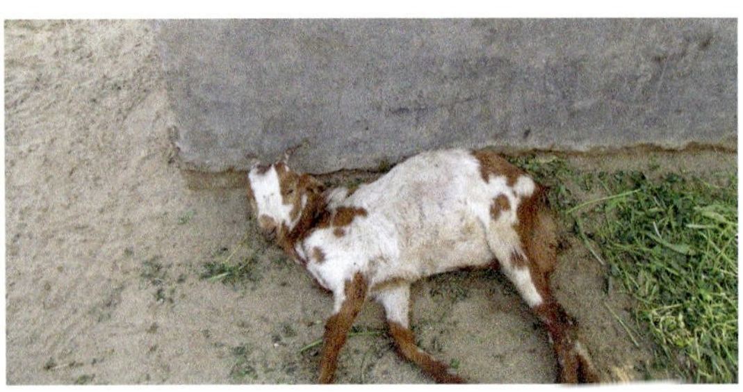

Fig. 5.5. An acute case of enterotoxaemia in a 4-month-old Barbari kid.

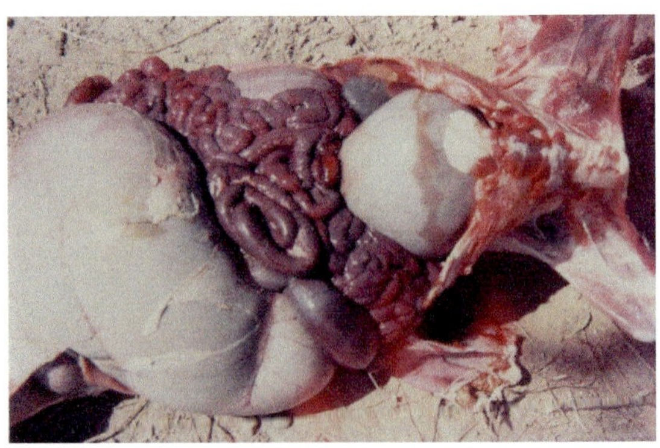

Fig. 5.6. Acute haemorrhagic enteritis in a kid with enterotoxaemia.

later. Any excessive feeding of carbohydrate should be curtailed immediately (Smith and Sherman, 2009). The initial vaccination should be followed by a booster vaccination 3–6 weeks later, and then booster vaccinations every 4–6 months. Vaccinations should be timed so that the last vaccination occurs 2–3 weeks before parturition in pregnant does to provide protection to new-born kids. Feeding should be carefully managed, which includes avoidance of sudden feed changes, vigilance against overeating, and guarding against accidental access to grains and other stored feeds. Other precautions include proper sanitation of the feeding trough by sprinkling lime and cleaning with disinfectant.

Brucellosis

Brucella melitensis is the aetiological agent of caprine brucellosis. It is an infectious zoonotic disease that has a significant economic impact on the livestock industry and public health. The disease is present in South and North America, Europe, Asia and Africa, and remains a major problem in the Mediterranean region, the Middle East, Central and Southeast Asia, sub-Saharan Africa, and parts of Latin America with the seropositivity of 2–10% in goats.

The serological incidence in goats in India is reported to be between 3% and 18% on the basis of samples collected and tested using different molecular diagnostic techniques (Gupta *et al.*, 2006). The disease causes abortion during late pregnancy (Fig. 5.7), retention of placenta, metritis and mastitis. It also affects males and symptoms include orchitis, osteoarthritis and synovitis. The disease is mainly transmitted through aborted foetal content and vaginal exudates. The major transmission of the disease in the flock occurs due to ingestion of infected colostrum and milk. After infection, the organism grows in the chorionic epithelium, spreading to the chorion and uterine mucous membranes.

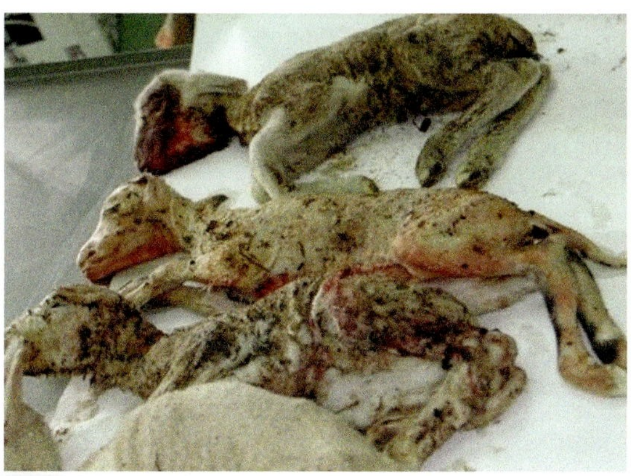

Fig. 5.7. Aborted foetuses in brucellosis that occurred at 4-months of pregnancy.

In post-mortem findings, the placenta appears focally or diffusely oedematous and covered with fibrin flakes. The cotyledons are dull in appearance and the chorion may be thickened. Beside isolation and characterization of the organism, molecular tests such as ELISA, PCR, LAMP and Taqman probes have been standardized to *B. melitensis* in goat serum and body fluids.

The treatment option is usually not economically feasible or therapeutically effective. No successful treatment can be prescribed, as the organism is an intracellular parasite. However, antibiotics like streptomycin, oxytetracycline or enrofloxacin are administered to control the disease. Long-acting tetracycline (25 mg/kg body weight on alternate days for 4 weeks) with streptomycin (20 mg/kg intramuscular on alternate days for 2 weeks) resulted in bacteriologic cure in all 36 goats in a study by Radwan *et al.* (1992). Vaccination with *B. melitensis* Rev-1 vaccine is recommended universally (Alton, 1987). The susceptible animals should be segregated from the healthy flock and slaughtered. A frequent seroscreening programme in the goat flock should be carried out, along with strict and proper disinfection of contaminated areas by fumigation or by sprinkling of lime. The culling of seroreactors from a healthy group should be implemented. A test-and-slaughter policy should be applied to control the infection in goat farms. *B. melitensis* is listed as an agent of bioterrorism and agroterrorism; it is thought that fewer than 10 CFU (colony-forming units) are capable of infecting humans via aerosols. This would require mass therapy of human populations and destruction of animal populations (Constable *et al.*, 2017).

Bacterial pneumonia

This disease is caused by infection with *Mannheimia haemolytica* or *Pasteurella multocida*. These organisms are normal inhabitants of the goat. Poor housing and ventilation conditions, extreme weather and overcrowding are predisposing factors for pneumonia. Parasitism, transportation stress, malnutrition, viral infections (e.g. parainfluenza-3 virus) and lung

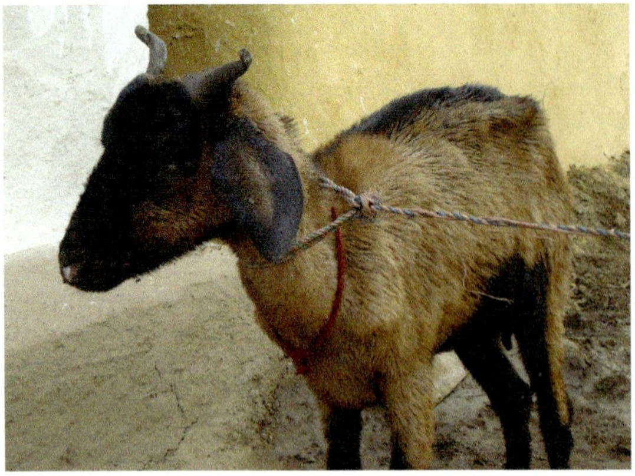

Fig. 5.8. Nasal discharge in a goat suffering from pneumonia.

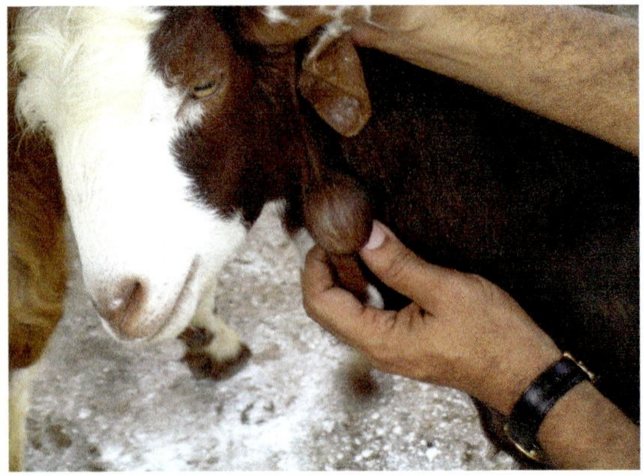

Fig. 5.9. Lymph node enlargement in a goat with caseous lymphadenitis.

parasites are also contributing factors for developing pneumonia. The disease affects both kids and adults. It has been observed that heat stress, sudden weather fluctuation and weaning stress act as predisposing factors in young animals to precipitate pneumonia. It is transmitted through inhalation of droplets and is clinically manifested by a high rise in temperature to 104–107°F/40–42.6°C, a mucopurulent nasal discharge, ocular discharge, anorexia (Fig. 5.8) and evidence of bronchopneumonia with moderate to high mortality. In severe cases, dyspnoea is marked with expiratory grunts.

Post-mortem findings can be used for a definitive diagnosis revealing varying degrees of bronchopneumonia. The effective treatment agents of pneumonia in goats include penicillin, ampicillin, sulphonamides, chloramphenicol, oxytetracycline, tylosin, florfenicol, enrofloxacin and ceftiofur. Antibacterial drugs should be given along with decongestant and antihistamine drugs to reduce lung congestion. Ceftiofur is the only antibiotic approved by the US Food and Drug Administration to treat caprine pneumonia. Inactivated vaccine and vaccine adjuvants may be used to prevent the outbreak of the disease at the age of 3–4 months and repeated annually or at 6-monthly intervals in regions that have a high occurrence rate.

Caseous lymphadenitis (CLA)

Caseous lymphadenitis is a highly contagious chronic bacterial disease in goats and sheep caused by *Corynebacterium pseudotuberculosis*. It mainly occurs in North and South America, Australia, New Zealand, Europe and South Africa.

This disease is clinically characterized by abscess formations in the skin, lymph nodes and internal organs (Fig. 5.9). Goats can pick up the infection from contaminated soil of the premises, water troughs, grazing on infected pastures or direct contact with a ruptured abscess. Ruptured abscess and nasal/oral secretions from animals with pulmonary abscesses are the major source of contamination. Thereafter, the organism multiplies and spreads throughout the body via the bloodstream and lymph nodes to the vital organs (e.g. lungs, kidney and liver) that also form abscesses. CLA abscesses are characteristically thick walled. At the early stage, the abscesses are filled with soft white to green exudate, but older abscesses display a very typical onion like appearance of rings of thick dry pus. The clinical signs and body condition depend upon the site of abscess formation in the vital internal organs or palpable lymph nodes. The most commonly affected lymph nodes are the submaxillary, prescapular, prefemoral, supramammary and popliteal nodes. The goats exhibit chronic weight loss, chronic pneumonia, pyelonephritis, ataxia and paraplegia. Sheep are more prone to internal abscesses and goats are more prone to external abscesses. Once infected, the animal is considered to be a carrier for life.

This is an occupational disease of shearers and abattoir workers with infection occurring through cuts that produce lymphadenitis. Serologic tests, radiography and ultrasonography may be useful diagnostic tests in goats with chronic weight loss.

Treatment with sensitive antibiotics is generally not rewarding, but subcutaneous abscesses can be managed with surgical drainage and an antiseptic dressing with iodine solution, with a precaution to dispose the pus to avoid soil contamination.

Vaccines formulated from concentrated, formalin-inactivated *C. pseudotuberculosis* are available in many countries and have been used with variable success. In comparison to sheep, vaccination appears less successful in goats in field trials. New goats introduced into a flock should be examined for evidence of the disease by meticulous clinical examination for enlarged palpable lymph nodes and serologic testing with a high sensitivity test.

Paratuberculosis (Johne's disease)

This disease is caused by *Mycobacterium avium* subspecies *paratuberculosis* (Map). It is one of the most important chronic wasting diseases of adult goats, causing significant economic loss. The susceptible animals pick up the infection through contaminated food and water. Milk is an important source of transmission of the disease in suckling animals. It is most common in goats at 2 to 3 years of age. The disease is associated with intensive management systems. The affected animals lose body weight continuously and become weak and emaciated (Figs 5.10 and 5.11) with a staggering gait, rough coat, anaemia, intermandibular oedema, intermittent diarrhoea followed by death. Persistent diarrhoea is not very common in goats except in the terminal stage of the disease.

No single test can identify animals with Johne's disease in a herd with sufficient specificity to avoid a false positive reaction (Smith and Sherman, 2009). Faecal culture is considered to be the 'gold standard' test for ante mortem diagnosis of this disease.

On necropsy there is focal or diffused thickening or oedema in the ileum, caecum and colon and enlargement of mesenteric lymph nodes. Common

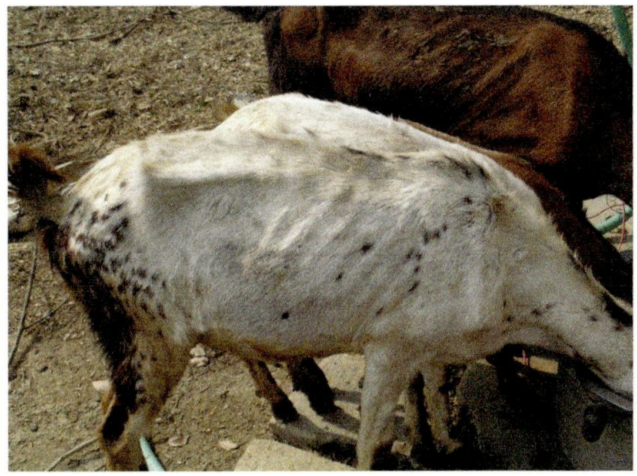

Fig. 5.10. Hide-bound condition in an adult Barbari goat with Johne's disease.

Fig. 5.11. A herd of Barbari goat in semi-arid region with Johne's disease.

lesions in cattle such as thickening and corrugation of mucosa of the ileo-caecal valve are not often observed in goats. Paratuberculosis is mostly a herd problem and can be minimized by identification and removal of infected individuals from a herd.

The demonstration of acid-fast bacilli within typical lesions and serodiagnosis by PCR using the IS900 sequence primers are used for confirmation of infection.

Treatment with antimicrobial and antitubercular drugs is neither effective nor economically feasible.

The control of Johne's disease is based on identification and elimination of infected animals and prevention of new infections by adopting best management practices and vaccination. Six-monthly or annual screening of the flock and subsequent, timely elimination of infected goats helps to reduce the prevalence rate.

Anthrax

Anthrax is caused by *Bacillus anthracis*, a large Gram-positive spore-forming bacteria. The disease is characterized by bloody discharge from orifices and by sudden death. The animals die due to septicaemia, respiratory distress and anoxia. The toxins released in the body cause generalized oedema and haemorrhage.

Anthrax is an important zoonotic disease, transmitted by direct handling of hair and skin, inhalation of spores from infected carcasses and contaminated animals. Therefore care must be taken when handling suspected or confirmed cases by using protective clothing, gloves and a mask.

The spores are highly resistant and often remain viable for long periods in contaminated soil or animal by-products. Goats are comparatively less susceptible than sheep and cattle. The animals get infected through contaminated feed and water. Anthrax has seasonal occurrence in endemic areas, with outbreaks often occurring after a period of drought or heavy rains. The disease has been reported in Nigeria, USA (Texas), China and Ethiopia (Okoh, 1981; Whitford, 1982; Shiferaw, 2004). Oedematous fluid or oozing blood from carcasses is a diagnostic feature. The organism can be easily demonstrable in a stained smear of oozing fluid. Penicillin, erythromycin, enrofloxacin and oxytetracycline are effective drugs for treatment. In addition, anthrax antiserum can also be given. An anthrax spore vaccine has been effectively and successfully used. Antibiotics should always be avoided in vaccinated animals.

It is advised not to conduct necropsy of the suspected animals. The carcass should be burned or buried deeply to avoid dissemination of anthrax spores into the environment.

Listeriosis (circling disease)

Listeria monocytogenes is the causative organism that is responsible for meningeoencephalitis, septicaemia and abortion in goats. It is a zoonotic disease and humans can get infected from exposure to milk and dairy products.

The infection spreads through an oral route. The disease is characterized by a rise in body temperature (106°F/41°C), abortions in the third trimester of gestation, dullness, depression, pressing of head against a wall or hard surface, unilateral facial paralysis, circling movements in either direction and conjunctivitis with corneal opacity. Septicaemia occurs in young animals and death results within 24 hours. In blood studies, leucocytosis, dominated by macrophages and lymphocytes, is observed.

Outbreak of the disease has been reported in North America, Europe, Spain, South Africa and Asia. Listeriosis is predisposed to sudden changes in the weather, stresses, parasitism and feeding of poor quality silage.

Suspected cases should be immediately isolated and treated early by using penicillin or tetracycline along with supporting fluid therapy and anti-inflammatory drugs and steroids for reducing brain oedema in the encephalitis form, particularly. The aborted foetus and discharges should be handled and disposed of carefully, and neonates should be given pasteurized milk to avoid the occurrence of neonatal septicaemia.

Mastitis

Mastitis is a common clinical condition of the udder. The disease is characterized by swelling of the udder, decreased milk production, discolouration of milk with the appearance of pus in late stages and atrophy of the udder in neglected cases. It is caused by many bacteria species (e.g. *Staphylococcus aureus*, *Escherichia coli*, *Pseudomonas* spp., *Klebsiella pneumoniae*, *C. pseudotuberculosis*, *M. haemolytica*), mycoplasma (contagious agalactia, caused by *Mycoplasma agalactiae* and *Mycoplasma mycoides* var. *mycoides*) and fungus. It is considered a disease of mismanagement, where unhygienic conditions increase the risk of occurrence. Gangrenous mastitis (blue bag) where the infected udder becomes cold and blue coloured and may slough off is also seen in goats. Treatment is often unsuccessful in such cases, but attempts may be made with antibiotic therapy. Hygiene, sanitation, cleanliness of farm premises and milk utensils, regular washing of the udder and milkers' hands are some important key factors to prevent mastitis.

The disease is observed in acute and subacute forms. In subacute cases, milk production declines and the milk is a yellow–green colour. In acute cases, the symptoms include fever, swelling of the udder, enlargement of the mammary lymph nodes and the development of multiple abscesses in the mammary gland, which may also lead to induration of the udder.

The mortality rate can be high (10–30%), and many adult females are culled because the udder is permanently damaged (Constable *et al.*, 2017).

The organism can be isolated from milk, ocular fluid, joint fluid aspirate, nasal swabs, ear swabs and lung lesions. Molecular detection from samples confirms the mycoplasma infection. The common effective antibiotics are enrofloxacin (fluoroquinolones), tetracycline, Tiamutin and macrolides. Flock biosecurity measures, milking process hygiene and test-and-slaughter should be employed. Vaccination with an attenuated-live vaccine or a killed-adjuvant vaccine of *M. agalactiae* gives variable results in reducing prevalence. This is a problem in dairy goats, and is endemic in many European countries, Africa, parts of Asia and the Indian subcontinent, the Middle East, and North and South America.

Dermatophilosis

This is a common skin infection of goats occurring worldwide caused by *Dermatophilus congolensis*, an anaerobic actinomycete. This is a soil-borne infection, entering through injured skin caused by tick infestation and thorny bushes during grazing. These skin lesions are intensified in humid seasons.

Clinical signs include papular and scab formation on the ear pinnae, the muzzle, nose, face, scrotum and feet. The scabby lesions on the skin surface are easily rubbed off, leaving a light-coloured area beneath (Scott, 2007). The lesions may be mistaken for contagious ecthyma and ringworm. The disease is confirmed by demonstration of Gram-positive mycelial organisms in the impression smear made from the under surface of the scab.

Heavy doses of penicillin with streptomycin and tetracycline are effective, with local application of chloramphenicol ointment. Cleaning the area with hydrogen peroxide is helpful in controlling infection. Control of tick infestation is desirable for the treatment and prevention of infection. Zinc sulphate solution (0.5–1.0%) has been found to be effective in reducing the spread and incidence of the disease.

Peste des petits ruminants

Peste des petits ruminants (PPR), also called goat plague, is a widespread viral disease caused by a Morbillivirus (Paramyxoviridae). The disease is considered to be a major constraint in the development of small ruminant productivity among smallholders, mainly by causing high mortality in young stock (Sumberg and Mack, 1985). There is a single serotype of PPR virus, but four distinct genetic lineages exist. Morbidity and mortality are high when occurring in native sheep and goat populations. The peracute and acute forms are seen mainly in goats. This disease is characterized by high fever, severe depression, decreased appetite, watery nasal discharge progressing to profuse muco-purulent discharge occluding the nostrils, respiratory distress and sign of bronchopneumonia, conjunctivitis, erosive and necrotic lesions in the mouth (dental pad, hard palate, cheeks) (Fig. 5.12), severe diarrhoea, dehydration and weight loss. Abortions have been reported during outbreaks. Death usually occurs within 1 week of the onset of illness and earlier in peracute cases.

This contagious disease of goats is now endemic in Western, Central, Eastern and Northern Africa, the Middle East and Asia. It is one of the most damaging animal diseases, affecting small ruminants in almost 70 countries in Africa, the Middle East and parts of Asia. It has been selected as one of the top priority diseases to be addressed by the World Organisation for Animal Health (OIE) because of high morbidity or incidence (80–90%) rate and mortality (50–80%) rate. It is more severe in young animals, those with poor nutrition and concurrent parasitic infections.

Spread of infection to new areas is largely by close contact between infected and susceptible animals, particularly during the movement of infected animals. PPR virus transmission requires close contact. Ocular, nasal and oral secretions and faeces are the sources of the virus. Contact infection occurs mainly through inhalation of aerosols produced by

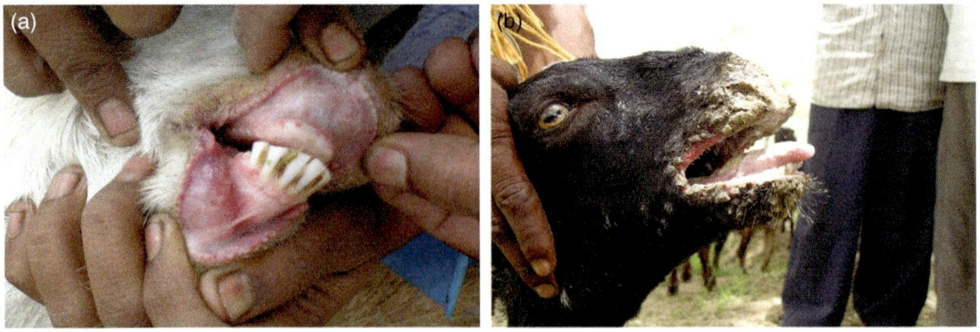

Fig. 5.12. (a) Buccal lesions and (b) pneumonia in PPR in an infected goat.

sneezing and coughing. Fomites such as bedding may also contribute to the onset of an outbreak. Infected animals may transmit the disease during the incubation period. Epizootics of PPR may be characterized by high mortality and poor prognosis. Young animals (4 to 8 months) are affected severely and morbidity and mortality rates are higher. The disease may be further complicated by secondary bacterial infections such as *E. coli* and *Pasteurella* spp. Coccidia have also been found to be associated with PPR to aggravate the condition. In typical cases of PPR, flocks have a history of stress-born predisposing factors. The risk of an outbreak is greatly increased when a new stock is introduced after transportation or when animals are returned unsold from the livestock markets where they may have contracted the disease (Kumar *et al.*, 2001).

Lesions are commonly observed in the digestive and respiratory tract. In acute cases, the main post-mortem findings include a dehydrated carcass with faecal soiling, necrotic lesions in the mouth/tongue and nose, congestion of the ileo-caecal valve, engorgement and blackening of the folds in the cecum, colon and rectum (zebra striping), an enlarged spleen, oedematous lymph nodes and pneumonia. For confirmatory diagnosis, samples include blood in EDTA, serum and biopsies of mesenteric lymph nodes, spleen, lung, tonsils, pieces of the ileum and large intestine. Swabs of serous nasal tract should be collected in the acute phase of the disease to be transported in cold storage.

PPR vaccine is now widely used for the control of PPR in Africa, the Middle East and Asia. Freeze-dried vaccine is reconstituted in sterile normal saline solution and given as a 1 ml subcutaneous injection to all animals above 3–4 months of age, which provides immunity for 3 years. If there is an outbreak, the diseased animals should immediately be separated from the healthy flock and all unaffected animals should be vaccinated. In established goat farms, introduction of new animals should be done only after PPR vaccination. Transportation stress precipitates PPR in goats, therefore, vaccination is advised before transportation (see Protocol 7 at the end of this chapter for more details on transportation). General management practices such as quarantine, proper disposal of carcasses and contact fomites, decontamination and restrictions on importation of goats from affected areas should be strictly implemented.

Foot-and-mouth disease (FMD)

FMD is one of the most regulated livestock diseases worldwide. The disease imposes considerable economic losses due to reduction of milk, meat and high kid mortality rates. When goats become infected with FMD virus, the disease may not be diagnosed for a considerable time because symptoms are very mild. However, some acute to subacute cases are characterized by vesicular eruptions in the epithelium of the buccal cavity, tongue, muzzle, feet, teats and udder, which is clinically characterized by lameness, fever and poor appetite. Vesicle formations in the feet are very fragile and mostly not observed (Fig. 5.13). In susceptible kids under 1 month of age, morbidity may approach 100% and mortality may be as high as 95% due to myocardial necrosis.

Infected animals are the main source of infection. It may spread either through direct contact or by contaminated feed and fodder, utensils and transportation.

There are seven distinct serotypes of foot-and-mouth disease viruses. Six of the seven serotypes (O, A, C, SAT-1, SAT-2 and SAT-3) are found in Africa; Asia has four (O, A, C, Asia-1); South America has three (O, A, C); and the Middle East periodically has incursions of two types (SAT-1, SAT-2) from Africa. There is no cross immunity between serotypes. In India, type O, A, C and Asia-1 serotypes are prevalent. It has been found that the serotype O (Kumar *et al.*, 2004) and the serotype A occur more frequently than the other serotypes. There has been no report or minor occurrence of type C outbreak in Indian goats.

A polyvalent vaccine containing A (22), O, C and Asia-1 strains of FMD virus is used for active immunization. To prevent antigen drift, vaccines comprised of more than one strain are advised. Initial vaccination is carried out at the age of 3–4 weeks and older; booster vaccination is given 2–4 weeks after the initial vaccination. Thereafter, re-vaccination is carried out at 6-monthly intervals. The essential features of a control and eradication programme include restriction of movement of animals and animal products, and disinfection of the infected area. Vaccination may be used to control the outbreak. FMD is controlled by restriction of livestock movement, isolation or slaughter of infected animals, ring vaccination around outbreak and regular vaccination. Immunity is broken not only by different serotypes but also by the variants of the same serotype. A 2–5% copper sulphate footbath can be used to control secondary bacterial infections and maggot infestations. Similarly, 2% alum solution is helpful to reduce the severity of buccal lesions.

Goat pox

Goat Pox is prevalent in North and Central Africa north of the equator, the Indian subcontinent, the Middle East, China and Southwest Asia. Goat pox is a serious disease with a high mortality rate of more than 50%. The disease is characterized by skin eruptions on the whole body. The clinical signs of goat pox virus infection vary with different hosts and in different geographical areas. Goats of all ages are affected but more severe effects are observed in young animals. Goat pox is an acute to chronic disease of goats characterized by generalized pox lesions throughout the skin (Fig. 5.14) and

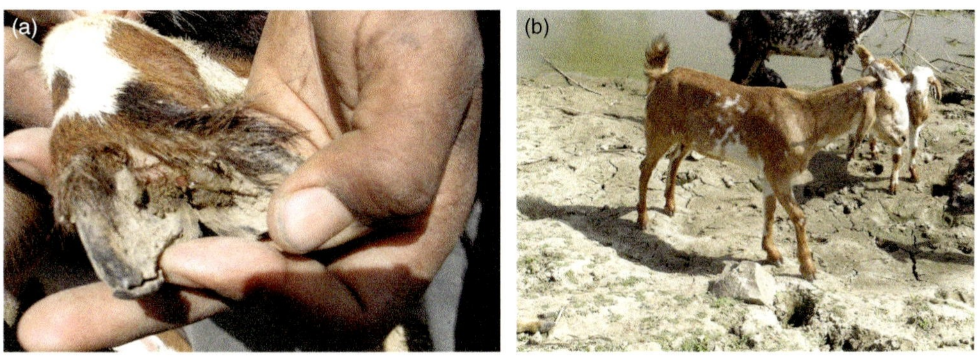

Fig. 5.13. (a) Vesicle formation in a foot and (b) lameness in a goat infected with foot-and-mouth virus.

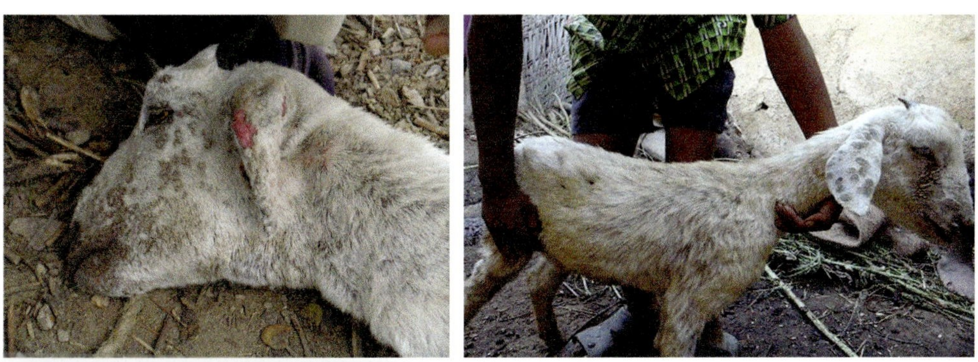

Fig. 5.14. Pox lesions on kids infected with goat pox.

mucous membranes, fever, rhinitis, conjunctivitis and lymphadenitis. The first signs may include fever, depression, conjunctivitis, lacrimation and rhinitis. Pox lesions are more easily observed on the hair-free parts of the body such as the perineum, inguinal area, scrotum, udder, axilla and muzzle. Lesions that present on the tongue and gums tend to ulcerate. The mortality rate becomes higher when lesions develop in the respiratory and alimentary tracts. Secondary bacterial pneumonia is the most common cause of death. In susceptible kids under 1 month of age, morbidity may approach 100% and mortality may be as high as 95%. The morbidity (67–75%) and mortality (45–58%) rates are very high in a susceptible goat flock in all age groups of animals (Barua *et al.*, 2005). In India, Black Bengal goats are highly susceptible to this disease.

Contact is the main means of transmitting the virus. Inhalation of aerosols from acutely infected animals and direct contact through skin abrasions are the natural means of transmission. Post-mortem examinations show pox lesions on all parts of the body. Skin nodules regress, forming reddish brown crusts that fall off to expose granulated tissue underneath. Multiple consolidated areas (0.5–2.0 cm in diameter) in the lungs are observed and similar types of lesions may found in the liver, kidneys, abomasum and other organs. Lesions are found in the trachea with accompanying catarrhal pneumonia. Immuno-capture ELISA can be used to detect the goat poxvirus in scab suspension. Indirect ELISA is employed to screen the serum samples for the presence of antibodies against goat pox virus in a flock. The attachment gene has been successfully adopted to detect the capri pox virus genomic DNA in skin scab samples. Infection results in solid immunity. There is no specific treatment to this disease, but antimicrobial therapy helps in recovery. Goat pox vaccine is recommended in prevalent areas of the country. Live-attenuated goat pox vaccine is also available, which confers solid protection. Restrictions on the movement of animals and animal products (meat, hair, wool and hides) are essential to prevent introduction of goat pox into uninfected regions. In endemic areas, vaccination is an effective means of controlling losses from goat pox. A few strains of goat pox virus are pathogenic for human beings, hence precaution in handling the infected cases is required.

Contagious ecthyma (contagious pustular dermatitis, orf, scabby mouth, sore mouth)

Contagious ecthyma is a viral disease of goats that is highly contagious in nature. Contagious ecthyma is caused by an epitheliotropic Parapoxvirus that infects goats through skin abrasions. The virus replicates in proliferating keratinocytes in the damaged epidermis. The distribution of the disease is worldwide. It occurs most commonly in young kids, but adult animals can also be severely affected. This disease can occur at any time, but outbreaks are most common during winter, especially in goats grazing on frozen grasses.

The disease appears in the form of papules, which progress rapidly to vesicles, pustules and scabs. It causes unthriftiness, associated with economic loss. Crusty proliferative lesions typically form on the lips (Fig. 5.15), but can also affect the face, ears, scrotum, teats or vulva. Large masses of granulation tissue develop under the scabs approximately 6 to 7 days after initial lesion formation. Bronchopneumonia and gastroenteritis have also been observed in kids. The skin crusts gradually dry and fall off, leaving areas of alopecia and depigmented skin.

Scabs that fall to the ground during resolution of lesions have been implicated as possible agents of transmission to other animals. Carrier goats may also be an important source of infection. The morbidity rate is comparatively high in kids. Local lesions should be dressed with stringent lotions or antiseptic agents; fly repellent cream should also be used to avoid maggot infestations. Antiviral drugs have been combined with emollients in gels and sprays (e.g. cidofovir and sucralfate) for recovery. Supportive treatment, such as providing soft,

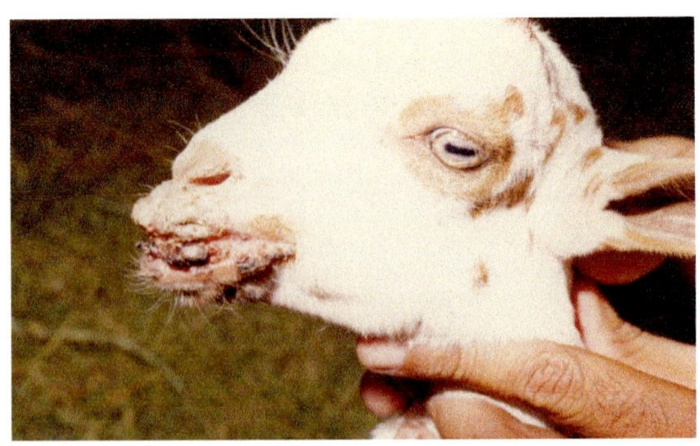

Fig. 5.15. Proliferative lesions on lips in a kid with contagious ecthyma.

palatable food, may be helpful. Affected kids should be isolated and applied ointment on lesion.

Bluetongue

Bluetongue is an infectious, arthropod-borne viral disease of ruminants caused by an Orbivirus transmitted by *Culicoides* biting midges, which severely affect sheep and cattle. Bluetongue infection of goats is typically an inapparent infection. Viruses from the western lineages circulate in Africa, the Caribbean and the Americas, whereas those from eastern lineages are endemic in Asia, Indonesia and Australia (Coetzee *et al.*, 2012). In endemic regions, cattle and goats are commonly considered to be amplifying hosts of bluetongue virus (BTV) due to a prolonged viraemia that usually occurs in the absence of clinical signs (Barratt-Boyes and Maclachlan, 1994). There are at least 24 different known serotypes of the BTV that share common complement fixing and precipitating antigens. Cross-protective immunity against heterologous serotypes is poor. However, only BTV serotypes 10, 11, 13, and 17 are currently active.

Depending on climatic factors affecting the vector population, the virus transmits seasonally to adjacent areas. Infection in goats is generally subclinical; however, more severe expression has also been reported (Sapre, 1964). Bluetongue infection in goats consists of mild depression, temporary loss of appetite, possibly a fever up to 105°F (40.5°C) and hyperaemia of the oral and nasal mucosa and excoriations of the tongue, lips, and gums that become ulcerative and necrotic. Coronitis also develops, with hyperaemia and swelling around the coronet joint. In affected goats, walnut-sized skin eruptions are distributed all over the body. Goats exhibit an acute drop in milk production, oedema of the lips and head, nasal discharge and erythema of the skin and udder (Dercksen *et al.*, 2007). Gross necropsy lesions in goats with full-blown clinical bluetongue are similar to the symptoms observed in sheep. The main post-mortem lesions are hyperaemia and oedema of the buccal mucosa and ulceration of the palate, lung, severe haemorrhagic gastroenteritis and haemorrhages and congestion in the spleen. The recommended tests for the detection of bluetongue serogroup-specific antibodies are agar-gel-immunodiffusion and competitive ELISA. The recommended test for the detection of serotype-specific antibodies is the virus neutralization test. Vaccination is the major means of control. Since, the various serotypes offer little cross immunity, multivalent vaccines are desirable.

Rift Valley fever (RFV)

This is caused by Rift Valley fever (RVF) virus, genus Phlebovirus, a member of the family Bunyaviridae, transmitted by an arthropod vector. This disease was first reported in the Rift Valley in Kenya in 1930; however, outbreaks occur throughout sub-Saharan Africa. This disease is now spreading into Egypt, Mauritania and Madagascar. The first RVF outbreaks outside Africa were recorded in 2000 in Yemen and Saudi Arabia (Constable *et al.*, 2017). Disease susceptibility in goats is lower compared to cattle and sheep. RVF is clinically characterized by abortion in pregnant does and death in young kids in the 1-week age group. Peracute death is reported in neonatal kids without any clinical signs. The acute form is clinically comprised of fever, vomiting, bloody diarrhoea, unsteady gait, stomatitis and death within 1–4 days (Smith and Sherman, 2009). The mortality rates are very low in adult goats, with the clinical signs of fever, anorexia and abortion. Extensive hepatic necrosis is the characteristic lesion in RVF. In neonates, the liver is enlarged and has a yellow–orange discolouration, whereas in older animals, pale foci of necrosis impart a mottled appearance to the organ.

Further spread of RVF into countries where the disease is not endemic may affect the economy and public health, and vaccination is an effective approach to prevent the spread of rift valley fever virus. The control of RVF is based on the restriction of animal movement to infected areas and mosquito-dominant areas, laricidal spray and vaccination. Live-attenuated vaccines (Smithburn strain) and mutagenized live virus vaccines confer immunity for at least 28 months. Females are not vaccinated because they are abortigenic. A live-attenuated MP-12 vaccine is one of the best-characterized RVF vaccines for safety and efficacy and is currently conditionally licensed for use for veterinary purposes in the United States (Ikegami *et al.*, 2015)

As a preventive measure, countries that are free from RVF should not import animals from Africa. Simultaneous control of mosquitoes, infected biological materials and prevention of human movement should be taken up.

Caprine arthritis encephalitis (CAE)

This disease, first diagnosed in 1974, is one of the most important diseases affecting the goat industry in the United States (Cork *et al.*, 1974). The highest prevalence occurs in Canada, France, Norway, Switzerland and the United States; the seroprevalence rate exceeded 65% in those countries. In countries actively importing dairy goats, such as Kenya, Mexico, New Zealand and Peru, the overall prevalence of CAE infection is often less than 10% (Smith and Sherman, 2009).

The disease is caused by caprine arthritis encephalitis virus (CAEV), a lentivirus belonging to the family Retroviridae. All breeds of goats are susceptible to CAE; however, serological surveys indicate that the disease is most common among the dairy goat breeds. CAE virus is transmitted naturally during the neonatal period from an infected adult goat to the kid through consumption of infected colostrum and milk. There is evidence to suggest that CAE can also be transmitted directly from goat-to-goat possibly through saliva, nasal secretions, urine or faeces.

It is manifested in two forms. The encephalitis form is most commonly encountered in kids of 2 to 4 months of age and is characterized by paralysis that may or may not progress to seizures or death. The arthritic form is most common and is seen in adult goats above 1 to 2 years of age. Affected goats gradually lose weight, develop a poor hair coat and enlarged joints, particularly the carpal, hocks and stifle. In the early stages, the affected animals may show a progressive leg lameness, and in terminal cases, goats may walk on their knees and refuse to rise. A tentative diagnosis is based on history and clinical findings, in particular the age of the animal and the diseases' manifestation pattern. Serological tests are commonly conducted for diagnosis and screening of the flock.

Treatments are generally not effective in any stage of CAE. Animals with mild cases of the arthritic form can be made more comfortable by providing regular, correct hoof trimming, providing easily accessible feed and water, and by long-term use of oral non-steroid anti-inflammatory drugs to relieve pain (aspirin at the dose of 10–20 mg/kg every 8 to 12 hours and phenylbutazone at a dose of 10 mg/kg once a day). Goats with advanced cases of the arthritic form, unable to extend their legs and forced to walk on their flexed knees, should be humanely euthanized (Smith and Sherman, 2009).

In control programmes, goats of breeding age and older should be regularly screened for detection of antibodies to the CAE virus. If a herd is negative for CAE, it can be kept free of CAE by managing it as a closed herd and only introducing new genetic stock that has been tested and confirmed to be free of CAE. Periodic herd testing for CAE should be carried out to monitor the herd's status. In an infected herd, culling should be considered. Kids should be removed from their dams before they are able to stand and suckle, and should be fed pasteurized goat colostrum and raised on pasteurized milk or milk replacer. Kids should also be kept separate to avoid contact with adults.

5.2.2 Parasitic diseases

A variety of internal parasites affect animal health and production. Some of the important infections are discussed in this section.

Gastrointestinal nematodiasis

Among nematodes, *Haemonchus contortus* is widely prevalent and highly pathogenic in goats. This infection causes severe diarrhoea and anaemia due to the voracious blood sucking nature of the parasite in the abomasum. *H. contortus* infection leads to gastric haemorrhages and sudden death.

The intestine can also be infected with *Ostertagia*, *Trichostrongylus columbriformis*, *Cooperia* spp., *Strongyloides papillosus*, *Nematodirus* spp., *Bunostomum trignocephalum* and *Gaigeria pachyscelis*, associated with acute to moderate green- or black-coloured foul-smelling diarrhoea with continuous soiling of the perineum and tail. All these worms are responsible for catarrhal or haemorrhagic enteritis and retarded growth of animals. Heavy infection may cause death. In prolonged infections, intermandibular oedema (bottle jaw) and anaemia may develop (Fig. 5.16). In chronic cases, there is dullness, loss of body condition and gradual weight loss, which affect production and performance.

Evidence of high faecal egg count and heavy worm load in necropsy are indicative of parasitic gastroenteritis. Therefore, periodical faecal examination is necessary for detection of egg loads to decide anthelmintic treatment (see Protocol 9 at the end of this chapter). Anthelmintic drugs used to treat and control infections include tetramisol (8–10 mg/kg body weight [BW]), albendazol/fenbendazole

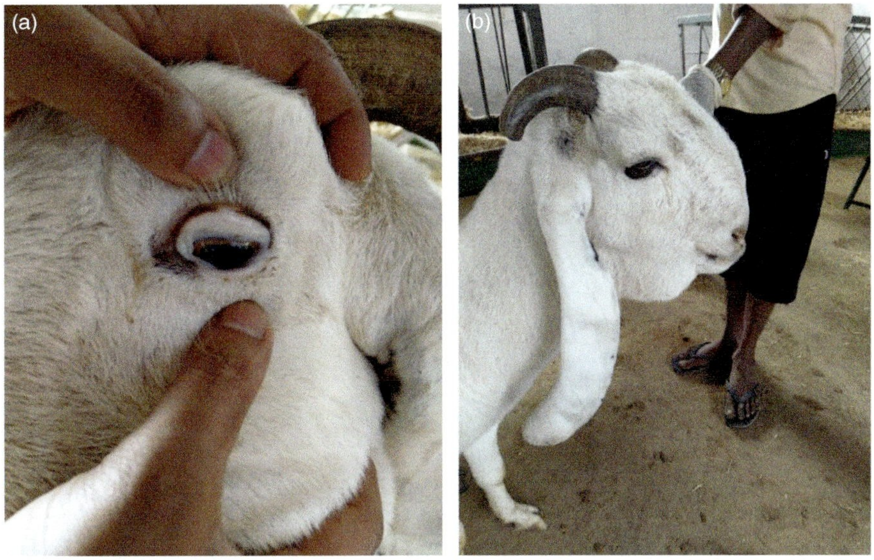

Fig. 5.16. (a) An anaemic mucous membrane and (b) bottle jaw in an adult Jamunapari goat with an *Haemonchus* infection.

(7.5–10 mg/kg BW), morantel citrate (10 mg/kg BW), doramectin/ivermectin/eprinomectin/moxidectin (0.2–0.4 mg/kg BW) and closental (7.5 mg/kg BW).

Immature paramphistomiasis

Clinical paramphistomiasis is associated with the immature form that infests the small intestine. This disease is commonly reported in Africa, Asia, Eastern Europe, Russia and Mediterranean countries (Katiyar and Varshney, 1963; Horak, 1971; Chhabra *et al.*, 1978). The outbreak mostly occurs in the wet season, when the intermediate host snail population is abundant in stagnant ponds and water resources. It has a morbidity rate of 35–79% (Katiyar and Varshney, 1963). The parasites cause severe irritation in the intestines, resulting in enteritis characterized by profuse fetid diarrhoea, depression, emaciation, anaemia and oedema in the submandibular space, with a high mortality rate. The presence of immature flukes in necropsy and faecal samples are diagnostic. Diseased animals should be removed from infected water sources and treatment should be given immediately. The effective anthelmintic drugs include oral administration of oxyclozanide (15 mg/ kg BW) and rafoxanide (7.5 mg/kg BW). For prevention, snail-infected water-logged areas should be avoided by grazing goats and sheep.

Fascioliasis

This is the most common liver fluke infestation worldwide. It is caused by *Fasciola hepatica* and *Fasciola gigantica*. The intermediate host breeds in stagnant water ponds that are the major source of infection in goats (Fig. 5.17). In acute forms it may cause sudden death in flocks, or depression, anorexia, progressive weakness and pain on deep palpation of the liver area. The chronic form is mostly observed in goats, characterized by diarrhoea, anaemia, rough hair coat, poor body condition score, anorexia, oedema in the submandibular space (bottle jaw) and the lower part of the abdomen. Bottle jaw is characteristic in long-standing cases. Diagnosis is by observation of fluke eggs on examination of the faecal matter. The infected flock may be treated with oxyclozanide (15 mg/kg BW orally), rafoxanide (7.5 mg/kg BW orally), triclabendazole (12 mg/kg BW orally/subcutaneously), nitroxynil (15 mg/kg BW subcutaneously), closentel (10–20 mg/kg BW orally) or clorsulon (7 mg/kg BW orally). For prevention and control, snail-infested water-logged areas should be avoided for grazing and flukicides should be strategically administered.

Cestodiasis (Monieziasis)

This is another type of parasitic enteritis caused by mainly *Moniezia expansa* (tapeworm) and sometimes

Fig. 5.17. Stagnant water pools are a source of infection of fascioliasis in goats.

Avitellina spp. The infection rate depends upon the presence of infective mites (the intermediate host) and requires a large number in the lumen of the intestine for clinical symptoms to be evident (Fig. 5.18). The infestation is clinically manifested by diarrhoea, malabsorption, dullness, depression, rough body coat, potbelly and stunted growth. It commonly occurs in young animals below the age of 6 months. Diagnosis is on the basis of clinical findings and the presence of the characteristic eggs/segments (proglottids) in the faeces. The effective and common anthelmintic drugs are niclosamide (50–100 mg/kg BW) and praziquantel (5 mg/kg BW). The benzimidazoles such as mebendazole (15 mg/kg BW), fenbendazole (15 mg/kg BW), cambendazole (20 mg/kg BW), oxfendazole (10 mg/kg BW) and albendazole (10 mg/kg BW) are also effective.

It is possible to control the disease by treating clinically infected kids and adult goats, and by restricting animals for grazing in fields, particularly avoiding those infested with grass mites. Ploughing and reseeding are considered to reduce mite populations at pasture.

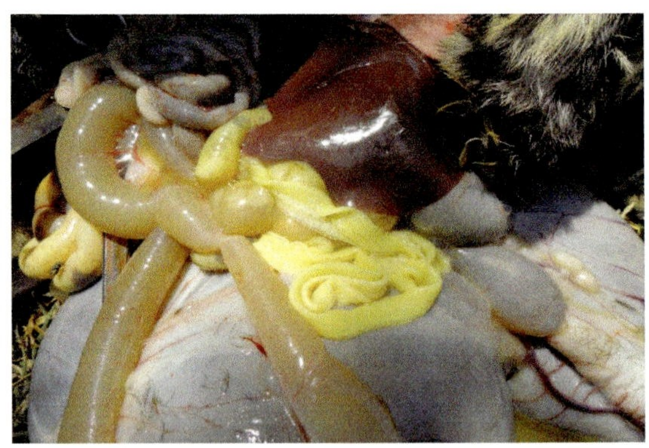

Fig. 5.18. Tapeworm (Moneizia) in loop of intestine in a 2-month-old goat kid.

Regular, timely mass medication with the correct dose of specific anthelmintic drugs should take place, timed to coincide with the particular meteorological conditions of the area. In general, at least two doses of anthelmintic drugs are drenched – before and after the monsoon season. Effective parasitic control cannot always be achieved by drugs alone.

Emphasis should be given to reduce contamination with maximum sanitation of the bedding and grazing areas, particularly in hot and humid rainy months when parasitic survival and infection rates are high.

Coenurus cerebralis cyst (gid, sturdy)

This parasitic disease is caused by the intermediate stage of *Taenia multiceps* through invasion of the brain and spinal cord and the development of localized and space-occupying lesions of the central nervous system. *Taenia multiceps* primarily inhabits the intestine of dogs and wild Canidae. The embryos, which hatch from eggs ingested in feed contaminated by faeces of infected dogs, reach the bloodstream and lodge in the brain and spinal cords of sheep and goats to cause clinical symptoms. The disease has been reported in Afro-Asian countries (Sharma and Chauhan, 2006).

The most obvious signs are slowly developing partial or complete unilateral blindness, dullness, head tilting, intermittent convulsions or intermittent circling, head-pressing, ataxia, loss of balance (Fig. 5.19) and incomplete mastication. The circling also depends on the location of the cyst in the brain – if the cyst is located in either the right or left parietal region, then the animal holds its head towards the affected side and moves in a circle towards that side. The other signs that develop as the disease progresses include grinding of the teeth, salivation, exophthalmos and, in the terminal phase usually 2–3 weeks later, recumbency, extensor rigidity and frequent paddling with the hind limbs. In some cases local and palpable softening of the cranium may also be noticed, particularly if the cyst is located in a superficial position. It may rupture, leading to spontaneous recovery. Death usually occurs after a long course of several months.

There are no known consistently effective anthelmintic therapies for elimination of the metacestode in clinically affected animals (Sharma and Chauhan, 2006); however, praziquantel administered orally at a dose of 100 mg/kg BW has been used in sheep with subsequent recovery (Verster and Tustin, 1990). Surgical removal of cysts is possible through the softened or perforated bone, with good recovery after surgery (Sharma and Tyagi, 1975). The occurrence of disease can be minimized if the dog population is controlled and given periodical deworming with effective anti-cestode drugs.

Coccidiosis

This is an important protozoan infection of the intestine that mostly affects kids in the 2- to 6-month age group. Severe enteritis is the main clinical sign in heavily infected kids. Several *Eimeria* spp. are responsible for coccidiosis in goats. In coccidiosis, the damage done to the host is essentially the result of intestinal cell destruction occurring when coccidian cysts are released from host cells. In young kids, coccidiosis is clinically manifested by the presence of diarrhoea. The faeces are loose, foul smelling and discoloured with or without blood. Animals may show abdominal pain, anaemia, inappetence and weight loss. The look of the animal is unthrifty with a rough coat. Death may occur in acute cases.

Fig. 5.19. Nervous system symptoms in an adult goat infected with GID.

The clinical disease is common after the stress of weaning, feed change or transportation. Kids that survive have poor appetites and poor body weight gain.

The clinical signs may be correlated with oocyst faecal counts. Though presence of oocysts does not indicate disease, the clinical disease is always associated with large numbers of oocysts in the faeces. The oocysts may be detected in faecal examination. Coccidiosis is self-limiting if not severe or acute, and clinical signs usually subside spontaneously. Coccidiosis is mainly a disease that results from poor management practices. It can be controlled by avoiding overcrowding, keeping younger stock separate from the adults, ensuring the sheds are cleaned and soil replaced at regular time intervals, and lime powder sprinkled inside the sheds. Medication may also be given to control the oocysts load.

Chemotherapeutic agents recommended for treatment and control of coccidiosis in kids are sulphadimidine (130–200 mg/kg BW for 4–5 days), nitrofurozone (15 mg/kg BW for 7 days), amprolium hydrochloride (10 mg/kg BW orally for 5 days) and toltrazuril (7–10 mg/kg BW for 2–3 days). The same drugs may also be used for prevention in lower doses for longer periods (Smith and Sherman, 1994). Monensin at the dose rate of 20–30 g/tonne feed may also be recommended for kids as a preventive measure during a period of susceptibility.

Filarial infection (setariasis)

Setaria digitata is a filarial worm of ungulates but it may also cause a disease in goats and sheep known as cerebrospinal nematodiasis, commonly referred to as lumbar paralysis or kumri. *S. digitata* is a non-pathogenic parasite of the peritoneal cavity of cattle and buffalo. Mosquitoes (*Anopheles* and *Aedes* spp.) serve as the intermediate hosts. Mosquitoes ingest the microfilaria along with blood during feeding from a definitive host. In unrelated hosts such as the goat, sheep and horse, the microfilaria migrates to the central nervous system, which causes the clinical condition. The incidence is correlated with the period of vector propagation. Morbidity rates can be high, with as many as 30–40% of susceptible animals.

It affects all age groups and may be acute or subacute, and mild to severe. Affected animals show mainly neurological signs such as ataxia, lack of coordination, nystagmus, circling and paresis or paralysis of the limbs, particularly the hind limbs (Constable *et al.*, 2017). The final outcome of the disease may be progressive worsening, acute death or, occasionally, spontaneous recovery. Treatment is effective if given early in the course of the disease. Diethylcarbamazine is given once at a dose of 100 mg/kg BW or at a dose of 40–60 mg/kg BW daily for 1 to 6 days. Ivermectin and doramectin at a dose of 100–200 μg/kg BW is also effective. Once the parasite has crossed the meninges, the response to therapy is poor. The use of diethylcarbamazine at an oral dose of 40 mg/kg given every 3 weeks before and during the mosquito season appears to reduce the number of clinical cases (Smith and Sherman, 2009). Reduction of mosquito populations could also reduce the transmission of the parasite from cattle to goats.

Ectoparasitic infestation

Goats can become infested with lice (*Bovicola caprae*, *Linognathus vituli*), mites (*Sarcoptes scabiei*, *Psoroptes caprae*, *Chorioptes caprae* and *Demodex caprae*) and ticks (*Ixodes ricinus*, *Rhipicephalus microplus* and *Rhipicephalus annulatus*). These ectoparasites cause cutaneous itching, irritation, restlessness, dermatotic lesions, anaemia and tick paralysis. Infestations are associated with poor production management. Tick infestations damage the skin and decrease the hide value. Other important consequences are blood loss and the transmission of very serious diseases such as anaplasmosis, babesiosis, heart water, theileriasis and tick-borne fever. Tick paralysis may also occur. Pituitary abscesses have been associated with attachment of ticks beneath the backswept horns of Boer goats in South Africa (Bath *et al.*, 2005). Infestations are common during hot and humid conditions. Many drugs may be used for the treatment and control of the infestation as a mass action programme. The chemical acaricides commonly used are deltamethrin (0.1–0.4% solution), cypermethrin (0.1% solution), amitraj (0.2–0.4% solution) and ivermectin/doramectin (1 ml/50 kg BW intramuscularly or subcutaneously). Timely mass dipping of animals along with using ectoparasiticide sprays at higher concentrations in areas frequented by goats helped to control ascariasis. A repetitive application – two to three times depending upon climate conditions and the size of the infestation – is required to decrease tick burdens (Harrison and Palmer, 1981). In certain circumstances, burning the pasture or cultivating the land may aid in tick control.

Mange

The sarcoptic mange (scabies) is caused by a goat-specific strain of *Sarcoptes scabiei*. It forms tunnels through the epidermis and feeds on tissue fluids. Human infection can occur from handling infected goats (Menzano *et al.*, 2007). Typical of scabies, lesions manifest with formation of crusts and intense pruritus. The lesion begins with a small area of skin with pruritus, which then spreads over a large area with severe pruritic dermatitis lesions around the eyes, ears and on the neck and thorax, inner thighs, udder and scrotum. Hyperkeratosis and alopecia are accompanied by self-excoriation and restlessness. In extensive cases, the skin becomes thickened, wrinkled and fissured (Abu-Samra *et al.*, 1981; Kambarage, 1992) and may harbour a secondary bacterial infection. Affected animals have decreased reproduction, meat gain and milk yield, and the infection may lead to death (Menzano *et al.*, 2007). *Chorioptes bovis* infests goats worldwide with papules and crusty lesions seen on the feet and legs.

Demodex caprae causing demodectic mange is relatively common in kids, pregnant does and dairy goats. In goats, non-pruritic papules and nodules develop, especially over the face, neck, shoulders and sides or udder. These nodules contain a thick, waxy, greyish material that can be easily expressed and mites can be found in this exudate.

Deep scrapings demonstrate the typical mites. The common ectoparasiticidal drugs such as amitraj (0.05%) and pyrethroids are used at weekly intervals for controlling scabies. Ivermectin with single or repeated doses at weekly intervals is recommended with clinical efficacy to avoid bathing in winter season, but the growing resistance and non-recommendation in lactating goats is of concern. For lactating animals, a single topical application of eprinomectin at 1 mg/kg (twice the cattle dose) might be preferable, because milk residues remain below the limit established for dairy cows (Dupuy *et al.*, 2001).

5.2.3 Mycoplasmal diseases

Caprine pleuropneumonia

Contagious caprine pleuropneumonia (CCPP) is a serious respiratory disease of goats in Africa and Asia (Thiaucourt and Bölske, 1996), associated with *Mycoplasma capricolum* subsp. *capripneumoniae* (previously known as mycoplasma strain F38) infection. It is also known as AbuNini in the Sudan. This is highly fatal with morbidity rates of 90% and case mortality rates of 60% to 100%. All age groups of animals are affected, but animals under 1 year of age are most susceptible. Animals may pick up the infection through the aerosol route or direct contact.

The clinical signs are restricted to the respiratory system, comprised of a high rise in body temperature (107°F/42.6°C), mucopurulent nasal discharge, coughing and sneezing, shallow and rapid respiration, dyspnoea with an extended neck, stiff gait, rough coat, dullness and animals laying down on the ground. In the terminal stages there is a rapid respiratory rate, mouth breathing, tongue protrusion and frothy salivation, followed by death in 2 or more days. The disease is clinically manifested by septicaemia, localization and inflammation in serosal cavities/joints and/or respiratory tract infections. The septic form usually occurs, causing acute fever that often leads to death within a few days. *M. mycoides* subsp. *mycoides* LC (MmLC) is mainly responsible for septicaemia in kids. The animals that survive may develop chronic arthritis. *M. capricolum* has also been reported from arthritis and mastitis. *Mycoplasma agalactaie* has also been isolated from vulvovaginitis cases in goats. These mycoplasmas are also responsible for agalactia and arthritis in goats.

Necropsy findings are characteristic in lungs showing congestion of one or more lobes, increased pleural fluid and fibrinous pleuritic lesions. Isolation of organisms by specific PCR, RT-PCR, ELISA and competitive ELISA using monoclonal antibodies is conducted with pleural fluid. Antibiotic therapy is effective by using intramuscular tylosin (10 mg/kg BW), oxytetracycline (15 mg/kg/day), tilmicosin, marbofloxacin and danofloxacin.

Effective biosecurity measures are needed to prevent the introduction of the disease into a flock via contact with infected carriers. Killed vaccines effectively reduce morbidity and mortality rates and are in practice in many countries with variable quality.

Contagious agalactia

This disease is mainly a clinical issue in dairy goats. It is endemic in most European countries, Africa and Asia, the Indian subcontinent, the Middle East, and North and South America. The mortality rate is 10–30%, and many adult females are culled because the udder is permanently damaged (Constable *et al.*, 2017). In acute cases the onset is sudden with no or very low milk production, fever, swelling of the

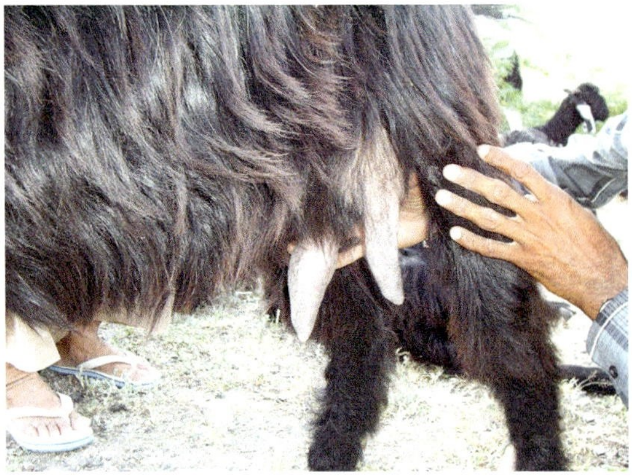

Fig. 5.20. Atrophy of the udder in chronic mycoplasmal mastitis in a goat.

udder, enlargement of the mammary lymph nodes and the development of multiple abscesses in the mammary gland, followed by induration and atrophy of the udder (Fig. 5.20). Other associated clinical symptoms include arthritis and conjunctivitis.

Diagnosis can be made by isolation of organisms from milk, ocular fluid, joint fluid aspirate, nasal swabs, ear swabs and lung lesions and molecular detection from samples. The common effective antibiotics are enrofloxacin (fluoroquinolones) tetracycline, tiamutin and macrolides. Instigating biosecurity measures at farm level, improving milking-time hygiene and test-and-slaughter policies may be adopted. Vaccination with an attenuated-live vaccine or a killed-adjuvant vaccine of *M. agalactiae* gives variable results in reducing prevalence.

5.2.4 Haemoparasitic diseases

Anaplasmosis

This is an arthropod-borne, rickettsial haemoparasitic disease caused by *Anaplasma ovis*. It is mostly subclinical in nature and often considered to be of minor economic importance in goats (Akerejola *et al.*, 1979). However, clinical disease due to *A. ovis* has been reported sporadically from Nigeria, India and Iraq (Kuil and Folkers, 1966; Mallick *et al.*, 1979; Yousif *et al.*, 1983). The most consistent clinical finding may be exercise intolerance, but other signs include a fever up to 107.5°F (41.9°C), anorexia, depression, weakness, pallor of mucous membranes, dyspnoea and increased heart rate. Oxytetracycline has been used successfully to treat clinically affected goats at an intramuscular dose of 10 mg/kg BW given once a day for 1 or 2 days (Smith and Sherman, 2009).

Babesiosis

This is a tick-borne blood protozoan infection caused by *Babesia motasi*. Babesiosis has been reported in Europe, the Middle East, India, Southeast Asia and parts of Africa (Purnell, 1981). An outbreak of naturally occurring disease in goats caused by *B. motasi* has been reported from India (Jagannath *et al.*, 1974). The clinical signs exhibited by affected goats include high fever (107°F/42.6°C), anorexia, weakness, anaemia, dyspnoea and tachycardia. Affected sheep showed icterus, haemoglobinuria and passing of coffee-coloured urine. Although these signs are observed less consistently in goats, they aid in the diagnosis when present (Jagannath *et al.*, 1974). Morbidity and mortality rates can be high. Diminazene (3 mg/kg) and imidocarb dipproprionate (1–2 mg/kg) are effective treatments. Diminazene at intramuscular doses of 12 mg/kg has been used in goats without adverse effects (Banerjee *et al.*, 1987). Chronic infections may lead to weakness and poor growth. The predominant findings at post mortem are icteric carcass, splenomegaly, lymphadenopathy and enlarged liver. Inoculation with indigenous strains of *Babesia* in conjunction with chemoprophylaxis (infection and treatment) was found to be effective in the control of babesiosis in enzootic areas (Smith and Sherman, 2009).

Theileriosis

Theileriosis is a tick-borne protozoal disease that has been given very little attention in goats when compared to sheep. Malignant ovine theileriosis, caused by *Theileria lestoquardi* and transmitted by *Rhipicephalus bursa*, is seen in sheep and goats. Malignant ovine theileriosis occurs in South-Eastern Europe and North Africa, the Near and Middle East, India and China (Smith and Sherman, 2009) and has been reported to be endemic in Nayla town, Sudan (Osman *et al.*, 2017). In milder cases, transient fever and mild lymph node swelling may occur and it may pass unnoticed by the farmers under field conditions. However, in some cases more acute signs include fever, anorexia, dullness, enlarged lymph nodes, reduced milk production, weakness, recumbency and death. Diagnosis is based on identification of Koch's blue bodies in lymph nodes or protozoa in erythrocytes. Specific information on the treatment of caprine theileriosis is lacking. In cattle, high doses of long-acting oxytetracycline (20 mg/kg) is given intramuscularly (Smith and Sherman, 2009).

Trypanosomiasis

Trypanosomiasis in goats produces acute, subacute, chronic or subclinical forms of the diseases. *Trypanosma vivax*, *Trypanosma congolense* and *Trypanosma evansi* are the most invasive trypanosomes in goats (Gutierrez *et al.*, 2006). Regional differences exist in the prevalence of caprine trypanosomiasis (Kramer, 1966; Griffin and Allonby, 1979a). Goats have also been implicated as a reservoir of *Trypanosma rhodesiense*, a nosodeme of *Trypanosoma brucei* transmissible to man (Robson and Rickman, 1973). The clinical symptoms vary depending on the specific species. Acute, subacute and chronic forms of the disease can occur in *T. vivax* and *T. congolense* infection. The chronic form of the disease is most common. The main signs include fluctuating fevers as high as 108°F/42.2°C, anaemia, enlarged lymph nodes and weight loss. Lymphadenopathy is a consistent finding and may be more pronounced with *T. vivax* or *T. congolense* infections (Smith and Sherman, 2009). Diminazene aceturate is given intramuscularly at a dose of 3.5 mg/kg BW and is considered effective against the three major trypanosomes.

5.2.5 Skin diseases

The health of the skin is important for the quality and quantity of fibre production (e.g. mohair, cashmere and pashmina). Goat skins are processed as leather, which is used for footwear, garments and luggage. The common diseases associated with skin are contagious ecthyma, capripox, *peste des petits ruminants*, bluetongue, staphylococcal folliculitis, dermatophilosis, dermatophytosis, sarcoptic mange, zinc deficiency, pemphigus foliaceus and protothecosis. All these diseases require systematic investigation for diagnosis and treatment. Hyperkeratosis and parakeratosis appear the same clinically and are consequences of ectoparasitisms, seborrhoea, zinc-responsive disease, dermatophytosis and dermatophilosis. Hair loss (alopecia) of different degrees may be due to nutritional deficiencies or imbalances (such as high calcium with low zinc) (van der Westhuysen *et al.*, 1988).

Skin diseases such as mange, grubs, tick infestations, goat pox, contagious ecthyma infections and dermatophilosis decrease the value of goat skins. The quality is also affected by maggot infestation or intense itching (pruritus), which frequently lead to excoriations and other secondary lesions associated with sarcoptic or chorioptic mange, lice and flea infections and hypersensitivity to other insects. Physical examination helps to identify most clinical conditions, but for long-standing, non-responding and non-specific lesions, systematic skin scrapping examination, bacteriological culture examination, histological biopsies and immunofluorescence testing (e.g. for autoimmune skin disorders such as pemphigus foliaceus) should be performed.

5.2.6 Reproductive diseases

The common reproductive ailments are associated with post kidding and need timely attention for maintaining reproductive efficiency.

Retained placenta and metritis

The placenta is normally passed within 3 or 4 hours. If the placenta has not been expelled within 12 hours after delivery of the last kid, it is then considered to be retained (Franklin, 1986). A retained placenta may occur in an infectious abortion caused by diseases such as toxoplasmosis (Calamel and Giauffret, 1975), brucellosis and chlamydiosis. Caesarean section is another risk factor for retained placenta (Brounts *et al.*, 2004). Metritis is an inflammation of the deeper layers of the uterus that is usually associated with inflammatory and infectious processes immediately after parturition. Metritis most commonly occurs when a retained placenta is present, which provides a good media for bacterial growth. It may be an important cause of infertility in some goats. Acute postpartum metritis is characterized by a malodorous, dark red uterine discharge. Does are febrile (104-106.7°F/40–41.5°C) and anorexic. Severe metritis may be complicated by peritonitis. Intrauterine antibiotics are commonly administered, but manual removal of the placenta should not be attempted. Instead, oxytocin (5 IU subcutaneously or intramuscularly) or prostaglandin may be administered several times a day, and systemic antibiotics are recommended to prevent the development of a life-threatening toxaemia or septicaemia (Smith and Sherman, 2009). Clostridial organisms, frequently *Clostridium tetani* and *C. perfringens*, may colonize the uterus and cause fatal toxaemia in affected goats. Post-parturient or genital gas gangrene caused by infection of the uterus with clostridial organisms has been reported in Angora goats in South Africa (Van Tonder, 1975). Tetanus and enterotoxaemia prophylaxis, as an adjunct therapy, may be essential. Chronic metritis or endometritis may occur after parturition or during the breeding season. Affected goats eat normally and are afebrile but have a whitish vaginal discharge. Chronic infections may be the result of postpartum metritis or the result of bacterial contamination during breeding. Systemic antibiotics and prostaglandin are the treatments of choice in these cases (William Braun Jr, 2007).

Prolapsed vagina and uterus

Eversion of the vagina or uterus occurs uncommonly in goats, but needs aseptic conditions to prevent outside contamination. The prolapsed part should be washed with copious amounts of water and a mild disinfectant, and then repositioned. An antibiotic powder is placed in the uterine cavity and oxytocin (5 IU) is given to contract the uterus. Systemic antibiotics may be given for several days, otherwise it may lead to metritis. In repeated prolapse, a stay suture should be placed in the vulva followed by the administration of antibiotic, anti-inflammatory and spasmolytic drugs.

5.3 Good Management Practices and Disease Prevention

5.3.1 Herd health management and skills

Goat production systems face many clinical challenges in different management practices, which reduces the economics of goat farming. Therefore, herd health management and preventive medicine programmes are formulated to reduce the risk of adverse effects. It can be accomplished by timely, cost effective and safe application of appropriate veterinary, nutritional, and management interventions. One protocol for prevention of clinical exigency is not perfectly suitable to all types of production systems; hence, it may be appropriate to modify protocols and preventive medicine programmes to suit the geographic, climatic, cultural and economic factors affecting herd health of specific regions (see Protocol 5 at the end of this chapter).

Body condition scoring

Body condition scoring is a management tool that can be used to evaluate the nutritional status of animals. Body condition, or fat cover, is an indication of the energy reserves in an animal. The fat reserve can be used by the animal during periods of high energy demand, stress and suboptimal nutrition. Thus, it is necessary to maintain goats with moderate body condition score, as a reflection of optimum production performance.

Generally, scoring is performed in goats using a range from 1.0 to 5.0 in 0.5 increments. Healthy goats should have a body condition scoring between 3.0 to 4.0. Goats with a body condition scoring of 1.0 or 2.0 indicate poor management or unhealthy conditions. A body condition scoring of 5.0 indicates an over conditioned (obese) body that is detrimental to the goat's health.

EVALUATION OF BODY CONDITION SCORE The lumbar, sternum area and intercostal (between the ribs) spaces are used for evaluation of the body condition score. Scoring in this area is based on determining the amount of muscle and fat over and around the vertebrae. Lumbar vertebrae have a vertical protrusion (spinous process) and two horizontal protrusions (transverse process). The observation is made by putting a hand over this area and trying to grasp these processes (spinous and transverse) with the fingertips and hand (Fig. 5.21). One has to feel the fat layer and predict the score. The second area is the sternum, or breast bone, where the amount of fat can be pinched (Fig. 5.22). Similarly, ribs and intercostal spaces are also used to feel the amount of fat cover. The details of the scoring should be carried out as described in Table 5.4.

Routes for administering medication

There are different approaches for the medication used in animals, depending upon the drug used.

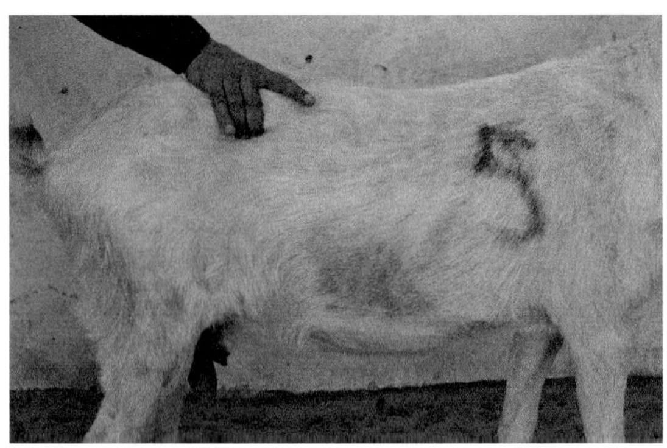

Fig. 5.21. Examination of lumbar region for body condition score.

INJECTION Careful administration of injectable medications is always necessary in herd health management. Strict hygiene and sanitation reduces the chance of introducing bacteria into medicine vials or the sick goat, thereby reducing injection reactions and abscesses. Disposable, single-use needles are commonly used to avoid injection reactions. However, multiple use of injections may be appropriate in healthy animals during vaccination programmes; the re-use should be restricted to 5–10 times and never re-used in sick animals. The appropriate needle size is also important for injection of different viscosities of the medicine. Most goat medications or vaccines can be given with 20 or 22 gauge (G) needles for thin, watery solutions or 18 to 20 G needles for thicker medications. In the case of subcutaneous injection, a 2 cm needle works well for subcutaneous injections, and for intramuscular injections a 2.5 or 3.5 cm needle works well.

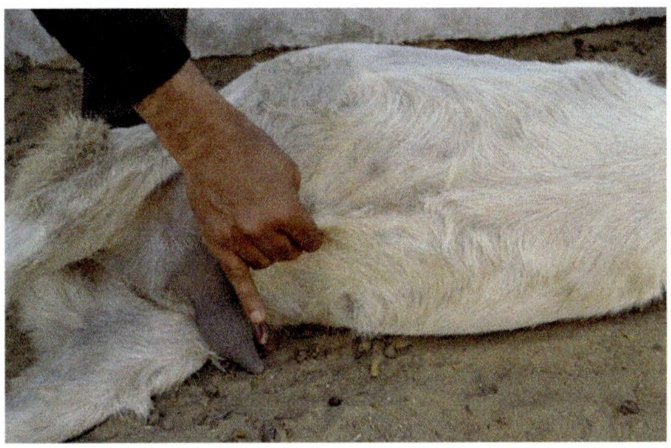

Fig. 5.22. Examination of sternum for body condition score.

Subcutaneous injections are normally administered in areas with loose skin folds such as the extra skin under the foreleg area or behind the elbow, over the shoulder blade, in the flank area and on the side of the neck. The selected site should be cleaned. The needle should be inserted facing bevel end downward to the ground, thus reducing the leaking of the drug after the needle is removed. Always ensure that the needle has not been inserted into a blood vessel. It is appropriate to inject 5 ml of any medication at one site. Intramuscular injections should be avoided in thigh muscles as they correspond to valuable cuts of meat such as the leg or loin, and lesions may develop and compromise meat quality, consumer confidence and producer profitability.

Sample collection, packaging and field necropsy

Livestock diseases are a major contribution to the economic loss incurred by farmers. Disease diagnosis plays an important role in identification through clinical examination or pathological changes during post-mortem examination. But accurate diagnosis always helps in finding the exact cause of the disease. Hence clinical or morbid samples need

Table 5.4. Body condition scoring in goats.

Score	Body condition	Interpretation
1	Emaciated and weak Backbone is highly visible and prominent Flank is hollow and ribs are clearly visible Fingers can easily penetrate into the intercostal spaces No fat cover	Extremely poor body condition
2	Backbone is still visible and prominent Some ribs can be seen and fingers can still pass under them with little pressure Minimum fat cover	Poor body condition
3	Backbone well covered, smooth and not prominent Ribs are barely discernible Intercostal spaces are felt using pressure Moderate fat cover	Moderate healthy body condition
4	Backbone and ribs not visible The side of the animal is sleek in appearance	Perfect healthy body condition
5	The backbone is buried in fat and the ribs are not visible The rib cage is covered with thick fat cover	Obese and over condition animal

Adapted from Detweiler *et al.* (2008).

to be collected from the site of the outbreak in an appropriate manner based on the clinical signs or post-mortem changes that would further aid in laboratory diagnosis (see Protocol 2 at the end of this chapter). Correct post-mortem techniques for the collection of samples should be followed (see Protocol 6 at the end of this chapter).

5.4 Strategies to Control Kid Mortality

5.4.1 Diseases of neonates

Low neonatal mortality rates are vital to economic goat husbandry and are critical for the fast turnover of flock strength and prolificacy. The losses caused by morbidity have been accounted for overall production losses and may sometimes exceed the losses caused by mortality. Diarrhoea and pneumonia are the top-listed diseases causing mortality in kids. Colibacillosis and septicaemia are generally the most common causes of mortality in juvenile kids, along with other infectious organisms such as *Cryptosporidium parvum*, rotavirus, coronavirus, *C. perfringens* and *Salmonella* spp. The dam and management practices have major roles in precipitating disease in young kids. There is a need to understand the common clinical ailments during kidding and kid management and accordingly develop preventive strategies to prevent kid mortality and economize goat rearing (see Protocol 3 at the end of this chapter).

Kid factors:

- Infectious diseases and parasitic diseases: colibacillosis (*E. coli*), PPR, salmonellosis, pneumonia, rotavirus, FMD, coronavirus, *C. perfringens*, contagious ecthyma, tetanus, *C. parvum*, coccidiosis, moniziasis.
- Non-infectious causes: low birth weight/premature birth (genetic reasons), hypothermia and hypoglycaemia syndrome, trace element deficiencies (copper, iodine, selenium).

Mother factors:

- Mother factors include low milk production syndrome associated with metabolic diseases (hypocalcaemia, pregnancy toxaemia, mastitis etc.). This factor contributes to poor nutrition for kids, who then become susceptible to infections. Underfeeding pregnant goats can have deleterious effects on foetal and/or new-born kids by adversely affecting placental size, foetal growth, deposition of foetal fat reserves for use after birth, maternal udder development and colostrum/milk production. A delay in the onset of suckling by such kids, who have low birth weight, is therefore more likely to lead to starvation-induced hypothermia than in normal weight kids.

Management factors:

- These include low milk production by the mother, sanitation and hygiene, rearing practices, shelter management, judicious and timely prophylactic measures, breeding management, nutrition and feeding practices that affect the neonate growth and disease resistance. Thus utmost care should be taken in rearing practice.

Colibacillosis

Diarrhoea is documented as a very frequent cause of kid mortality throughout the world. The bacterium *E. coli* is a common inhabitant of the gastrointestinal tract in both non-pathogenic and pathogenic forms. Pathogenic serotypes are capable of gaining entrance to the body (septicaemic form) through the intestinal wall, while others remain in the intestinal tract and liberate a toxin that is absorbed by the body and causes generalized disease and diarrhoea (enterotoxogenic form). The common signs are anorexia, loss or decreased appetite, off feed, bloody stools (haematochezia), ataxia, incoordination, staggering and falling over, and dehydration and fever. Haemorrhagic enteritis is a common post-mortem lesion found in kids (Fig. 5.23). Colibacillosis is managed by administration of antibiotics, preferably both oral and systemic, and rehydration therapy, along with a hygienic drinking water supply and environmental sanitation. Damhood vaccination with a prevalent *E. coli* strain is an effective alternative to reduce the economic losses. Vaccination of

Fig. 5.23. Haemorrhagic enteritis in a kid with colibacillosis.

goats with single and polyvalent strains of bacteria could provide sufficient immunity for preventing colibacillosis in new-born kids.

Salmonellosis

The bacterium *Salmonella typhimurium* is the most common serovar of the genus causing diarrhoea in goats. Some of the affected neonates die even without clinical signs. In general, *Salmonella* can cause diarrhoea in kids of any age. *Salmonella* also causes outbreaks of diarrhoea in late gestation and is frequently associated with abortion. Vaccine efficacy is questionable, and vaccination has not been thoroughly evaluated in sheep and goats.

Cryptosporidiosis

Cryptosporidiosis caused by *C. parvum* is primarily a disease in kids below 30 days of age and is usually a milder disease than coccidiosis. Infective oocysts are passed in faeces and are transmitted by oral ingestion. *C. parvum* is not host-specific and infection from other species (e.g. rodents, farm cats) via contamination of feed is also possible. Cryptosporidiosis is characterized by acute gastrointestinal disturbances, haemorrhagic or mucoid diarrhoea, fever, lethargy, anorexia and loss of body condition leading to significant economic loss. The faeces are yellow or pale, watery, contain mucus and have a strong odour. Concurrent infections with other enteric pathogens, especially rotavirus and coronavirus, are common, and epidemiologic studies suggest that diarrhoea is more severe in mixed infections. Large numbers of oocysts (108 to 1010 oocysts/g) are present in the diarrhoeic stool. Recovered kids serve as carriers and become a potential source of infection to susceptible populations.

Diagnosis is based on detection of oocysts by examination of faecal smears with Ziehl-Neelson stain or by faecal flotation. The oocysts are small (5–6 mm in diameter) and relatively non-refractive. Oocyst are difficult to detect by normal light microscopy but are readily detected by phase-contrast microscopy. Treatment with halofuginone and paromomycin sulphate have proved successful in preventing natural disease in kids.

Rotavirus gastroenteritis

Kids are infected with group B rotaviruses, whereas most other animals and humans are infected with group A rotaviruses. Rotavirus generally causes diarrhoea in kids between 2 and 14 days of age. Young animals become very depressed and dehydrated. Rotavirus is treated with supportive care. Vaccinating does with bovine rotavirus vaccine during pregnancy will increase passive immunity.

C. perfringens infection

C. perfringens is a known cause of diarrhoea in older kids along with *E. coli* and viral infections. For diagnosis in dead animals, a loop of bowel can be ligated, removed and submitted to the lab unopened under refrigeration. Formalinized tissues from major organs and various portions of the gastrointestinal tract should also be submitted for histopathologic evaluation and toxin typing. Vaccination (*C. perfringens* types C and D) of the doe in the last month of pregnancy may improve the concentration and specificity of protective antibodies available in the colostrum.

Floppy kid syndrome

The causative agent has not been identified. The affected kid is normal at birth and develops a sudden onset of muscular weakness (flaccid paresis or paralysis) or ataxia between 3 and 10 days of age. Affected kids are depressed, cannot use their tongues to suckle but can swallow, and have marked paradoxical metabolic acidosis. There are no signs of diarrhoea, respiratory disease or other signs. The clinical signs of paresis/paralysis/ataxia in 3- to 10-day-old kids and supporting blood chemistry values are diagnostic features. Differential diagnoses include white muscle disease, abomasum bloat, colibacillosis, septicaemia or enterotoxaemia. Early detection and correction of a base deficit as well as good supportive care are important. A mixture of 2 teaspoons of baking soda and half a teaspoon of salt in a quart of water has been used successfully. Kids may need to be fed milk by stomach tube. More severe cases may require biochemical analysis of blood and intravenous fluid administration.

Tetanus

Tetanus is caused by *C. tetani*, which persists in the soil of most farms. The bacteria are anaerobic, and wounds where air contact is limited are the most susceptible to tetanus. Castration with rubber rings increases the incidence of infection. Infected animals become stiff (Fig. 5.24), move with a straddled gait

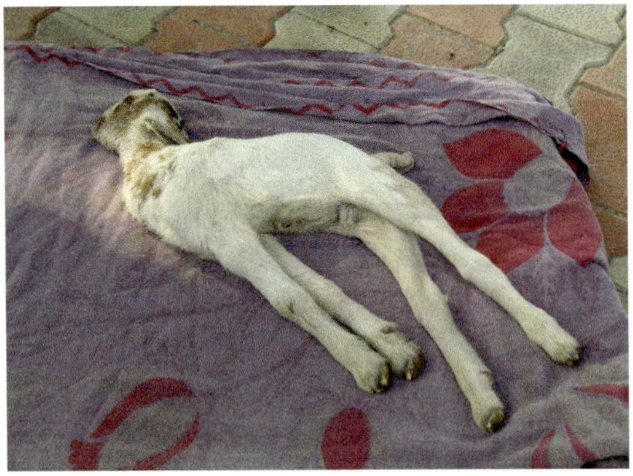

Fig. 5.24. A one-week old kid with tetanus exhibiting stiffness of the body.

and usually die. Antisepsis and disinfection before castrating wounds will minimize the infection. Vaccinating with tetanus toxoid and antitoxin prior to operation will also prevent infection.

Pneumonia

Pneumonia is a clinical condition that is associated with poor management of farms, such as dampness and high ammonia concentrations in the premises. Environmental stress is also a precipitating factor of pneumonia in kids. Pneumonic signs such as nasal discharge, respiratory distress and fever are diagnostic. It is important to detect the disease as early as possible and to start antibiotic treatment as soon as possible. Broad-spectrum antibacterial drugs used to treat pneumonia include tetracycline, penicillin, streptomycin, sulfamethazine, enrofloxacin, marbofloxacin and cephalosporin. Adequate selenium and vitamin E help the kid withstand pneumonia.

Foot-and-mouth disease

FMD is a highly contagious viral disease characterized by fever and formation of vesicles in the mouth and feet. In kids, mortality occurs due to acute myocarditis, the fatal form of FMD without classic vesicular lesions. A sudden death of a kid without any apparent signs may be due to FMD. On post-mortem examination, there is greyish or yellowish streaking in the myocardium with degeneration and necrosis described as 'tiger' stripes, particularly in the left ventricle and interventricular septum (tiger heart). A serum troponin I assay and ECG can be used for diagnosis of myocarditis and prognosis of affected kids during FMD outbreak.

Orf

Contagious ecthyma is characterized by scabby lesions on the lips. It is not generally harmful to the host; however, it may lead to maggot infestation and death by starvation in neglected cases (see Section 5.2.1).

White muscle disease

The cause of white muscle disease (muscular dystrophy) in young kids is a deficiency of selenium or vitamin E or both, although lack of vitamin E is more likely to be the cause. The clinical signs are dead or weak new-born kids, or a kid unable to rise or walk with stiffness. It may affect 6- to 8-week-old kids, particularly, fast-gaining kids (see details in Section 5.6.1)

Hypothermia and hypoglycaemia

Hypothermia occurs when body heat loss exceeds heat production, and may result in neonatal death in kids. Hypoglycaemia relates to starvation associated with insufficient intake of colostrum/milk, weakness and inadequate mothering. Starvation also deprives new-born kids of immunoglobulins that are present in the colostrum in high concentrations, which may result in kid mortality.

5.4.2 Diseases in growing kids up to 3 months of age

In growing kids, coccidiosis is an important protozoan infection of the intestine usually noticed during weaning, feed change or transport. The severity of the infection depends on the number of oocysts ingested. Heavily infected kids suffer from severe enteritis. The different life cycle stages in the intestine result in intestinal cell destruction. In young kids, signs of coccidiosis include diarrhoea with or without blood, and foul-smelling stools. Animals may show abdominal pain, anaemia, inappetence and weight loss (Fig. 5.25). Later on, the animal appears unthrifty with a rough skin coat. Death may occur in severe cases due to dehydration and electrolyte imbalance. Kids that survive have poor appetites and poor body weight gain. The oocysts

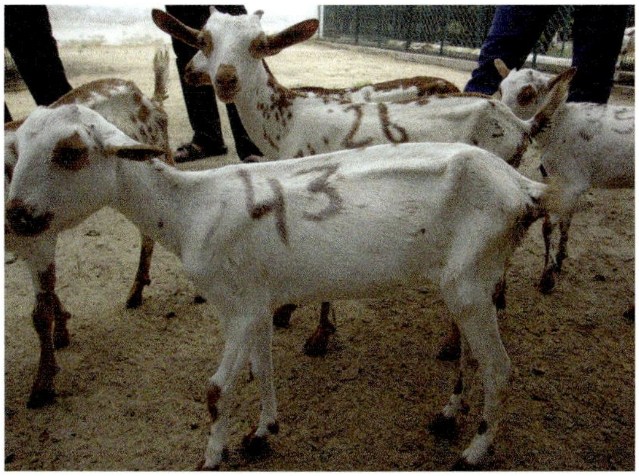

Fig. 5.25. Stunted growth in 3-month-old kids with coccidian infections.

may be detected on faecal examination and help to assess the load of infection. In most cases, the self-limiting and clinical signs usually subside spontaneously. Coccidiostats are recommended for treatment and control of coccidiosis in kids (see Protocol 4 at the end of this chapter).

5.5 Disease Genetics

Efficient animal production systems are often hindered by disease and environmental stress. Economic losses from diseases of farm animals have been estimated to amount to 10–20% of the total production costs in developed countries and 35–50% in developing countries (Muller and Brem, 1991). Mortality is the major cause of loss; however, subclinical infection hampers production efficiency and increases the veterinary costs and condemnation of products. Despite traditional disease control measures, losses attributable to infectious diseases continue to impede the livestock industry. Moreover, consumers are conscious about drug residues and agro-chemicals in milk, meat and other products. Therefore, it is necessary to devise more effective approaches to combat diseases. For alternative approaches to disease control, it is necessary to understand the disease and the animal's defence system. The onset of disease is often the result of the complex interaction between an individual the animal's genotype and the environment to which the animal is being exposed. Diseases also poses threats across species barriers as those in one species may act as reservoirs for infections in others. In particular, animal infections pose zoonotic threats to human health. The control or management of infectious disease is currently achieved by several mechanisms: (i) chemical intervention, notable examples being anthelmintics for nematode parasite control, acaricides for tick control and antibiotics for the control of many bacterial diseases; (ii) vaccination; (iii) sanitation and disinfection; and (iv) culling, isolation and control of the movements of animal and/or animal products.

Disease control or management using genetics (i.e. exploiting genetic variation among hosts in disease resistance) is increasingly looked upon as an additional way of controlling many diseases, complementing or sometimes replacing existing strategies. Breeders and agricultural industries now face considerable pressures to select animals for enhanced disease resistance. The pressure to breed for disease resistance is the evolution of resistance in parasites to chemical or vaccine control measures. There are a number of examples such as resistance to anthelmintic drugs by nematodes in all major sheep-producing countries, resistance to antibiotics by bacteria and decreases in the efficacy of antibiotics in many production systems, resistance to anti-protozoal drugs, and vaccine failure in Marek's disease in poultry.

5.5.1 Basic definitions and concepts for disease biology

The background information and terminology need to be understood to conceptualize disease genetics and epidemiology. Many of the basic definitions come from Grenfell and Dobson (1995); however additional context has been added (Bishop, 2007).

Infection versus disease

Parasite is a general term describing an organism with a dependence upon a host. Parasites include many organisms, including viruses, bacteria, protozoa, helminths, flies and ticks. Disease is often used to cover two distinct concepts: infection and disease. Infection is the colonization of a host animal by a parasite. Disease describes the side effects of infection by a parasite and may take several forms including acute, subacute, chronic and subclinical, which may or may not be debilitating.

Resistance versus tolerance

Resistance is the ability of the individual host to resist infection or control the parasite lifecycle.

Tolerance is the ability of a host to tolerate infection and show little or no measurable detriment, i.e. minimal effects of disease.

Horizontal versus vertical transmission

Horizontal transmission describes the transmission of infection between infected and susceptible individuals, or between disease vectors (an animal or object that transmits parasites) and susceptible individuals. **Vertical transmission** is from parent to unborn offspring. Sources of horizontal transmission include ingestions of contaminated food or drink, inhalation of contaminated air droplets, direct contact, injection via an animal's saliva or bite, invasion via open wounds or penetration of the host by an actively searching parasite.

Epidemic versus endemic

An **epidemic** is a sudden, rapid spread or increase in the prevalence or intensity of a parasite or disease. An epidemic is often the result of a change in circumstances that favour parasite transmission such as a rapid increase in host population density, climatic changes that favour the parasite, the introduction of a new parasite or strain of parasite, or the introduction of a new and susceptible host population. A widely distributed epidemic is known as a **pandemic. Endemic** describes a parasite or disease that does not exhibit wide fluctuations through time in a defined location.

Immunity and herd immunity

Immunity is the ability of the individual host to combat infection or disease due to the presence of antibodies or activated cells. It may be divided into three types: (i) acquired immunity which is conferred on an individual after recovery from a disease; (ii) natural or innate immunity which is inherited from parents, or in some cases antibodies may be passed across the placenta and therefore present in the blood at birth; and (iii) artificial immunity that may be induced by the administration of a vaccine, denatured antigens of a parasite, or antiserum that contains antibodies. Immunity is often not complete and may vary between individual animals within a population. **Herd immunity** describes the immunological status of a population of hosts with respect to a given parasite. The level of herd immunity is determined by the net rates at which individual animals acquire and lose their immunity. A feature of herd immunity is that it is sometimes not necessary for all animals in the population to be completely immune for the population as a whole to be free of infection. This is the basis of public health vaccination programmes where (for example) not all children are vaccinated against a particular disease, but the mass of vaccinated children effectively protect the unvaccinated children.

Infectious period, latent period and carrier state

The **infectious period** is the period during which an infected animal is able to transmit an infection to a susceptible host or to a vector. The infectious period may or may not coincide with the disease. The **latent period** is the period during which the individual is infected but is not yet capable of transmitting the infection. The **carrier state** is a state in which an infected individual shows no clinical signs but is capable of transmitting the infection. Carrier state occurs with many bacterial infections such as tuberculosis.

Prevalence and incidence

Prevalence is the proportion of the host population that is infected or diseased at a specific point in time. It is often expressed as a percentage, and it is a measure of how widespread the infection or disease is. **Incidence** is the number of new cases that arise in a population over a specified time period (e.g. per year). Therefore, **incidence** is a rate parameter. It does not describe the proportion of infected or diseased animals.

5.5.2 Mechanisms of disease resistance

For almost every disease that has been intensively and carefully investigated, evidence for host genetic variation in either resistance or tolerance has been found. However, it is often not clear whether the observed genetic variation is for resistance, tolerance or a combination of both. Clunies-Ross (1932) was the first to recognize the need for the distinction between resistance to infection and resistance to the effects of infestation. Generally, two types of resistance have been described in all the bacterial, viral and parasitic diseases. Resistance to infection means the mechanism that prevents the organisms penetrating into the living host cell (i.e. first line of defence or cellular resistance). The second type of resistance/second line of defence is the resistance to the disease process or effects of the infestation in the host in response to pathogen infection.

- The animal may have an appropriately targeted immune response against the pathogen. This may enable the animal to successfully combat the infection or avoid pathogenic effects of the disease.
- The animal may have non-immune response genes that limit infection in target organs, for example, the binding-protein genes that allow specific strains of *E. coli* with K88 or F18 fimbriae, to attach to the gut of the pig resulting in pre-weaning or post-weaning diarrhoea. The absence of appropriate binding proteins results in resistance to infection.
- The animal may have physical attributes that make infection by the pathogen difficult. An example is the role of skin thickness in conferring resistance to ticks.
- The animal may have behavioural attributes that enable it to avoid infection. An example is the hygienic behaviour of honeybees. Behavioural attributes are also very important for understanding nematode resistance in small ruminants. Goats tend to be browsers rather than grazers like sheep; therefore there is less chance for infection by contaminated pastures. Hence goats are likely to have evolved less resistance.
- Genetic resistance to infection or tolerance of infection may be due to a single gene. Examples include resistance to *E. coli* diarrhoea in pigs and the PrP gene, which is associated with resistance to scrapie in sheep. In other cases, resistance or tolerance is due to the combined effects of several or many genes. Examples of this include nematode resistance in ruminants, mastitis resistance in dairy cattle and sheep, and trypanosome tolerance in locally adapted breeds of cattle. Although resistance and tolerance are sometimes qualitative phenomena, as in the case of *E. coli* diarrhoea in pigs, more often they are quantitative phenomena. That is, they show continuous variation from one extreme to the other. Continuous variation is seen when resistance or tolerance is due to the combined effects of several genes.

5.5.3 Host resistance phenotypes and approaches for selection for disease resistance

Exploiting genetic differences in resistance (i.e. using host genetic variation to reduce the transmission of infection) has the benefit of reducing the incidence of infection in the target population. This reduces the effects of disease and may also reduce transmission of infections to other populations. Using genetics to assist in the management of the clinical signs of disease, i.e. genetically increasing the **tolerance** of a population to infection, will be effective in reducing the incidence or the effect of disease in the target population. However, it may not decrease the prevalence of the parasite. Hence, the disease incidence or prevalence in other populations in the same environment will not be affected.

Resistance and susceptibility are generally measured on a quantitative scale and various criteria are used to define the phenotypic expression of disease resistance. They are generally measured as a differential response such as mortality or morbidity, or as other expressions of infection such as antibody response, tumour production, and so on, or expressed in quantitative scales such as survival time and density of pathogen in target organs (virus production, oocyst production, etc.). Pock count/pock forming unit is used to assess the susceptibility of cells to transformation by oncogenic viruses (Prince, 1958). Faecal egg count is a useful indicator trait for studying resistance of gastrointestinal nematodes (Baker *et al.*, 1998; Baker, 1999). The difference in expression may be seen between breeds, strains, families and individuals, or at different loci within an individual. Examples of intraspecific disease resistance in different domestic animals include Marek's disease and leucosis in poultry (Hutt, 1958; Payne, 1985; Freeman and Bumstead, 1987); atrophic rhinitis and respiratory disease in pigs (Lundeheim, 1979; Kennedy and Moxely, 1980); mastitis and ketosis in cattle (Philipsson *et al.*, 1980); and parasite infestation and foot rot in sheep (Albers *et al.*, 1987; Bulgin *et al.*, 1988).

To breed a resistant stock, it is essential to know the mechanism by which resistance is mediated and this knowledge helps to control the disease effectively. The studies in mice have highlighted the mechanism of control of different bacterial, viral and parasitic diseases. The approaches for selection for disease resistance can be achieved by direct and indirect methods. Direct methods for selection for disease resistance are based on monitoring disease status of animals after challenge and include observing breeding stock, challenging breeding stock, challenging progeny/siblings of breeding stock and challenging clones. In farm animals, the simplest and most widely used approach is the observation of breeding stocks and selection of breeding stock for disease resistance under normal production conditions. Challenging the breeding stock with infectious agents may have negative effects on production traits; however, the expression of disease resistance is good. This method would not be advised if death usually results from a particular disease. Testing of progeny/siblings response to a disease agent is an excellent way to measure disease resistance. Indirect methods are to identify resistant animals on the basis of indicator traits without the need for direct challenge and disease expression, and may include vaccine challenge, *in vitro* tests and genetic markers (Gavora and Spencer, 1983). Indirect methods of selection for disease resistance are an ideal approach for the improvement of disease resistance since it does not require exposure to pathogens. The indirect methods are vaccine challenge, *in vitro* tests and genetic markers. These methods have no negative effect on production, and expression of disease resistance is also good. The disease resistance gene variants or quantitative trait loci (QTL) (or even genomic predictors) have the benefit of enabling the selection of animals without exposing them to infection. For many diseases, molecular techniques may be the only way of enabling effective selection for disease resistance. For others, such as mastitis or nematode parasites, a combination of phenotypic measurement and genetic markers may be the optimal approach.

5.5.4 Genetic variation to diseases in goats

Genetic variation in resistance to viral diseases is found in all farm animals and provides an idea for selection for improved resistance. Laboratory studies on susceptibility of mice to infection with viruses such as FMD, blue tongue, RVF, capripox and rinderpest over many years strongly suggest that breeding for resistance to these and other viral diseases is feasible (Kitching, 1991; Schat and Davies, 2000). Capripox challenge in British goat breeds caused several generalized skin and internal lesions and death on day 10–14 after infection. But the same isolate in North Yemen susceptible black goats did not develop generalized lesions and the animals fully recovered (Kitching *et al.*, 1987).

The genetic variation in host resistance for major bacterial diseases in farm animals has been reported. The major bacterial diseases for which genetic variation have been described worldwide are tuberculosis, mastitis, brucellosis, salmonellosis, pasteurollosis, mycoplasmosis and ovine foot rot. The breed differences in susceptibility to Johne's disease in goats have been reported by Singh *et al.* (1990) and Molina *et al.* (1991).

Certain breeds of sheep and goat are more resistant to helminth parasites than others, as first documented by Stewart *et al.* (1937). Since then numerous studies have reported the response to gastrointestinal nematode and trypanosome infections (Dargie, 1982). In goats, breeding for disease resistance to helminths has been reported in Africa and India (Rout *et al.*, 2011). Dalliance and Red-Maasai sheep breeds of Africa and dwarf goats of West and East Africa are more resistant to trypanosomiasis (Griffin and Allanby, 1979b). The genetic resistance to heart water in indigenous goats was reported in South Africa (Matheron *et al.*, 1987).

A case study of nematode infections in goats

Helminthosis, which includes nematode infections, is by far the most important disease problem affecting the livelihoods of livestock producers in developing countries. Although helminth infections affect all livestock species (including monogastrics), they are a particular problem for ruminants where obligate grazing invariably brings the animal into contact with nematode parasites.

Gastrointestinal (GI) parasitism presents a major challenge to livestock productivity worldwide (Perry *et al.*, 2002), and more specifically to small ruminant production in the tropics. Control strategies include anthelmintic treatment, grazing management and nutritional supplementation; however, these standard control measures are not fully effective in preventing losses due to parasitic infection in different production systems. Anthelmintic resistance, treatment costs and consumers' awareness of drug residues in milk and meat have pressed the scientific community to search for viable alternatives to control parasites in both smallholder and intensive production systems. Enhanced genetic resistance of hosts to parasites contributes to sustainable control strategies in smallholder production systems as it will have a direct impact in terms of reducing concerns about food safety, animal welfare and environmental pollution (Waller, 1999; Bishop and Morris, 2007). There is increasing evidence for genetic variation in resistance to nematodes in sheep and goats (Baker *et al.*, 1998, 2001; Bishop and Morris, 2007). In goats, evidence for genetic variation for resistance to GI parasites has been established between breeds (Pralomkarn *et al.*, 1997; Baker *et al.*, 1998, 2001) and within breeds (Morris *et al.*, 1997; Mandonnet *et al.*, 2001). If selective breeding for nematode resistance is to be implemented as an alternative or complementary nematode control strategy, then it is necessary to quantify the variation in resistance and establish the relationship between indicator traits and performance traits. Faecal egg counts (FEC) and packed cell volume (PCV) are the indicator traits commonly used to assess the level and impact of infection. For effective use of selection to assist in parasite control in goats, it is essential to assess the ability of indicator traits to aid selection for resistance to nematode infections. Published heritability estimates for nematode resistance, as assessed by FEC, have generally been reported to be low to moderate in goats in tropical conditions (e.g. Woolaston *et al.*, 1992; Baker *et al.*, 2001; Mandonnet *et al.*, 2001; Vagenas, 2002). The heritability of PCV followed a similar trend to FEC, maximizing at about 0.24 (standard error = 0.09) at 9 months of age. The heritability of PCV was 0.32 in African goats (Baker *et al.*, 1998, 2001). In contrast, under temperate conditions the heritability of FEC and PCV in Australian cashmere goats was higher at 5 months of age (RIRDC, 2005). The genetic correlation between FEC and PCV from 6 months of age was moderate to strong and negative, and similar observations in Galla and Small East African goats have been reported by Baker *et al.* (2001). Estimated genetic and phenotypic correlations between FEC and PCV are consistently negative (and therefore favourable) in goats when they are infected with *H. contortus* (Albers *et al.*, 1987; Baker *et al.*, 1998). Effective disease control will require integrated approaches, combining varied strategies including improved management, nutrition, judicious use of chemicals and genetic control options.

Sustainable disease control

Sustainable disease control is characterized by avoiding reliance upon a single control strategy and attempting to combine complementary approaches for controlling the transmission of infection. A strategy that is reliant upon a single mechanism (e.g. using a chemical that kills the parasite in a specific way) essentially gives the parasite a single target against which to mutate. The parasite that has a mutation which allows it to avoid that control strategy will have a huge selective advantage over the rest of the parasite population; hence that genotype will propagate and become dominant, leading to the often-observed breakdown in the control strategy. Disease control strategies that combine different approaches will generally be more sustainable, as parasites with the mutation allowing them to escape one strategy will still be susceptible to other strategies. Genetic and non-genetic strategies may be combined complementarily, and this would be advocated under most circumstances. Moreover, genetic strategies depending upon more than one gene may also be considered to be a form of combined control strategy.

5.6 Fundamentals of Clinical Practice

5.6.1 Prevention of intensive farming and production associated diseases

Intensive methods of management and feeding may result in environmental stress on the metabolic health of the animals, rendering them more prone to metabolic disorders. Production/metabolic diseases are becoming increasingly serious problems with intensive husbandry/practices. Metabolic diseases such as pregnancy toxaemia, hypomagnesaemia and hypocalcaemia are attributable to an imbalance between the rates of input of dietary nutrients and the output of production. If intake of any given nutrient is significantly below daily needs for a sufficient period of time, then nutritional deficiency disease can occur.

Pregnancy toxaemia, ketosis or fatty liver disease (pregnancy disease, pregnancy ketosis or twin kid disease)

Pregnancy toxaemia is a common metabolic disease occurring in the last 6 weeks of gestation in does with multiple foetuses, undernourishment during late pregnancy, and precipitated in severe weather, sudden changes in feed or other concurrent disease or after transportation. Mortality may also occur. The deficiency of glucose as an energy source is the main cause. Blood sugar levels decrease as pregnancy progresses (hypoglycaemia) from a normal 35–45 mg per 100 ml blood to 20–25 mg per 100 ml blood in late pregnancy. In such conditions, body fat is utilized as an energy source, which results in excessive ketone bodies (acetone, acetoacetate and beta-hydroxybutyrate) and fatty infiltration of the liver (fatty liver disease). The affected pregnant does show inappetence, aimless walking, muscle twitching, opisthotonos and grinding of the teeth, progressing to blindness, ataxia and, finally, sternal recumbency, coma and possibly death. The levels of low glucose and high ketones may be detected in the urine using diagnostic strips. Hypoglycaemia can be treated by a single intravenous injection of 50–100 ml 50% dextrose, followed by a 5% dextrose and electrolyte solution. Corticosteroids (dexamethasone and triamcinolone) are effective in management. The insulin dose (20–40 IU protamine zinc insulin) may be standardized and can be administered intramuscularly every other day (Matthews, 1999). Oral administration of glycerol or propylene glycol may be effective in the early stages of the disease. Ketoacidosis can be managed by administering isotonic or hypertonic sodium bicarbonate solution intravenously and balanced electrolyte solution appropriately. An adequate nutritional level throughout the pregnancy will prevent pregnancy toxaemia.

Lactational ketosis/fatty liver disease

This disease is clinically similar to pregnancy toxaemia, but occurs in the lactational stage. The clinical condition is manifested by reduction of milk production, inappetence and poor body condition. In severe cases, neurologic signs (nervous ketosis) may also appear as in pregnancy toxaemia. Like pregnancy ketosis, good supportive care following kidding and appropriate dietary management of the early lactation doe are important.

Periparturient hypocalcaemia (milk fever)

Classical milk fever is commonly seen in dairy cattle after calving; however, dairy goats are susceptible to hypocalcaemia. The initial symptoms are characterized by nervous signs including hyperactivity, twitching of the lips, eyelids and ears, and the body is cold to the touch. The doe quickly becomes sternally recumbent and laterally recumbent in the final stages. The head may be turned back to the flank (opisthotonos posture). In less severe cases, animals are lethargic, anorexic with poor milk production. Death may also occur in critical cases. Dramatic recovery occurs after careful intravenous calcium borogluconate solution infusions, followed by subcutaneous injections of calcium solutions at a maintenance dose. Oral calcium therapy is usually helpful in follow-up after intravenous therapy. Less severely affected does can be managed by subcutaneous solutions only. It is necessary to maintain appropriate dietary calcium and phosphorus content in late pregnancy to support foetal bone development.

Hypomagnesaemia (grass tetany, lactation tetany, milk tetany)

Lactating goats uncommonly suffer from grass tetany or lactation tetany, and growing kids reared on milk replacer may also be affected (milk tetany). In acute cases, animals suffer with ataxia, stiffness and hyper-excitability, recumbency and paddling, and possible death. The condition can be managed by administering calcium and magnesium solutions, followed by subcutaneous injections of magnesium sulphate solutions as well as oral magnesium supplementation. Appropriate dietary supplementation of magnesium from late pregnancy through early lactation is needed.

Nutritional deficiency diseases

Nutritional deficiency slows or blocks the related metabolic process resulting in deficiency symptoms. Vitamin A deficiency results in poor appetite, weight loss, unthrifty appearance with a poor hair coat, night blindness and a thick nasal discharge in goats, especially in young animals. Deficiency also predisposes animals to diarrheal and respiratory disease and to parasitism. Vitamin A deficiency may promote desquamation of urinary epithelium and nidus formation with subsequent urolithiasis (Schmidt, 1941). Vitamin E deficiency alone can cause nutritional muscular dystrophy (white muscle disease). Kids may have muscle disease at birth and be too weak to suckle. Sudden death related to Zenker's necrosis of the heart muscle or diaphragm may occur. Inhalation pneumonia can also result from muscle weakness of the muscle of the larynx and pharynx. Affected kids are sometimes noted to cough or have milk run out the nose after drinking (Smith and Sherman, 2009). Kids may also develop muscle stiffness after exercise (Byrne, 1992). Vitamin E and selenium are closely interrelated; a deficiency of one can be at least partially offset by increasing the intake of the other (Smith and Sherman, 2009). It is advised to add 25–50 mg of vitamin E/kg of concentrate for adults and 50–100 mg/kg of concentrate for kids (Ferrando and Barlet, 1978). Oral and injectable vitamin E are also prescribed.

Calcium and phosphorus deficiency is associated with retarded growth and rickets in growing kids. Approximately 99% of the calcium in the body is found in the skeleton in combination with phosphorus (Ca:P ratio of 2.2:1). A severe imbalance of calcium and phosphorus (relatively too much phosphorus from a mostly grain diet) can cause osteodystrophia fibrosa in goats, in which the lower jaws are enlarged but soft and rubbery, with medial angulation of the crowns of the cheek teeth (Andrews *et al.*, 1983). Supplementation of these minerals should be maintained as per the requirement.

Rumen acidosis

Rumen acidosis is a well-known management disease in intensive goat rearing systems. Rumen acidosis develops due to excess feeding on carbohydrate feed in growing kids and adult goats. Rumen acidosis lowers the gastric pH and creates an environment that favours the proliferation of harmful bacteria, resulting in potentially fatal enterotoxaemia both in kids and adult goats.

Major symptoms include anorexia, apathy, ruminal impaction, ruminal atony, watery faeces, haemoconcentration, metabolic acidosis, prostration and death. The ruminal fluid changes to a milky colour with a sour odour, low pH and predominance of Gram-positive bacteria in smears. Goats are highly susceptible to acute ruminal acidosis induced by sorghum at a dose rate of 50–100 g/kg BW, whereas wheat at a dose rate of 100 g/kg BW is required to produce the classical signs in goats. Management should focus on the removal of rumen contents, and replacement with ingesta taken from healthy animals is necessary. Rumen lavage may be accomplished with a stomach tube. Initially, 5% sodium bicarbonate solution should be given intravenously (20–400 ml). During the next 6–12 hours, a balanced electrolyte solution or a 1.3% solution of sodium bicarbonate in saline, may be given intravenously.

In less severe cases, magnesium hydroxide (50 g) and sodium bicarbonate (50–100 g) should be given orally to neutralize lactic acid in the rumen. In addition, oral antibacterial drugs (oxytetracycline) are advisable to control Gram-positive pathogenic bacteria. Antihistaminics are recommended to check histaminosis, and oral rumeotorics are suggested to regain rumen function and appetite. As a preventive strategy, good feeding management is the key to reducing the mortality rate due to rumen acidosis. Feedlot animals should be introduced gradually to concentrated rations over a period of time. Lactic acidosis can be reduced by immunization against *Streptococcus bovis*.

Urea toxicity

This happens due to the consumption of urea in feed because of over mixing or improper mixing of urea in feed. Excess urea is degraded into ammonia and absorbed systemically, rendering the rumen pH to alkaline. The main clinical signs are muscle and skin tremors, salivation, frequent urination and defecation, incoordination, bloating and death. Urea toxicity can be best managed by administration of acetic acid (vinegar) (200–500 ml) by a stomach tube, which reduces the rumen pH and retards the absorption of urea.

Urolithiasis

Clinical obstructive urolithiasis is most frequently noted in young, castrated males. Small calculi become lodged in the male urethra and the condition becomes more severe by early castration of male ruminants. Goats kept on an excessive grain diet become prone to calculi formation. The predisposing factors are increased urine concentration, decreased water intake or increased water loss, urine stasis and increases in urine pH. Forage high in oxalates promote oxalate calculi. The concentrated grain rations are commonly high in phosphorus and relatively low in calcium. This imbalance promotes the formation of phosphate calculi, as frequently seen in feedlot kids. Vitamin A deficiency may also initiate urolithiasis in goats. The signs depend on the degree and place of obstruction. Initially, male goats show restlessness, anxiety, tail twitching and vocalizing, dysuria and arching of the back. Animals react to pain on palpation at sigmoid flexure. Drops of bloody urine and/or crystals may be seen attached to preputial hairs. Rupture of the bladder or urethra usually results within 24 to 48 hours in untreated cases. In a state of rupture of the bladder, the discomfort alleviates and animals resume a more normal attitude, but uraemia (characterized by anorexia, weakness, depression) follows, resulting in markedly elevated blood urea nitrogen (11.7 to 47.5 mmol/l [70 to 285 mg/dl]) and creatinine levels. Urine sediment should be examined (Fig. 5.26). Marked crystalluria supports the diagnosis of obstructive urolithiasis. An odour of urine may be helpful in field diagnosis. Contrast radiography and ultrasonography help in assessing the extent and location of calculi.

The management depends on the nature and extent of the calculi present. In acute cases and rupture of the bladder, surgical intervention becomes urgent. Diazepam, at a dose range of 0.1–0.5 mg/kg and

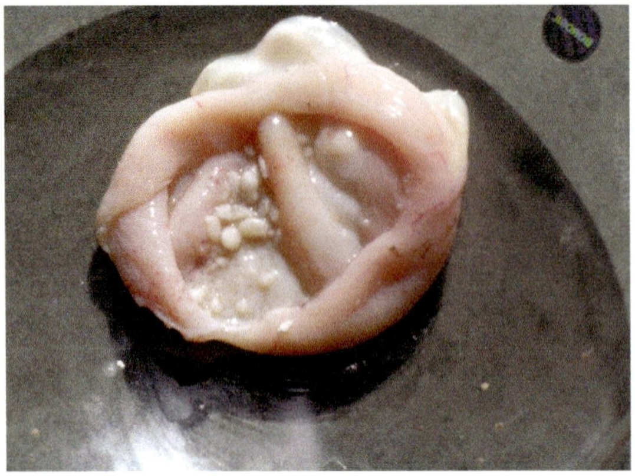

Fig. 5.26. Calculi in the urinary bladder of a 4-month-old kid.

antispasmodic drugs (anafortan) may help to promote urethral relaxation and facilitate natural expulsion of the calculus by the pressure of attempted urination. Dietary management is the key to the control and prevention of urolithiasis. The calcium to phosphorus ratio of the overall diet should be in the range of 2:1. Heavy grain feeding should be restricted and supplemented with balanced mineral mixture. Prophylactic use of urinary acidifiers is advocated. Continuous administration of ammonium chloride at a dose of 10 g per head per day or at a level of 2% in the concentrate ration has been recommended in goats. Delaying castration may be another management tool for reducing the prevalence of obstructive urolithiasis.

5.6.2 Farm health management and application of annual health calendar

Shifting from extensive to semi-intensive and intensive systems of management has led to increased mortality of kids and adults due to diseases. The prevention of common goat diseases has been standardized and an annual goat health calendar has been developed for goat flocks maintained in the tropics. Commercial goat farmers should follow the schedule as it helps in the prevention of goat diseases under different agro-climatic zones of tropical countries (see Protocol 5 at the end of this chapter).

5.6.3 Alternative treatment strategies for disease management

Synthetic drugs can cause toxicity affecting various physiological functions. Plant medicines are relatively non-toxic, cheaper and are eco-friendly and have been in use in different communities worldwide for many years. Herbal medicine has been used in day-to-day problems of healthcare in animals. Seventy-five per cent of medicinal plants grow naturally and are known to cure many ailments in animals in Asia and Africa.

Collection of medicinal plants

The desired plant materials are collected from medicinal gardens and dried in a shady area. This plant material is prepared using different methods.

MACERATION Whole or coarsely powdered crude drug is placed in a stoppered container with the solvent (organic solvent or water) and allowed to stand at room temperature for a period of at least 3 days with frequent agitation until the soluble matter has dissolved. The macerated mixture is strained and filtered. The filtered liquid can be used orally.

DECOCTION This is the most commonly used method for preparation of extract. This procedure is suitable for extracting water-soluble and heat-stable constituents. The dried coarse plant material is boiled in a specified volume of water (1:4 or 1:16) for sufficient period of time until it becomes reduced to one-fourth of its original volume. It is then cooled and strained, and may be used as such or processed further.

PERCOLATION This method can be used by farmers with percolation equipment. The procedure is used most frequently to extract active ingredients from a tincture of plant material. The plant material is kept in a glass percolator and allowed to stand for 3–4 hours, after which time the liquid slowly filters through a powdered substance that absorbs certain constituents. The extract drips out the bottom of the percolator.

HOT CONTINUOUS EXTRACTION (SOXHLET) In this method, the finely ground crude drug is placed in a porous bag or 'thimble' made of strong filter paper, which is placed in Soxhlet apparatus. The extracting solvent in the flask is heated, and its vapours condense in a condenser. The condensed extract drips into the thimble containing the crude drug, and extracts it by contact. The extraction process is continuous and is carried out until a drop of solvent from the siphon tube does not leave residue when evaporated. The advantage of this method, compared to previously described methods, is that large amounts of the drug can be extracted using a much smaller quantity of solvent. This has a tremendous effect on the economy of production in terms of time, energy and, consequently, financial inputs. On a small scale, it is employed as a batch process only, but it becomes much more economical and viable when converted into a continuous extraction procedure on a medium or large scale.

5.7 Zoonotic Diseases

Zoonoses are among the most common potential risks to which humankind are exposed. The close association between humans and animals can lead to a serious risk to public health with severe economic consequences in both developing and developed countries. All major zoonotic diseases hamper animal production, including goats, and create obstacles to international trade in animals and animal products.

5.7.1 Contagious ecthyma (Orf, contagious pustular dermatitis)

Contagious ecthyma is a highly contagious, zoonotic viral skin disease in goats, exhibiting painful skin lesions, often on the mouth and muzzle leading to anorexia and starvation. The morbidity in young kids may be as high as 90–100%. In humans, the infection is localized and self-limiting. Scabs/crusts can act as the source of infection to other animals and humans. Non-porous gloves must be worn while handling the infected animals. Scabs and crusts must be burned and the area must be disinfected.

5.7.2 Rabies

Rabies affects all warm-blooded animals, and is transmitted to humans by canines and vampire bats. Rabies-infected goats may pass the infection onto the human population through saliva or neurological tissues. Rabies virus is inactivated by heat; however, as a precaution, the US National Association of State Public Health Veterinarians recommends against consuming tissues and milk from rabid animals. Post-exposure prophylaxis consists of immediate wound cleaning. An inactivated human vaccine is available for veterinarians, animal handlers, wildlife officers, laboratory workers and others at a high risk of exposure.

5.7.3 Brucellosis

Brucellosis is also called undulant fever, Malta fever, Gibraltar fever, Mediterranean fever and Mediterranean remittent fever in human infections. Brucellosis in goats is caused by *B. melitensis*, and rarely by *Brucella abortus* or *Brucella suis*. Most cases are caused by occupational exposure to infected goats. Other sources of infection for humans include contact with animal abortion products, ingestion of unpasteurized dairy products or ingestion of undercooked meat, bone marrow or other uncooked meat products. It may also spread by contact with laboratory cultures and tissue samples, accidental injection of live brucellosis vaccines, blood transfusion, bone marrow transplantation or sexual intercourse.

Human brucellosis is usually prevented by controlling the infection in animals. Pasteurization of dairy products and proper cooking of animal products are important safety measures. Good hygiene and protective clothing/equipment are very important in preventing occupational exposure. Precautions should be taken to avoid contamination of the skin as well as inhalation or accidental ingestion of organisms when assisting kidding, performing a necropsy or butchering the animal for consumption. Particular care should be taken when handling an aborted foetus or its membranes and fluids. The *B. abortus* strain 19 vaccine and *B. melitensis* Rev-1 vaccine must be handled with caution to avoid accidental injection or exposure. In the laboratory, *Brucella* spp. should be handled under biosafety level 3 conditions or higher. Frequent seroscreening programmes in goat herds should be implemented and seropositive goats must be slaughtered immediately as per a 'test-and-slaughter' policy.

5.7.4 Listeriosis (circling disease/silage disease/caprine bacterial encephalitis)

Listeriosis is an important infectious disease of goats associated with neurological signs, septicaemia, mastitis and abortion. Milk and meat are the potential sources of the infection for humans. Chevon should be cooked properly and unpasteurized milk products should not be consumed. All cooking utensils as well as the hands should be washed after their contact with raw meat and milk. Sanitation and hygiene during kidding and postmortem examination decreases the risk of contracting the disease in veterinarians and farmers who are occupationally exposed.

5.7.5 Tuberculosis

Tuberculosis is a chronic bacterial disease of cattle that occasionally affects other species of mammals including goats. This disease can spread to humans typically by the inhalation of aerosols or the ingestion of unpasteurized milk. Raw or undercooked meat can also be a source of the organism. Only test-and-slaughter techniques are successful to eradicate tuberculosis from domesticated animals

5.7.6 Anthrax

Anthrax is an infectious and contagious disease with zoonotic potential. Transmission of the disease from animals to humans is usually by direct handling of hair, wool, skin or carcasses of infected animals. It is caused by *B. anthracis*, a large Gram-positive aerobic spore-forming bacteria. The spores of this bacterium can survive in extreme environmental conditions for years. The infection spreads through feed and water contaminated with spores. Sudden death with bleeding from all natural orifices (especially nostrils and mouths) are characteristic of the disease. The carcasses of the infected/suspected goats must not be opened, because this will initiate the formation of spores. In dead goats, excessive splenomegaly, no clotting of blood and no rigor mortis are observed, which are salient post-mortem findings of this disease. Carcasses of the disease-affected goats should be burned or buried deeply to avoid dissemination and persistence of spores in the environment.

5.7.7 Leptospirosis

The other names of this disease are autumnal fever, Cane cutter's disease, Canicola fever, harvest fever, infectious jaundice, 'Lepro', mud fever, rice field worker's disease, seven-day fever, swamp fever, swine herd's disease, Walter fever and Weil's disease. Leptospirosis is caused by *Leptospira interrogans*. The disease is prevalent in almost all species of domestic animals, including goats and several rodent species. Leptospirosis is primarily a disease of animals, and man is considered as a 'dead-end' host that, at times, may serve as a reservoir. The infected rodents (mainly) and mammals, usually herbivores, serve as natural reservoirs of infection. In the carrier animals the bacteria survive well and shed in urine. Leptospires can survive for weeks in soil and water that is frequently contaminated by urine of carriers, particularly if the pH is alkaline. Fish, reptiles and birds have also been implicated as carriers. Goats may remain symptomless but shed leptospires in the urine for short periods or may develop jaundice, haemoglobinuria and abort their foetuses.

All measures should be adopted to control the rodent population. In order to avoid leptospiral infection, the advice is to consume boiled or chlorinated water, wash hands with disinfectants, drink pasteurized or boiled milk, protect food articles and utensils from contamination with rodent urine, encourage use of protective clothing in farmworkers, particularly when working in water-logged areas or handling animals/animal products during slaughtering and parturition, etc., and avoid swimming and wading in lakes, ponds and swimming pods contaminated with rodent and livestock urine.

5.7.8 Campylobacteriosis (vibriosis)

Vibrionic abortion in goats is commonly caused by *Camphylobacter jejuni* and less frequently by *Camphylobacter fetus*. *Campylobacter* spp. are widely distributed in nature, with most of the species adapted to the intestinal tract of warm-blooded animals and birds. The infection can be transmitted by ingestion of contaminated water and feed, direct contact with carriers (human and animals) and sexual contact. Aborting does usually show no clinical signs but may have mild diarrhoea and mucopurulent vaginal discharge. The placenta is often oedematous with necrosis of cotyledons. Sanitation at farms is the necessary step for prevention and control of *Campylobacter* infections. Faecal contamination of feed and water should be avoided and the aborting does should be isolated. Placentas, body fluids and foetuses should be burned or buried deeply.

5.7.9 *Taenia* infections (taeniasis, cysticercosis, neurocysticercosis, coenurosis, neurocoenurosis)

Taenia spp. are long, segmented parasitic tapeworms and have an indirect life cycle between a definitive and an intermediate host. Some *Taenia* species are zoonotic, with humans serving as the definitive host, the intermediate host, or both. Carnivores are usually the definitive hosts for *Taenia* spp. A definitive host becomes infected when it eats tissues from the intermediate host that contain larvae. Human coenurosis is much less common than cysticercosis. Taeniasis is usually mild or asymptomatic. Morbidity and mortality are mainly related to the central nervous system, ocular or cardiac disease. Permanent brain damage or blindness is possible. Cysticercosis often causes epilepsy in endemic areas. Prevention of taeniasis depends on avoiding raw or undercooked meat and meat products. Prevention of cysticercosis or coenurosis also depends on good hygiene, including hand washing. In endemic areas, people should avoid eating raw fruits and vegetables that cannot be peeled. Only bottled water, filtered water or water that has been boiled for at least 1 minute should be drunk. Human sewage can contain *Taenia* eggs and should not be used to irrigate crops. People carrying tapeworms in the intestines should be treated as soon as possible.

5.8 Protocols

Protocol 1: Placenta disposal

Material kit: gum boots, face mask, hand gloves, disinfectant, lime powder, spade.
Precaution: Person handling the aborted foetus and placenta should wear a facemask, hand gloves and gum boots to avoid any possible infection.

Procedure

1. Do not handle infected material with bare hands since the disease may be zoonotic.
2. The aborted foetus, placenta, contaminated bedding, feed, etc., should be buried at least 4 feet deep after a liberal sprinkling of lime.
3. Disinfect the shed after isolating the doe from the flock.
4. Disinfect the lochial discharges daily with 1–2% NaOH or 5% sodium hypochlorite (bleach) solution till the discharges cease, usually by day 10–15.
5. The person handling the placenta must use antiseptic cream as a safety measure to prevent infection.
6. Isolate the affected animal immediately for a minimum of 20 days after kidding/abortion, for detailed investigation.

Protocol 2: Sample collection, packaging and field necropsy

Sample collection requires proper logistics for collection, preservation and dispatch of morbid samples to the diagnostic laboratory. Toxicity-related mortalities require prior knowledge for collection and preservation of viscera or other organs (bone, skin, hair, etc.) based on the suspected toxic agent. The samples collected should be carefully packed, labelled and transported to the laboratory by the fastest practicable method with the appropriate storage conditions. All samples should be accompanied by a letter or submission form, which includes the name and address of the submitter, the origin of the material, relevant history, animal identification and corresponding specimens.
Material kit: Different sized syringes and needles, vacutainer, sample collection bottles, 50% glycerol saline, ice, 10% formal saline, sterile swab, EDTA, phosphate-buffered saline (PBS), merthiolate (1:1000), specimen collection kits, glass slides, pre-filled custom shipping labels, etc.
Precaution: Before taking samples, careful consideration should be given to the purpose for which they are required. This will determine the type and number of samples needed to provide the valid result. When samples are taken from live animals, care should be taken to avoid injury or distress to the animal or danger to the operator and attendants. Whenever handling biological material, from either live or dead animals, the risk of zoonotic disease should be kept in mind and precautions should be taken to avoid human infection.

Collection of samples

ORGANS AND TISSUES Tissues can be collected for microbiological culture, parasitology, biochemistry, histopathology and immunohistochemistry. The affected organ should be excised in such a manner that the sample should have the affected part along with some normal organ. The collected samples/organ should be kept in a clean, sterilized glass vial neatly labelled, within normal saline or transport medium for bacteria or viruses. Organs collected for histopathological examination should be stored in a vial containing 10% formal saline.

BLOOD Blood should be collected from an external jugular vein in goats with a sterile needle and syringe. The blood should be transferred carefully into a sterilized glass vial containing an anticoagulant solution such as sodium citrate, sodium EDTA or heparin. The vial should be rotated from the base for proper mixing of the anticoagulant with the blood, which otherwise may lead to clotting. The vial should be dispatched in an icepack to the laboratory.

SERUM The blood should be collected from the jugular vein and placed in a sterile glass vial without anticoagulant and allowed to clot with the tubes placed in a slanting position. After some time, the serum is collected by sterilized Pasteur pipette and transferred into another sterilized glass vial or Eppendorf tube. Merthiolate (1:1000 strength) may be used as a preservative. The serum samples should also be sent in an ice pack to the laboratory.

FAECES/DUNG SAMPLES A minimum of 10 g of freshly voided faeces should be collected. Faeces samples collected for parasitology examination should be sent to arrive at the laboratory within 24 hours. If transport times are likely to be longer than 24 hours, the sample should be sent on ice to prevent the hatching of parasite eggs. Alternatively, commercially available swabs may be used. Swabs should be transported in the appropriate transport medium. Faeces are best stored and transported at 4°C.

PUS AND NASAL DISCHARGE Pus and nasal discharge are collected directly on a sterilized cotton swab placed in a sterile glass vial. Later, the glass vial can be preserved in an ice pack and dispatched for further laboratory examination.

BODY FLUIDS AND EXUDATES The samples are collected from the body cavities of dead animals. They are collected by a sterile syringe or sterile Pasteur pipette and are dispatched in a sterile glass vial under refrigeration conditions.

BLOOD AND TISSUE IMPRESSION SLIDES Thin smears of blood should be made by pressing a clean, non-greasy slide against the cut part of the organ/tissue after blotting with filter paper. The slides are dried and fixed with methyl alcohol. These slides should be dispatched in a slide box.

SKIN AND EPITHELIAL TISSUES Plucked hair or wool samples are useful for diagnosing surface feeding mites, lice and fungal infections. Deep skin scrapings using the edge of a scalpel blade are useful for burrowing mites. In disease-producing vesicular lesions, if possible 2 g of affected epithelial tissue may be collected under aseptic conditions and put it in 5 ml phosphate-buffered glycerine or tris-buffered tryptose broth virus transport medium. The vesicular fluid can be aspirated using a sterile syringe and collected in a sterile glass vial.

GENITAL TRACT AND SEMEN SAMPLES Samples may be taken by vaginal or preputial washing using suitable sterile swabs. The cervix or the urethra may be sampled by swabbing. Samples of semen are best collected using an artificial vagina or by the extrusion of the penis and artificial stimulation. The samples collected in a sterile tube can be sent on ice to the nearest laboratory.

EYES A sample from the conjunctiva can be taken by holding the palpebra and gently swabbing its surface. The swab is then put into a transport medium. Mucopurulent lacrimal discharges are also useful in some cases.

PIECES OF INTESTINE AND THEIR CONTENTS The intestine is cut from both sides of the affected area and tied at the ends with a clean thread. These pieces are then put in an open-mouthed glass bottle and dispatched in ice. It is necessary to dispatch the samples as early as possible to the laboratory after collection.

MILK Milk samples should be collected after cleansing and drying the tip of the teat. Milk samples are collected usually for detection of subclinical mastitis (California mastitis test or somatic cell count), brucellosis and for bacteriological examination in milking does. Milk samples can be collected in glass or plastic vials and preserved in ice for transport to the laboratory. Milk for bacteriological testing should be transported as frozen samples.

Disease-wise sample collection

PESTE DES PETITS RUMINANTS This is a viral disease. In live animals, swabs are made of the conjunctival discharges, nasal and buccal mucosae. During the very early stage of the disease, whole blood is also collected in anticoagulant for virus isolation, PCR and haematology. At post mortem, lymph nodes (especially the mesenteric and bronchial nodes), lungs, spleen and intestinal mucosae should be collected aseptically, chilled on ice and transported under refrigeration. Fragments of organs collected for histopathology are placed in 10% formalin.

CONTAGIOUS CAPRINE PLEURO PNEUMONIA This is a severe disease of goats caused by *M. capricolum* subspecies *capri pneumoniae* (Mccp). Lung lesions (particularly from the interface between consolidated and unconsolidated areas), pleural fluid and mediastinal lymph nodes are important samples during necropsy. Samples for microbiological examination should be transported under refrigeration or cool arrangements, as mycoplasma viability diminishes rapidly with increasing temperature. Pleural fluid and tissue suspension swab should also be collected and suspended in 2–3 ml of culture medium and sent on ice.

GOAT POX This is a viral disease characterized by fever, generalized papules or nodules, vesicles and internal lesions particularly in the lungs. Material for virus isolation and antigen detection should be collected by biopsy or at post mortem from skin papules, lung lesions or lymph nodes. Samples for virus isolation should be collected within the first week of the occurrence of clinical signs, before the development of neutralizing antibodies. Buffy coat can also be used for virus isolation. Samples for histology should include tissue from the surrounding area in 10% formalin.

BLUE TONGUE Whole blood can be collected during acute pyrexia without anticoagulants and transported to the laboratory on ice. Serum can be collected for screening of antibodies in the goats.

FOOT-AND-MOUTH DISEASE The tissue of choice in this disease is epithelium or vesicular fluid. Ideally at least 1 g of epithelial tissue should be collected from an unruptured vesicle, usually from the tongue, buccal mucosa or feet. Epithelial samples should be placed in transport medium composed of glycerol phosphate buffer, preferably with added antibiotics (1000 IU penicillin, 100 IU neomycin sulphate, 50 IU polymixin sulphate and 100 IU mycostatin) and should be sent to the laboratory as soon as possible.

BRUCELLOSIS Serum for by standard tube agglutination test (SAT) should be harvested. Titres of 1:40 and above is considered positive. The aborted contents should be handled carefully, as *B. melitensis* is zoonotic in nature. Use of gloves and masks is recommended with safety precautions for handling the aborted contents. The abomasal fluid of the aborted foetus can be collected using a sterile syringe and packed and sent on ice to the laboratory for bacteriological examination.

ENTEROTOXAEMIA In dead animals, the intestine can be cut and ligated at both the ends and sent on ice to the laboratory for identification of possible toxins produced by *C. perfringens* type D.

Table 5.5. Morbid sample preservation for laboratory diagnosis of important infectious diseases.

Name of the disease	Samples to be collected	Preservative
Foot-and-mouth disease	Tongue epithelium, vesicular fluid, buccal mucosa	50% glycerol saline with 1000 IU penicillin
	Serum – for ELISA	Ice
Peste des petits ruminants	Blood, lymph nodes, spleen, nasal swabs, buccal swabs, serum (ELISA)	Ice
	Histopathology: intestine, spleen, lungs	10% formal saline
Blue tongue	Blood in EDTA (during viremia), serum (ELISA, AGPT)	Ice
Goat pox	Pock lesions, dried scabs, lungs, lymph nodes (virus isolation)	Ice (within 1 week of onset)
Haemorrhagic septicaemia	Blood smear for staining, whole blood, visceral organs culture	Packed in butter paper Ice
Enterotoxaemia	Whole intestinal loop, tied at both ends – for toxin estimation	Ice
Brucellosis	Aborted contents for culture	Ice
	Serum, milk– STAT, ABRT	
Listeriosis	Brain tissue for culture, PCR	Ice
Rabies	Brain hippocampus smear, saliva	Ice for isolation
	Brain tissue: histopathology	10% formal saline
Mastitis	Milk for mastitis tests: CMT, NAGase	Formalin Ice (4°C)
	Milk: for microbiology/ABST	

Continued

Table 5.5. Continued.

Name of the disease	Samples to be collected	Preservative
Parasitological examination	Dung or faecal sample for eggs of nematodes, tapeworm, oocyst of coccidia, cryptosporidia	Packed in sealed polythene covers 4°C (<24 h)
CCPP	Lungs, pleural fluid	Ice
Chronic diarrhoea/J.D.	Faecal sample, mesenteric lymph node	Packed in sealed polythene covers sent through ice
Dermatitis/Mange	Deep skin scrapings, hair	Packed in sealed covers or butter paper
Tumours/ neoplasms	Affected tissue/organ for histopathology	10% formal saline
Neonatal diarrhoea	Faecal swab or faeces for microbiological examination	Swab dipped in PBS and packed with ice packs
Endometritis/pyometra	Discharge or swab for ABST	Collected in sealed vials, packed in ice

ABST, antibacterial sensitivity test; AGPT, agar gel precipitation test; CMT, ELISA, enzyme-linked immunosorbent assay; PCR, polymerase chain reaction; STAT, standard tube agglutination test ; NAGase, N-acetyl-β-D-glucosaminidase.

Table 5.6. Important information to be submitted with morbid sample.

Information required	Details
Name of the owner/sender	
Address and telephone number of the owner/sender	
Animal description (species, breed, sex, age and identity of animals sampled)	
Date of sample collection and submission	
List of samples submitted	
History of the case	
Length of time of sickness of animals	
Recent arrival history and place	
Post-mortem findings, if any	
The date of the first cases and of subsequent cases or losses, with any appropriate previous submission reference numbers	
A description of the spread of infection in the herd or flock	
Mortality rate, morbidity rate and their age, sex and breed	
Clinical signs and their duration	
The type and standard of husbandry, including the type of feed available, possible contact with poison or poisonous plants	
Any treatment history	
Any vaccination history	
Other observations about the disease, and other disease conditions	

Protocol 3: Control and prevention of neonatal mortality in kids

Material kit: Tincture iodine or povidone iodine, soft clean cloths, scissors, cotton thread/dental floss, disinfectant scrub, thermometer, disposable plastic sleeves or gloves, clean bucket and hot water, collar, lead rope, plastic-coated wire to assist kidding during a difficult birth, baby nasal syringe, flashlight, 20–50% dextrose or glucose solution, stomach tube kit, dosing syringe, colostrum (frozen colostrum), warming box/heating pad/blow dryer to warm up hypothermic kids, plastic bottles with a variety of nipples for bottle feeding, baking soda to treat acidosis.

Procedure

- The new-born kid should be born in clean, dry and soft bedding. The new-born kid should get up after birth and suckle the dam.
- Kidding should be in separate clean and well-bedded pens.
- Apply tincture iodine on the navel to prevent the entry of infection.

- Kids should be provided with sufficient amounts of colostrum in a timely manner. In the case of insufficient colostrum, frozen colostrum should be provided.
- New-born kids may be transferred to a clean pen separate from the main kidding ground.
- Diseased new-born kids should be transferred with their dam to clinical wards for the period of treatment and convalescence.

Table 5.7. Common clinical ailments during kidding time and management.

Clinical condition	Identification/symptoms	Management/therapy
Management in kidded does		
Mastitis	Swollen, hot, painful udder with blood, flaked milk, fever in acute mastitis; ice-cold udder in gangrene mastitis	Clean the udder before and after milking with antiseptic water Frequent milking Used dip teats after milking Inject effective antibiotics Teat infusions and parentally
Ketosis	Inappetence to anorexia, urine and breath may smell like bitter almond, loss of muscle control, recumbency, convulsions and eventually death	60 ml of propylene glycol orally two to three times daily along with a tablespoon of baking soda to counteract acidosis until appetite returns Jaggery feeding provides instant energy Intravenous dextrose or fluids
Milk fever	Observed in dairy goats, due to calcium deficiency, difficulty in standing, subnormal temperature, coma, death	Calcium gluconate as slow infusion (about 60 ml or more)
Prolapsed uterus	May be after kidding, uterus comes out and looks like a mass of tissue	Recovery good if undamaged and repositioned within 24 hours Administer calcium gluconate Administer anti-inflammatory drugs and antibiotics, check for further complications
Prolapsed vagina	Pregnant doe's vagina protrudes from under her tail like a small red handball	Prolapsed vagina appears in lying position and encourages doe to stand much of the time, wash vagina with mild soap and push it back
Retained placenta	Placenta is not passed within 24 hours after kidding	Administer effective antibiotic Monitor temperature for possible infection Do not pull hard on placenta Call veterinarian for removal and further management
Septicaemia	High fever, poor appetite, depression, and foul-smelling discharge, eventually death in untreated cases	Immediately administer effective antibiotic, may be intravenous along with fluids and anti-inflammatory drugs
Management in new-born kids		
Diarrhoea	White/yellow liquid faeces, foul smelling, dehydration, sunken eyes May be infectious or occur after overfeeding of milk	Recovered in mild cases associated with overfeeding Treatment with suitable antibacterial/ chemotherapy Rehydration therapy, Ringers solution IV and subcutaneous route
Floppy kid syndrome (acidosis)	Neonate kid (3–10 days old) Sudden difficulty in standing or walking, Kid becomes loose, flaccid with watery stomach. Mainly due to acidosis, occurs after drinking milk	Provide immediately with half a teaspoon of baking soda (sodium bicarbonate) Commercial electrolyte solution for 12 hours rather than milk Antibiotics and/or sodium bicarbonate IV may also be used

Continued

Table 5.7. Continued.

Clinical condition	Identification/symptoms	Management/therapy
Hypothermia and/or hypoglycaemia	Kid exhausts body reserves and is unable to maintain body heat or activity. May have abnormally low temperature, mouth will often be cold to touch, belly is not well distended, poor appetite	20% warmed dextrose IV Warm kid up to 99^{0}F/37.2°C if necessary using water bath, heating pad, etc. Ensure colostrum or milk feeding Keep watch on proper timely suckling to does
Navel ill	Hot, pus-filled navel stump, swollen joints, fever, convulsions, rigid movements, death	Prompt early treatment with antibiotics
White muscle disease	Muscle degeneration and progressive paralysis; Arched back, stiff hind end, respiratory distress, Kid may appear healthy and suddenly die after exercise or stress due to cardiac arrest, caused due to deficiencies in selenium or vitamin E	Prompt early treatment with vitamin E capsules and injection of selenium and Vit E supplement are effective

Adapted from: Kidding with confidence (2008) Cornell University Cooperative Extension and the Empire State Meat Goat Producers Association's Education Committee.

Protocol 4: Common drugs and dosages used in goats in clinical practice

Precaution: The preventive chemicals/drugs/feed mix should be followed as per the guidance of the veterinary physician, recommendation of the pharmaceutical company and animal husbandry experts.

Table 5.8. Common drug and dosages used in clinical practice.

Drug	Dose (per kg BW)	Route of administration
Sedatives and anaesthetics		
Diazepam	0.25–0.5 mg/kg, 2–4 ml/40 kg	Slow IV
Acepromazine maleate	0.5–0.1 mg/kg, 1–2 ml/40kg	Slow IV
Medetomidine	25–35 µg /kg, 1–1.5 ml/40 kg	IM
Xylazine	0.05 mg/kg, 0.1 ml/40 kg slowly, 0.1 mg/kg, 0.2 ml/40 kg	IV IM
Detomidine	20–40 µg/kg, 0.08–0.16 ml/40 kg	IV or IM
Alphaxolone/alphadolone	3 mg/kg, 10 ml/40 kg	Slow IV
Etorphine hydrochloride/acepromazine maleate	0.01 ml/kg	IV or IM
Ketamine	10–15 mg/kg, 4–6 ml/40 kg	Slow IV
Methohexital sodium	4 mg/kg, 6.5 ml/40 kg	Orally
Pentobarbital sodium	10–15 mg/kg, 40–60 ml/kg	IV
Propofol	3–4 mg/kg, 12–16 ml/40 kg	Slow IV
Thiopental sodium	10–15 mg/kg, 8–12 ml/ 40 kg	IV
Tiletamine-zolazepam	5.5 mg/kg	IV
Anthelmintics for gastrointestinal nematodes		
Albendazole, febental, fenbendazole, oxfendazole	7.5 mg/kg	Orally
Mebendazole	22.5 mg/kg	Orally
Netobimin	11.25 mg/kg	Orally
Levamisol	12 mg/kg orally (toxic in goats at higher dose rates at 20 mg/kg)	Orally
Ivermectin	200 µg/kg,2.5 ml/10 kg	Orally
Thiophanate	9 kg/tonne feed as single dose or 2.8 kg/tonne feed daily for 5 days in meat producing animals	Feed mix/orally

Continued

Table 5.8. Continued.

Drug	Dose (per kg BW)	Route of administration
Oxyclozanide	15 mg/kg	Orally
Triclabendazole	10 mg/kg	Orally
Closantel	10 mg/kg	Orally
Nitroxynil	10 mg/kg	Subcutaneous injection
Niclosamide	100 mg/kg	Orally
Praziquantel	5 mg/kg	Orally
Analgesics and anti-inflammatory drugs (non-steroidal)		
Aspirin	50–100 mg/kg	Orally every 12 hours
Carprofen	1.4 mg/kg, 1 ml/35 kg	SC or IV every 36 to 48 hours
Flunixin meglumine	2 mg/kg, 2 ml/45 kg	IV or IM
Meloxicam	0.5 mg/kg, 1 ml /10 kg	IV or SC
Phenylbutazone	4 mg/kg, 1 ml/50 kg	IV
	10 mg/kg	Orally
Analgesics and anti-inflammatory drugs (steroidal)		
Betamethasone	0.04–0.08 mg/kg, 1 ml/30 kg	IM
Dexamethasone	0.1 mg/kg, 1 ml/20 kg	IV
Methylprednisolone	10–30 mg/kg	IV
Anticonvulsants		
Diazepam	0.25–0.5 mg/kg, 2–4 ml/40 kg	IV to effect
Phenobarbitone	0.04 ml/kg	IV to effect then 0.22ml/kg IV every 8 to 12 hours as required
Coccidiostat		
Sulphadimidine	200 mg/kg initial dose, then 100 mg/kg	Orally
Sulphadimidine	200 mg/kg, 6 ml/10 kg initial dose, then 100 mg/kg, 3 ml/10 kg	SC or IV
Decoquinate	100 g/tonne feed or 1 mg/kg for 28 days	Feed mix
Sulphamethoxypyridazine	20 mg/kg, 1 ml per 12.5 kg	SC
Diclazuril	1 mg/kg, 1 ml/2.5 kg as a single dose	Orally
Amprolium	5–10 mg/kg orally daily for 3 to 5 days	Orally
Toltrazuril	20 mg/kg once every 3 to 4 weeks	Orally
Drugs acting on the digestive tract		
Loperamide hydrochloride	100–200 μg/kg, 0.5–1 ml/kg every 8 or 12 hours	Orally to reduce intestinal motility
Metamizole, hyscine butylbromide	0.5–5 ml	IV as spasmolytics
Metaclopramide,	0.5–1 mg/kg, 0.5–1 ml/5kg IV	IV, restores abomasal tone and promote emptying
Chlorpromazine	0.5 mg/kg	IV, appetite stimulant
Diazepam	0.04 mg/kg	IV, appetite stimulant
Drugs used for external parasites		
Amitraz	0.025% solution, by diluting 1 volume in 250 volumes of water	Spray or wash
Cypermethrin	1.25% solution, ' pour-on' 0.25 ml/kg or as dip	Spray or dip
Deltamethrin	1% solution	'Spot-on' or as dip
Permethrin	4% solution	'Spot-on' or as dip
Doramectin,	10 mg/50 kg, 1 ml 50 kg	SC
Abamectin	10 mg/50 kg, 1 ml/50 kg	SC
Ivermectin	10 mg/50 kg, 1 ml/50 kg	SC
Moxidectin	10 mg/50 kg, 1 ml/50 kg	SC
Eprinomectin	500 mg/kg, 1 ml/10kg	Orally
Fertility		
Dinaprost	2 ml	IM or SC

Continued

Table 5.8. Continued.

Drug	Dose (per kg BW)	Route of administration
Clorprostenol	0.5 ml	IM or SC
Oxytocin	2–10 U, 0.2–1 ml	IM or SC
Chorulon	500 U	IM
Gonadorellin	2.5 ml	IM (at time of service or to prevent luteolysis)
Buseerelin	2.5 ml	IM or SC or IV
Folligon	Dose depends on time of year, weight of goat and milk yield	SC or IM
Myometrial relaxants		
Clenbutarol hydrochloride	0.8 µg/kg, 1.25 ml/50 kg	Slowly IV or IM
Vetrabutine hydrochloride	2 mg/kg , 1 ml/50 kg	IM
Topical fungicides		
Eniliconazole	0.2% solution	By wash or spray, every 3 days for three or four applications
Natamycin	0.01% solution	Locally, repeat after 4 or 5 days and again after 14 days
Griseofulvin	7.5 mg/kg	Orally for 7 days

Adapted from Matthews, 1999. IM, intramuscular; IV, intravenous; SC, subcutaneous.

Protocol 5: Preventive health calendar for goats

Table 5.9. Annual vaccination, deworming and screening schedule for goats.

A. Vaccination			
	Initial vaccination and booster dose		
Name of disease/Core vaccine	Initial dose	Booster dose	Annual vaccination/repeat vaccination
Peste des petits ruminants	At 3–4 months age	–	Provide protection up to 3 years and repeat after 3 years
Foot-and-mouth disease	At 3 months age	3–4 weeks after of first injection	Repeat every 6 months interval
Goat pox	At 3–5 months age	3–4 weeks after of first injection	Repeat annually
Enterotoxaemia	At 3–4 months age	3–4 weeks after of first injection	Repeat every 6 months or early in high occurrence
Haemorrhagic septicaemia	At 3–4 months age	3–4 weeks after of first injection	Repeat annually
Rift valley Fever (live attenuated)	6 months onward	–	Repeat annually
Optional vaccine			
Anthrax	At 3–4 months age	–	Repeat after 9–12 months
Chlamydophila abortus	At 3–4 months age	–	Repeat after 12 months
Botulism	At 3–4 months age	3–7 weeks after first injection	Repeat after 9–12 months
Brucellosis (Rev.1)	3–5 months	–	–

Continued

Table 5.9. Continued.

B. Deworming Schedule		
Name of Infestation	Age	Period of administration
Coccidiosis	2 – 6 months	Any coccidiostat for a week period or provide as feed mix
Endoparasitic infection	After 4 months	Pre and Post monsoon, drug selected on type of parasitic infection
Lice infestation	Sizeable load of infection	Pre and Post winter
Tick infestation	Sizeable load of infection	Monsoon months

C. Regular screening		
Diseases	Period	Recommendations
Brucellosis	Once in a year	Positive cases are to be slaughtered
Johne's Disease	Once in a year, preferably after parturition	Positive cases are to be removed
Mycoplasmosis	Once in a year	Treatment with specific drugs

Protocol 6: How to conduct a post-mortem

Material kit: Mayo scissors (straight), bowel scissors, dissection scissors (straight), boning knife (sharp), skinning knife (curved), skinning knife (curved), autopsy knife (curved), chopper stainless, bone saw, post-mortem hammer, scalpel handles of different sizes, 2 m measuring tape, spine wrench, brain knife, dressing forceps (16 cm), bone-cutting forceps, bone chisel, Kelly forceps (14 cm, straight), Kelly forceps (14 cm curved), tissue forceps, metal box, Pasteur pipettes, syringes, spirit lamps, disinfectants to be used at the site of post mortem, cotton, antiseptic solution.

Precaution: Post-mortem examination is a very important step in disease diagnosis. It should be done carefully with all precautions as it involves a great potential to spread infection. The necessary precautions should be followed during transportation of the carcass, post-mortem examination, disposal of the carcass and disinfection of the site and handlers.

- The post mortem should be carried out in sufficient light.
- The post mortem should be conducted as early as possible after death.
- Protective coverings like gum boots, apron, gloves, face mask and head cover are mandatory for all personnel to avoid exposure to zoonotic infection.
- One person should be available to record the observed changes in different organs/tissues/body cavities/body fluids.

Procedure

Post mortem or necropsy is a systematic and scientific examination of the dead body/carcass/cadaver with the objective of ascertaining the cause of death by examining gross pathological changes in different vital organs and also facilitating the collection of biological materials leading to confirmed diagnosis (Pawaiya and Shivasharanappa, 2014).

Post-mortem procedure

1. Verification of the carcass: The details such as species, breed, sex, age, identification number, colour, etc. must be verified.

2. History: History regarding the clinical symptoms, duration of the disease, mortality and morbidity, age group and sex, vaccination history, any sudden change in feed or source of feed, introduction of new animals in the herd, change of climate, involvement of more than one species of animals in the area with a similar disease manifestation and any other relevant information regarding the disease should be obtained. It is important to record the date and time of death and when the post mortem is performed.

3. External examination of the carcass: Includes general body condition, post-mortem stage (algor mortis, rigor mortis or livor mortis); condition of the natural orifices (any discharge, prolapse), position of the lesions, visible mucous membranes, marks of injury (fresh or old) or any other pathological lesions. In cases where the animal has been suspected to have died of anthrax, the carcass should not be opened. The approximate age of the animal can be assessed by body appearance, horns and dentition.

4. Examination of carcass: The carcass should be examined for body fat and the presence of fluid in the abdominal cavity.

- Material for bacterial or viral isolation should be collected before opening the cadaver further.
- For opening the thorax, the sternum is removed by cutting the ribs at costochondral joints. The neck and thoracic organs are then removed together. The diaphragm is taken out by cutting at its insertion at xyphoid cartilage, following on either side along the costal attachments and finally the sub-lumbar insertion.
- Collect the thoracic fluid for cultural examination at this stage otherwise it will get contaminated. Any fluid or abnormal contents in the pericardial sac is observed and collected. The structure of the oral cavity and neck are then removed and the tongue examined. Follow this with the pharynx, oesophagus, salivary glands and trachea backward along the neck where the previous incision was made.
- The body lymph nodes are examined by dissecting longitudinally (forelimbs: prescapular and axillary; hind limbs: prefemoral and superficial and deep-inguinal and popliteal). When intracranial lesions are suspected, the brain as a whole is taken out by carefully removing the top of the skull and it is then systematically examined for pathological changes in brain parenchyma.

5. Examination of Gastrointestinal system: Starting with the buccal cavity, the entire mucosal surface of the cheeks, gums, tongue and pharynx along with the lips, commissures and teeth should be examined for any pathological lesions.

- The oesophagus is examined for changes in the mucosa (inflammation, erosion, ulceration, etc.), obstruction, stricture, stenosis, tumour growth, parasite lesions, etc.
- The rumen is examined for the quantity and consistency of its contents (thin, doughy, impaction, presence of free gas or gas bubbles trapped in ingesta – tympanites), pH, condition of mucosa, rumen pillars, foreign bodies and parasites (mature amphistome), etc.
- The reticulum is examined for the presence of any foreign body and adhesions with the diaphragm or liver.
- The omasum is examined for impaction, inflammation or ulceration of mucosal folds.
- The abomasum is opened from its greater curvature and examined for distension, type and consistency of contents, parasites, foreign objects (hair balls, etc.) and condition of the mucosa (swollen, oedematous, inflamed, haemorrhage, ulceration, perforation, rupture, nodule, timorous growth, etc.).
- The intestines (small: duodenum, jejunum, ileum; and large: caecum, colon, rectum) are examined for malpositions, obstruction, type and consistency of contents, parasites, inflammation, haemorrhage, mucosal thickness, corrugations, etc.
- Mesenteric lymph nodes should invariably be examined along with intestines.
- The liver is examined for size, consistency, colour, cut surfaces by slicing the organ, condition of intrahepatic bile ducts, parasites, etc.
- Check the gallbladder for the amount, colour, consistency and transparency of bile, the presence of parasites, calculi, mucosal condition (smooth, rough, inflamed, oedematous or timorous), patency of bile duct, etc.
- The spleen should be examined for size, shape, consistency, colour, presence of infarcts or haemorrhages on the surface, etc.

6. Examination of the **urogenital system**: It should be examined systematically.

- The kidneys are examined for location, size (atrophy or hypertrophy), colour, condition of perirenal and pelvic fat, cut surface cortex, medulla, renal pelvis for dilatation (hydronephrosis), infarcts, calculi, pus (pyelonephritis), etc.
- Ureters are checked for dilatation in case of hydronephrosis, pyelonephritis or calculi.
- The urinary bladder is checked for patency of the urethra by pressing the distended bladder and observing the passing of urine and its colour; it is checked for inflammation, haemorrhage or tumorous growths in the bladder.

- The urethra is examined for stenosis and lodgement of calculi.
- Accessory sex glands (seminal vesicles, prostate and bulbo-urethral), gonads (testes and ovaries), uterus and other genital organs are similarly examined for gross pathological changes.

7. Respiratory system: Examination includes external nares, nasal and paranasal sinuses, larynx, trachea, bronchi, lungs and associated lymph nodes (bronchial and mediastinal).

- The air passage and sinuses are examined for presence of exudates, inflammation, haemorrhage, foreign bodies, parasites or growth.
- In lungs, congestion, haemorrhage, consolidation, emphysema, atelectasis, mucus plugging, presence of parasites in terminal bronchioles, cysts, nodules or tumours may be encountered on gross examination. The changes may be described in terms of size, shape, area and involvement of lobe, type of exudates, etc.

8. Cardiovascular system: The heart is examined for condition of the pericardium, presence of fluid/exudate (colour, consistency, transparency, amount, etc.) in the pericardial sac, epicardial and endocardial surfaces (petechiae, echymoses, growth, etc.), myocardial thickness, ventricles and auricles and large blood vessels (aorta, vena cava, pulmonary vein and artery) for the presence of worms, growth, thrombus, intimal thickening, etc.

9. Nervous system: The brain is examined for condition of meninges and blood vessels, surface corrugations, ventricles, presence of parasitic cysts, malacic foci, etc. Similarly, the spinal cord is also examined for any pathological malformations, constriction, injury especially when peripheral nervous signs are indicative of spinal cord involvement.

10. Disposal of the carcass: The disposal of the carcass and disinfection of the site as well as of the persons involved in the handling of the carcass during necropsy is one of the most important tasks after post-mortem examination and collection of morbid material for laboratory investigations is over.

11. Disinfection of handlers: The persons involved should remove the protective coverings and dispose of them with the carcass. They should wash their hands thoroughly with soap and disinfectant lotion and finally rinse with clean hot water with a handful of starch or flour added to deodorize the hands.

12. Sterilization of the instruments: The instruments used for necropsy should be cleaned thoroughly with hot water and detergent and then sterilized by autoclaving.

Protocol 7: Transportation of goats

Transportation: Goats are transported from one place to another for various reasons including marketing, slaughter, re-stocking/destocking and from drought-affected areas to better grazing sites, etc. In general, the goats are traditionally transported by foot (trekking). However, with increasing urbanization and commercialization, transport by road is becoming increasingly frequent and more important. Rail transport is rarely used to transport animals. Major animal losses and damage result from inadequate care during the transport of goats using all three means of transport (hoof, road and rail). In transportation, animals should be maintained in good condition and with good animal welfare standards.

Poor transportation can cause various problems such as stress, predisposition to diseases especially PPR, respiratory diseases, bruising, trampling, suffocation, heart failure, heat stroke, sun burn, bloat, predation, dehydration, exhaustion, injuries and fighting leading to huge losses. The following guidelines should be followed to avoid poor transportation conditions.

Pre-transport activities

1. Ensure that all animals to be transported are healthy and strictly avoid transportation of sick, weak or extremely weak goats.

2. It has been noted that PPR occurs in unvaccinated goats after transportation. Avoid vaccinating immediately preceding or following transportation, as this could add to the stress.

3. The goat should be provided with sufficient space in the vehicle.

Table 5.10. Space in vehicle based on body weight of animal.

Approximate weight (kg)	Area (m^2/animal)
<35	0.20–0.30
35–55	0.30–0.40
>55	0.40–0.75

4. Sheds and premises should be ready for the arrival of the animals.

5. It is desirable to carry sufficient feed and fodder on a long journey in case of any unforeseen delays, and to ensure a gradual change-over to any new diet at the destination over a period of 3 to 5 days.
6. Adult males should be transported separately to female goats.
7. Large- and small-sized goats should be kept separated in the vehicle to prevent injury.
8. Where possible, goats should be acclimatized to the type and size of the pen, crate or box in which they will be transported.
9. Each animal must have a legible identification number that must be recorded accurately on any accompanying documentation.
10. Goats in the last term of pregnancy (approximately 15 days before expected date of parturition), or that have given birth in the last week, are not fit for transportation.
11. It is not advised to provide feed to the animals before loading and during the transportation, but water should be provided *ad-libtum*.
12. It is important that the transport unit should be cleaned and disinfected before loading and after unloading.
13. Diseased animals, whenever transported for treatment, should not be mixed with healthy animals.
14. If the animals are planned for trekking then the journey should be planned, paying attention to the distance to be travelled, opportunities for grazing, watering and overnight rest.

In-Transit activities

1. The interior of the transport unit should be specially designed for animal transportation and should be well lit.
2. Ensure adequate bedding over a non-skid floor in the vehicle to avoid body injury.
3. Use a loading dock that allows vehicles to drive away from it with minimal turns, stops and starts.
4. Drive cautiously to reduce the chance of falling and related damages.
5. The optimum temperature inside the vehicle should be managed by providing proper covering to protect the goats from extreme external temperatures.
6. The animals should be checked for any discomfort during transportation.
7. If the transport unit is not equipped for feeding and watering the goats, provision should be made for unloading and reloading en route for that purpose when required.
8. Lactating goats must be milked at intervals of no longer than 12 hours.
9. An attendant should accompany the animals and be responsible for feeding and watering.

Post-transport activities

1. Noise and harassment during loading and unloading should be avoided.
2. After prolonged periods of transport (36 hours or more), the health of the animals should be monitored.
3. Water should be provided.
4. The animals should be examined in a quarantine compound for at least 3–4 weeks, and should be vaccinated as per the area-specific requirement if animals are not vaccinated before transportation.

Protocol 8: Spot diagnosis of clinical conditions and diseases

Examination of blood, faeces, urine and milk provide a preliminary indication of clinical condition in the animals. Some details are given below.

Examination of faeces

BLOOD IN FAECES

1. Prepare a saturated solution of benzidine in glacial acetic acid (2 ml).
2. Add equal amounts of hydrogen peroxide solution to it.
3. Take 2 ml of the above mixture in a separate tube.
4. Add 3–4 drops of the faecal suspension prepared in distilled water.
5. Development of a blue colour indicates the presence of blood in the faecal material.

Examination of urine

REACTION OR PH The pH is tested by dipping a red or blue strip of litmus paper into the urine sample. A change from red to blue suggests an alkaline reaction and a change from blue to red indicates an acid reaction.

ALBUMIN: ROBERT'S REAGENT METHOD

1. Place 2 ml of Robert's reagent in a test tube.
2. Stratify 2 ml of urine on the reagent with the help of a medicine dropper or pipette.

3. If the sample is positive for albumin, a white ring appears at the point of contact.
4. Report the findings as follows:

–	No ring
+	A definite narrow ring
++	A definite wider ring
+++	Heavy ring
++++	Very heavy ring

GLUCOSE: BENEDICT'S METHOD The cupric-oxide is reduced to cuprous oxide which gives a green, yellow, orange or red colour in the presence of sugar.

1. Take 0.5 ml of urine in a test tube.
2. Add 5 ml of Benedict's qualitative reagent and mix thoroughly.
3. Place in a boiling water bath for 5 minutes or heat on flame.
4. Place the tube in the test tube stand and allow it to stand for a few minutes.
5. Note as follows:

–	Clear blue
+	Green to yellow sediment
++	Orange with heavy sediment
+++	Brick red with heavy sediment

ACETONE: ROTHERA'S TEST

1. Put 3 ml of urine in a test tube.
2. Add 1 ml of Rothera's reagent.
3. Add 1 ml of concentrated ammonium hydroxide on the top of the mixture.
4. A dark purple colour indicates the presence of acetone in the sample.

BILE: HAY'S TEST

1. Put about 5 ml of urine in a test tube.
2. Put a small amount of Flowers of sulphur on the surface of the sample.
3. The sample is negative if they float and positive if they sink.

BLOOD

1. Prepare a saturated solution of benzidine base in glacial acetic acid (2 ml).
2. Add an equal amount of hydrogen peroxide solution to it (2 ml).
3. Add 2 ml of urine to the tube containing the reagent and mix well.
4. The sample is positive for blood if a blue colour appears.

CALCIUM

1. Place 5 ml of urine in a test tube and add an equal amount of distilled water to it. Use it as a control.
2. In a second tube take 5 ml of urine and add an equal amount of Sulkowitch reagent.
3. Let the tube stand and read at intervals of 2 and 10 minutes and report as follows:

–	No visible precipitate
+	Mild precipitate
++	Lightly cloudy precipitate
+++	Light to moderate milky white precipitate
++++	Heavy chalky precipitate which settles down rapidly

Examination of milk for mastitis

MODIFIED WHITE SIDE TEST

1. Place 5 drops of milk in a watch glass or glass slide.
2. Add one drop of 4% NaOH and stir it for 20 seconds with a glass rod.
3. If the sample is positive for mastitis, there will be flaking and thickening.

Examination of skin samples for mites and fungus

DIRECT SMEAR METHOD

1. Treat the scraping with 10% KOH or NaOH for one hour at room temperature.
2. Transfer it with the help of a glass slide in moist conditions.
3. Triturate it with the help of a spatula or knife to get it well macerated.
4. Cover it with coverslip.
5. Examine it under low power with a microscope.

SEDIMENTATION METHOD

1. Keep the scraping in 10% NaOH or KOH solution for a sufficient time so that debris is digested or heat it gently for some time, but avoid damaging the parasites.

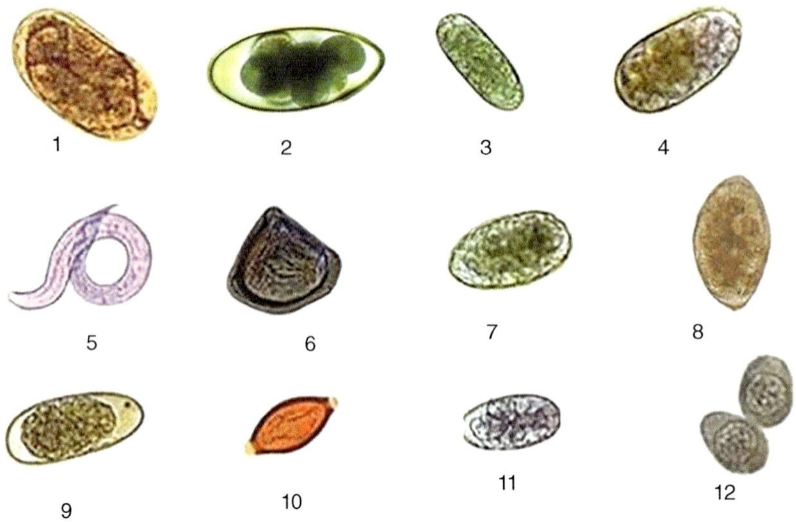

Fig. 5.27. Important parasitic egg/oocyst in faecal samples. 1. *Bunostomum* (hook worm); 2. *Nematodirus* (thread neck worm); 3. *Cooperia* (small intestine worm); 4. *Oesophagostomum* (nodular worm); 5. *Dictyocaulus* (lung worm); 6. *Moniezia* (tapeworm); 7. *Haemonchus* (barble pole worm), 8. *Fasciola hepatica* (liver fluke); 9. *Trichostrongylus*; 10. *Trichuris* (whip worm); 11. *Ostertagia* (brown stomach worm); 12. Coccidia oocyst.

2. Transfer it to a centrifuge tube and centrifuge it until the sediment settles at the bottom.
3. Discard the supernatant and make smear from the sediment and cover it with cover slip.
4. Examine it under low power with a microscope for mites and fungal spores.

Examination of rumen fluid for hydrocyanic acid

Suspected plants or ruminal contents can be tested in the field by the Henrici (picric acid) test.

TEST FOR PLANT MATERIAL

1. Plant material is placed in a test tube containing a little water and a few drops of chloroform and heated very gently in the presence of sodium picrate paper.
2. The tube should be corked and reagent paper hung from the top without touching the test material.
3. Colour of the reagent paper from yellow to red indicates the presence of free HCN.
4. Reagent papers are easily prepared by mixing 0.5 g picric acid and 5 g sodium carbonate in 100 ml water. Filter paper is dipped in the reagent and allowed to dry in a dark place.

TEST FOR HYDROCYANIC ACID IN RUMEN FLUID

1. Place a drop of ruminal fluid on a test paper.
2. Red discolouration is a positive reaction.
3. It only detects free HCN.

Protocol 9: Faecal egg count, faecal culture and parasite DNA isolation

Identification of eggs

A total of 2 g of faeces is dissolved in 20 ml of tap water and strained through an ordinary kitchen sieve. The filtrate is centrifuged at about 3000 rpm for 10 minutes. The supernatant fluid is removed and the sediment portion is shaken with saturated salt solution (sodium chloride) with a cover slip on the top. The tube is centrifuged at 2000 rpm for 10 minutes. The cover slip is gently picked up and examined under the microscope for parasitic eggs over a glass slide. The eggs can be identified by their morphological shape and size (Fig. 5.27), which will help to decide the type of anthelmintic to be administered to the goats.

METHOD FOR FAECAL EGG COUNT To obtain accurate information with regards to the severity of an infection, egg counting methods have been devised in order to determine the number of eggs per gram of faeces.

McMaster egg counting technique Fresh samples work best, but accurate results can be obtained if the sample is refrigerated during the interim period. If samples are not refrigerated, the eggs will hatch and develop larvae and the eggs will disappear from the faecal samples.

The McMaster counting chamber is made of two glass slides separated by three or four narrow transversely placed strips of glasses 1.5 mm thick, so that two or three spaces of 1.5 mm depth are obtained between the two slides.

Two grams of faeces from the collected samples are soaked or triturated in 30 ml of water until it is sufficiently soft. Then 30 ml of saturated salt solution is added. After thorough shaking, the sample is taken by means of a wide (8 mm) pipette or dropper and run in to the counting chamber, so as to fill all the spaces. The eggs are floated to the under surface of the upper slide in such a way that they are all in compound microscope focus against the ruled area. The number of eggs within the ruled area are counted and multiplied by 200, representing the number of eggs per gram of the original sample.

Precaution: The chamber should not be soaked in soapy water for long periods as this will cause the chamber to become cloudy. If the chamber gets dirty, it should be soaked with dish soap and then rinsed thoroughly with tap water.

Faecal culture Copro culture is a method to identify the type of parasitic infection in animals by culturing faecal samples. The faeces is put in a medium-sized Petri dish up to the brim level. Then each Petri dish is placed in a bigger Petri dish that is filled with a sufficient quantity of water. A few drops of 0.01% sodium azide is added to check for any fungal growth. The faeces is kept moist by sprinkling water from time to time, avoiding excessive water in it. After a week or so larvae hatch and start migrating into the water of the bigger Petri dish. The presence of a large number of larvae is observed by microscopy. For storage purposes, larvae are transferred to a 50 ml centrifuge tube and centrifuged at 2500 rpm for 10 minutes (two or three times). Larvae so procured are examined morphologically under a microscope for specific identification. The larvae collected in the form of a pellet are transferred to Eppendorf tubes containing 1 ml Tris EDTA (TE) buffer and stored at –20°C for further use.

DNA extraction:

Collection of adult worms from the abomasum

Gastrointestinal worms of *Haemonchus contortus* are collected from slaughtered or dead animals post mortem. The abomasum of the animal is opened in an enamel tray, washed with PBS solution and the content is collected from the abomasal wall. The live/dead worms are collected manually with the help of a tearing needle. The collected worms are placed in ice-cold normal saline. They are washed thoroughly three to four times by centrifugation. Lastly, the worms are kept in TE buffer at –20°C. They are kept in fraction in Eppendorf tubes for further use in DNA extraction.

Extraction of genomic DNA from Haemonchus contortus worms

1. Use worms/larvae stored –20°C.
2. Quickly transfer to a previously cooled and autoclaved mortar for grinding.
3. Grind the worms thoroughly but gently with an equal volume of lysis buffer (1 ml) with a pestle and mortar until homogenized.
4. The homogenate is transferred to a new Eppendorf tube for further processing.
5. Lysis buffer with 10 μl proteinase K enzyme (5mg/ml) is added to the Eppendorf tube.
6. Add 100 μl SDS (sodium dodecyl sulphate).
7. Incubate the Eppendorf tube at 41°C overnight to facilitate lysis.
8. After incubation, divide the homogenate into two equal volumes for further processing.
9. Add saturated phenol–chloroform in equal volume (about 0.65 ml) and centrifuge at 11 000 rpm for 10 minutes.
10. Using a micropipette, gently collect the upper layer of the homogenate containing DNA and transfer it to a new Eppendorf of tube.
11. Add equal volumes of phenol–chloroform–isoamyl alcohol (25:24:1; about 0.65 ml) to the original tubes, and centrifuge again at 11 000 rpm for 10 minutes.
12. Once again, gently collect the upper layer of the homogenate and transferred it to new Eppendorf tubes.
13. Add an equal volume (0.65 ml) of chloroform to each Eppendorf tube and centrifuge at 11 000 rpm for 10 minutes.
14. Transfer the supernatant to a new Eppendorf tube, and repeat the process of chloroform extraction and centrifugation.

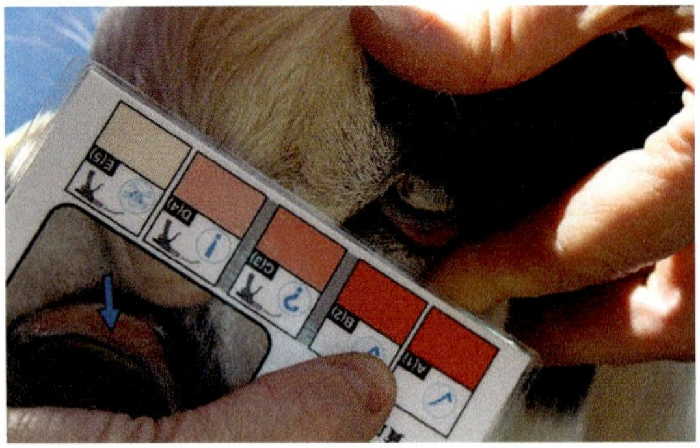

Fig. 5.28. FAMACHA card matching with eye mucous membrane colours. Source: http://www.acsrpc.org/resources/famacha.html.

15. Collect the supernatant and add chloroform and isoamyl alcohol (24:1) and centrifuge at 11 000 rpm for 10 minutes.
16. Using a micropipette, gently collect the upper layer of the homogenate containing DNA and transfer it to a new Eppendorf tube.
17. Add equal volumes of chilled propanol-2 to the Eppendorf tubes and mix gently to precipitate the DNA.
18. Centrifuge the tubes at 14 000 rpm for 10 minutes.
19. Discard the supernatant and keep the resultant pellet.
20. Wash the pellet with 70% alcohol (~200μl) and centrifuge at 11 000 rpm for 10 minutes.
21. Carefully discard the alcohol.
22. Dry the pellet for 5 minutes with the help of a DNA Speed Vac.
23. Finally, add TE buffer (~100μl) and store the DNA sample at −20°C.

Protocol 10: Eye colour test for *Haemonchus* infection

Gastrointestinal nematode (GIN) infections are an important cause of low productivity in small ruminants. Different groups of anthelmintic drugs are used in goats two to three times a year. This may lead to development of anthelmintic-resistant strains of parasites and thus limit the effectiveness of worm control programmes. It was opined that strategies that do not rely heavily on chemical treatment of entire flocks or herds are a better approach. One possibility is the use of clinical anaemia as a determinant, with subsequent selection and de-worming of only those animals that seem unable to cope with the infection. The FAffa Malan CHArt (FAMACHA) was developed with this concept in mind. It is a plastic card with five high-resolution photos showing different colours of eye mucous membranes correlating with different stages of barber pole worm infection resistance (Wyk and Bath, 2002).

Table 5.11. FAMACHA score card.

Clinical category	Colour of conjunctiva	Interpretation	Treatment guidelines
1	Red	Optimal	No
2	Pinkish-red	Acceptable	No
3	Pink	Borderline	Maybe
4	Pinkish-white	Dangerous	Yes
5	White	Fatal	Yes

How to use FAMACHA

The FAMACHA system was developed in South Africa by Dr Francois 'Faffa' Malan and is an assessment method for worm burden in livestock and the need for treatment. FAMACHA-based deworming is a low-cost, targeted treatment method that identifies and treats only those individual animals that require supportive treatment, while leaving the rest untreated.

This system helps in categorizing the anaemic status of goats and sheep based on the conjunctivae colours on a scale from 1 (red/non-anaemic) of healthy animal through shades of pink to 5 (white/severely anaemic) of an unhealthy animal. The anthelmintic treatment is recommended for goats categorized in groups 4 and 5 only, while the animals falling in categories 1 to 3 are considered normal. According to the FAMACHA system, the colour of the conjunctiva of each animal is classified into five categories and treatment guidelines are based on the colour of the conjunctiva.

The experiment that led to the development of the first FAMACHA card was conducted on a sheep farm near Badplaas, South Africa (Vatta *et al.*, 2007), in a climatic zone characterized by hot, wet summers and mild winters. Later, the system was tested with indigenous goats in South Africa, and subsequently tested and adopted in several other countries, including India (Singh and Swarnkar, 2012), Brazil, Morocco, Kenya, Italy and Switzerland. Faecal examination of both field- and farm-based herds revealed that, even during a worm-favourable monsoon season, almost 20–50% of animals had very low faecal egg count and did not require anthelmintic treatment. This is in contrast to the normal practice where the anthelmintic drug is administered to all the animals in a herd, resulting not only in a high cost of drugs and drenching, but also contributing to the emergence of anthelmintic-resistant strains of nematodes.

The FAMACHA system can be effectively implemented by smallholder livestock farmers even with low literacy levels. Further, application of the FAMACHA tool leads to a large reduction in anthelmintic use when compared to conventional drenching approaches and slows down the development of anthelmintic resistance (Wyk and Bath, 2002).

Some of the recommendations for use of the FAMACHA system are listed as follows (Kaplan *et al.*, 2004).

1. Examine the animals at least every 2–3 weeks at the beginning of the expected period of occurrence of nematode infections. During the critical period the animals must be examined on a weekly basis.
2. Adult animals in good health and body condition must be treated for the 4th and 5th category only. However, in the case of animals that are not in good health, treatment must be provided at the 3rd category.
3. Anorexic and weaker looking livestock and the ones with a submandibular oedema (bottle jaw condition) must be specifically identified and given treatment.
4. If the number of anaemic animals in a flock or herd is greater than 10%, the animals in the 3rd scale must also be treated. It is also advisable to change the pasture used by an anaemic flock.
5. Animals should be observed in direct sunlight and matched to the chart, which should always be used for matching colours. If the animal's eye colour is in between two categories, score as the lighter category (higher number).
6. Pregnant and lactating animals and animals under 1-year of age should be dewormed when they are in the 3rd category as their immune systems are not fully functional.
7. Store in a dark place when not in use because the card will fade with time.
8. Replace the card after 12 to 24 months of use (varies depending upon use and storage conditions).
9. Keep a spare card in a location protected from light (compare with the card in use).

Protocol 11: Morbidity and mortality recording formats

Format for collecting information when attending a disease outbreak

DETAILS OF FARMER:

Name:

Village:

District:

Mobile phone no.:

A. INFORMATION ABOUT FLOCK (NUMBER OF ANIMALS)

1. Total no. of goats in farm:
2. Male adult:
3. Female adult:
4. Kids less than 1 month: 1–2 months: 2–3 month: 3–6 months:

B. MANAGEMENT SYSTEM (√) TICK ONLY ONE IN GIVEN OPTION

1. Grazing only / grazing + feed supplement / fully stall fed
2. Floor type: bricks / cemented / plastic / others
3. Animals kept: all together / age-wise / sex-wise / age- and sex-wise
4. Location of animal house: near to water logging area / away from water logging area
5. Direction (long axis of animal hose): east–west / north–south
6. Total covered area: length ft / width ft
7. Timing and interval of cleaning the shelter: morning / noon / afternoon / evening / two times in a day / daily / once in two days / once in three days / others
8. Status of floor of house is wet / dry / semi-dry
9. Cleanliness of premises / house
10. Farmer obtained training for goat rearing?

Table 5.12. Morbidity and mortality pattern recording during outbreak (format).

Age of animals	Sex	No. of animals at farm	No. of sick animals	No. of animals died	Main symptoms	Duration of sickness	Treatment given	Animals recovered or still under treatment
0–1 month								
1–3 months								
3–6 months								
6–12 months								
Adult (non-Pregnant)								
Adult (pregnant)								
Adult (milking)								
Adult								
Breeding buck								
Total number of animals								

Table 5.13. Mortality recording in different age groups of goat (format).

Diseases clinical condition	0–1 month	1–3 months	3–6 months	6–12 months	Above 1 year
Diarrhoea					
Neonatal septicaemia/toxaemia					
Pneumonia					
Cold shock/hypothermia					
Weakness/inanition					
Stomatitis					
Joint ill					
Predation					
Tetanus					
Rumen acidosis					
Acute tympany/bloat					
Coccidiosis					
Foot-and-mouth disease					
Enterotoxaemia/toxaemia					
PPR					
Goat pox					
Gastrointestinal parasitic infestation					
Fascioliasis					
Ectoparasitism					
Debility /JD					
Heat stroke/hyperthermia					
Snake bite					
Plant poisoning (specify if possible)					
Chemical poisoning (specify if possible)					
Others (area-specific clinical problem)					

C. FEEDING MANAGEMENT

1. Type of feed offered to animals; the timing and amount.
2. Udder washes before suckling to kid?
3. If yes, only with water / with any antiseptic.
4. Source of drinking water to animals: river / pond / well / tube well / tank water / other (specify).

D. HEALTH MANAGEMENT

1. Recommended annual health calendar used at your farm? (Yes or No)
2. Deworming schedule followed in the last year.
3. Vaccination schedule followed in the last year.
4. Treatment given and response.

E. TRANSPORTATION HISTORY

1. Have you transported the animals in the last 30 days? (Yes or No)
2. If yes,
   - Date of transportation:
   - Number of animals:
   - Type of vehicle used:
   - Distance covered:
   - Duration of transportation:
   - Journey starting time:
   - Journey completion time:
   - Vaccinated before/after the transportation: (yes or no)
   - What feed you offered after arrival:
   - Any previous major disease during last one year? If yes, please gives details:

Format for survey on neonatal mortality (0 days to 3 months)

Name and address of farmer:
Age:
Sex:
Any special training in goat production: Yes/No
Education:
Working family member:
Primary occupation: Secondary occupation:
Land holding: Yes/No
Annual income:
Livestock reared: cattle: buffalo: sheep: goat: poultry: other:
Farming system: intensive / semi-intensive / extensive
Date of survey:

1. No. of goats reared: Breed:
2. No. of kids born in a year:
3. No. of kids died in a year:
4. No. of deaths (season wise):
   Summer: Winter: Spring: Rain:
5. No. of deaths (age-wise):
   Neonate (1–7 days): 7–30 days: 1–3 months:
6. No. of deaths (diseases wise):
   Diarrhoea: Pneumonia: Weakness: Tetanus:
   Goat pox: Poor mothering: Other:
7. Any other remarks:

Protocol 12: Standard operating procedures for biosecurity measures at goat farms

Biosecurity focuses on routine farm activities to protect herd health by limiting the transmission of infectious agents responsible for diseases. Biosecurity practices improve food safety potential and animal welfare and minimize the risk of zoonotic disease exposure. Guidelines/standard operations procedures must be followed to prevent the transmission of infections to farm animals.

Farm personnel/visitors

1. Ensure that people entering the farm, family members, visitors and service providers do not compromise the animal and human health.
2. Ensure that all farm workers are aware and understand the biosecurity practices on the farm.
3. All visitors and service providers are advised to follow biosecurity practices and should be briefed about practices to follow.
4. Ensure that visitors and service providers access only areas of the farm that are necessary.
5. Permit visitors contact with animals only when necessary.
6. Hands should be washed or sanitized before and after any contact with animals, and entry to specially identified risk areas of the farm premises such as isolation/sick pens, etc., should be restricted.
7. Hands should also be washed or sanitized before and after any contact with animals, especially those that are diseased or of unknown health status, following contact with any potentially contaminated material, such as deadstock, aborted foetuses, placentas or manure.
8. Explain and encourage the use of personal protective equipment when visitors move onto farm property. Personal cleanliness and general hygiene should be encouraged among the visitors and service providers.
9. Meet with farm workers and their family members at least twice yearly to discuss the usefulness and effectiveness of each of the practices in the biosecurity plan. Basics such as not leaving the animal area without cleaning any contaminated items such as animal excreta from their clothes, or not leaving the animal area without cleansing and disinfecting their shoes may be highlighted in such sessions.
10. Maintain a visitors' register with all required information duly entered on entry and exit from the farm.

Movement of livestock and people

1. Animals maintained at the farm should be protected by timely vaccination and other applicable preventive measures.
2. Check animals for health status before purchasing.
3. Quarantine (isolation in separate pens) for 3 weeks to ensure animals are free from underlying diseases.
4. Newly arrived goats should be routinely dewormed using an effective anthelmintic drug.
5. Complete all disease testing, treatments, procedures and vaccinations before animals are released from isolation to the farm premises.
6. Test animals returning from exhibition to determine disease status by performing serology, culture, and faecal egg counts.
7. Do not use feeding or watering equipment, penning, handling facilities or equipment from isolated animals for resident animals.
8. Ensure that a mandatory veterinary inspection of all animals attending any show/fair is conducted prior to unloading, and that any animal with evidence of disease is not unloaded.
9. Ensure that animals that have recently kidded/aborted (i.e. within the last 2 weeks) or near to kidding at the time of the show are excluded due to risk of transmission of infectious abortion diseases.
10. Transport animals in a vehicle that has been cleaned and disinfected prior to use. Ideally, the vehicle should be dedicated exclusively to the farm's use.
11. Animals that are sold and moved out of the farm should be transported as per the transportation rules, taking into account their welfare, as per country guidelines.

Feed and fodders

1. Feed and feed resources should be examined for contamination, infection or infestation, chemical residues, weed seeds and/or pests.
2. Ensure proper storage of feed, as incorrectly stored feed can deteriorate and develop fungal toxins and other unwanted disease agents.

3. Transportation of feed should be in a clean carrier, with a record of quality and related details.

Vehicles, tools and equipment

1. Restrict access of vehicles that travel from farm to farm. Vehicles should not be allowed to enter the farm from other unidentified places, unless they have been cleaned and disinfected. Cleaning and disinfection is the principle biosecurity tool for reducing vehicle-related disease risk.
2. Animal transportation vehicles should be loaded and unloaded at the periphery of the farm and animals can then be led to the isolation area.
3. Maintain a vehicle register with all required information duly entered on entry and exit from the farm.
4. Clean and disinfect tools and equipment before and after use on livestock or between different batches or herds of animals.

5.9 References

Abu-Samra, M.T., Imbabi, S.E. and Mahgoub, E.S. (1981) Mange in domestic animals in the Sudan. *Annals of Tropical Medicine Parasitology* 75, 627–637.

Akerejola, O.O., Schillhorn van Veen, T.W. and Njoku, C.O. (1979) Ovine and caprine diseases in Nigeria: A review of economic losses. *Bulletin of Animal Health Production in Africa* 27, 65–70.

Albers, G.A.A., Gray, D., Piper, L.R., Barker, J.S.F., Le Jambre, L.F. *et al.* (1987) The genetics of resistance and resilience to *Haemonchus contortus* infection in young Merino sheep. *International Journal for Parasitology* 17, 1355–1363.

Alton, G.G. (1987) Control of *Brucella melitensis* in sheep and goats: a review. *Tropical Animal Health and Production* 19, 65–74.

Andrews, A.H., Ingram, P.L. and Longstaffe, J.A. (1983) Osteodystrophia fibrosa in young goats. *Veterinary Record* 112, 404–406.

Baker, R.L. (1999) Genetic resistance to endoparasites in sheep and goats. A review of genetic resistance to gastrointestinal nematode parasites in sheep and goats in the tropics and evidence for resistance in some sheep and goat breeds in sub-humid coastal Kenya. *Animal Genetic Resources Information Bulletin* 24, 13–30.

Baker, R.L., Rege, J.E.O., Tembely, S., Mukasa-Mugerwa, E., Anindo, D. *et al.* (1998) Genetic resistance to gastrointestinal nematode parasites in some indigenous breeds of sheep and goats in East Africa. *Proceedings of the 16th World Congress on Genetics Applied to Livestock Production* 25, 269–272.

Baker, R.L., Audho, J.O., Aduda, E.O. and Thorpe, W. (2001) Genetic resistance to gastro-intestinal nematode parasites in Galla and Small East African goats in the sub-humid tropics. *Animal Science* 73, 61–70.

Banerjee, P.K., Guha, C. and Gupta, R. (1987) A note on incidence of *Babesia motasi* infection in a goat in West Bengal. *Indian Veterinary Journal* 64, 71–73.

Barratt-Boyes, S.M. and MacLachlan, N.J. (1994) Dynamics of viral spread in bluetongue virus infected calves. *Veterinary Microbiology* 40(3–4), 361–371.

Barua, S., Vaid, R.K., Rana, R. and Kumar, A. (2005) Pox in sheep and goats: pock marks in small ruminant health. Book on Infectious diseases of domestic animals and zoonoses in India. *National Academy Sciences India* – special issue 2005, 41–54.

Bath, G.F., van Wyk, J.A. and Pettey, K.P. (2005) Control measures for some important and unusual goat diseases in southern Africa. *Small Ruminant Research* 60, 127–140.

Bishop, S.C. and Morris, C.A. (2007) Genetics of disease resistance in sheep and goats. *Small Ruminant Research* 70, 48–59.

Brounts, S.H., *et al.* (2004) Outcome and subsequent fertility of sheep and goats undergoing cesarean section because of dystocia: 110 cases (1981–2001). *Journal of the American Veterinary Medicine Association* 224, 275–279.

Bulgin, M.S., Lincoln, S.D., Parker, C.E., South, P.J., Dahmen, J.J. *et al.* (1988) Genetic associated resistance to foot rot in selected Targhee sheep. *Journal of the American Veterinary Medical Association* 192, 512–515.

Byrne, D.J. (1992) Vitamin E deficiency in a flock of Angora goats. *Goat Veterinary Society Journal* 13, 55–62.

Calamel, M.M. and Giauffret, A. (1975) An eUne enzootie de toxoplasmose caprine abortive. *Bulletin de l'Academie Véterinaire de France* 48, 41–45

Chhabra, R.C., Gill, B.S. and Dutt, S.C. (1978) Paramphistomiasis of sheep and goats in the Punjab state and its treatment. *Indian Journal of Parasitology* 2, 43–45

Clunies-Ross, I. (1932) Observations on the resistance of sheep to infestation by the stomach worm (*Haemonchus contortus*). *Journal Council for Scientific and Industrial Research (Australia)* 5, 73–80.

Coetzee, P., Stokstad, M., Venter, E.H., Myrmel, M. and Vuuren, M.V. (2012) Bluetongue: a historical and epidemiological perspective with the emphasis on South Africa. *Virology Journal* 9, 198.

Constable, P.D., Hinchcliff, K.W., Done, S.H. and Grunberg, W. (2017) *Veterinary Medicine: a textbook of the diseases of cattle, horses, sheep, pigs, and goats*, Eleventh edition. Elsevier, St. Louis, Missouri, USA.

Cork, L.C., Hadlow, W.J., Crawford, T.B., Gorham, J.R. and Piper, R.C. (1974) Infectious leukoencephalomyelitis of young goats. *The Journal of Infectious Disease* 129, 134–141.

Dargie, J.D. (1982) The influence of genetic factors on the resistance of ruminants to gastro–intestinal nematode and trypanosome infections. In: Owens, D.G. (ed.) *Animal Models in Parasitology*. Macmillan Education, London, 17–51.

Dercksen, D., Groot Nibbelink, N., Paauwe, R., Backx, A., van Rijn, P.A. *et al.* (2007) First outbreak of bluetongue in goats in the Netherlands. *Tijdschrift Voor Diergeneeskunde* 132, 786–790.

Detweiler, G., Gipson, T., Merkel, R.C., Goetsch, A. and Sahlu, T. (2008) Body Condition Scores in Goats. Proceedings of the 23rd Annual Goat Field Day, Langston University, Langston, Oklahoma, USA, pp. 127–133.

Dupuy, J., Chartier, C., Sutra, J.F. and Alvinerie, M. (2001) Eprinomectin in dairy goats: dose influence on plasma levels and excretion in milk. *Parasitology Research* 87(4), 294–298.

Ferrando, R. and Barlet, J.-P. (1978) Vitamines. In: Jarrige, R. (ed.) *Alimentation des Ruminants*. INRA Publications, Versailles, France, pp. 161–176.

Franklin, J.S. (1986) Retained placenta, metritis and pyometra. In: Morrow, D.A. (ed.) *Current Therapy in Theriogenology*, vol. 2., W.B. Saunders Co., Philadelphia, 595 pp.

Freeman, B.M. and Bumstead, N. (1987) Breeding for disease resistance – the perspective role of genetic manipulation. *Avian Pathology* 16, 353–365.

Gavora, J.S. and Spencer, J.L. (1983) Breeding for immune responsiveness and disease resistance. *Animal blood groups and biochemical genetics* 14(3), 159–180.

Grenfell, B.T. and Dobson, A.P. (1995) *Ecology of Infectious Diseases in Natural Populations*. Cambridge University Press, UK.

Griffin, L. and Allonby, E.W. (1979a) Studies on the epidemiology of trypanosomiasis in sheep and goats in Kenya. *Tropical Animal Health and Production* 11, 133–142.

Griffin, L. and Allonby, E.W. (1979b) Trypanotolerance in breeds of sheep and goats with an experimental infection *Trypanosoma congolense*. *Veterinary Parasitology* 5, 97–105.

Gupta, V.K., Verma, D.K., Rout, P.K. and Singh, S.V. (2006) Polymerase chain reaction (PCR) for detection of Brucella melitensis in goat milk. *Small Ruminant Research* 65, 79–84.

Gutierrez, C., Corbera, J.A., Morales, M. and Büscher, P. (2006) Trypanosomosis in goats: current status. *Annals of the New York Academy of Sciences* 1081, 300–310.

Harrison, R. and Palmer, B.H. (1981) Further studies on amitraz as a veterinary acaricide. *Journal of Pesticide Science* 12, 467–474.

Horak, I.G. (1971) Paramphistomiasis of domestic ruminants. *Advances in Parasitology* 9, 33–72.

Hutt, F.B. (1958) *Genetic Resistance to Disease in Domestic Animals*. Comstock Publishing Associates, Ithaca, New York.

Ikegami, T., Hill, T.E., Smith, J.K., Zhang, L., Juelich, T.L., Gong, B., Slack, O.A.L., Ly, H.J., Lokugamage, N. and Freiberg, A.N. (2015) Rift Valley fever virus MP-12 vaccine is fully attenuated by a combination of partial attenuations in the S, M, and L segments. *Journal of Virology* 89, 7262–7276.

Jagannath, M.S., Hegde, K.S., Shivaram, K. and Nagaraja, K.V. (1974) An outbreak of babesiosis in sheep and goats and its control. *Mysore Journal of Agricultural Sciences* 8(3), 441–443.

Kambarage, D.M. (1992) Sarcoptic mange infestation in goats. *Bulletin of Animal Health and Production in Africa* 40, 239–244.

Kaplan, R.M., Burke, J.M., Terrill, T.H., Miller, J.E., Getz, W.R., Mobini, S., *et al.* (2004) Validation of the FAMACHA eye color chart for detecting clinical anemia in sheep and goats on farms in the southern United States. *Veterinary Parasitology* 123, 105–120.

Katiyar, R.D. and Varshney, T.R. (1963) Amphistomiasis in sheep and goats in Uttar Pradesh. *Indian Journal of Veterinary Science and Animal Husbandry* 33, 94–98.

Kennedy, B.W. and Moxley, J.E. (1980) Genetic factors influencing atrophic rhinitis in the pig. *Animal Production* 30, 277–283.

Kimberling, C.V. (1988) *Jensen and Swift's Diseases of Sheep*, 3rd edn. Lea & Febiger, Philadelphia, Pennsylvania, USA.

King, N.B. (1980) Pregnancy toxemia and enterotoxemia. In: *Proceedings, Refresher Course for Veterinarians, Post-Graduate Committee in Veterinary Science*, University of Sydney, vol. 52, pp. 209–212.

Kitching, R.P. (1991) Genetic control of acute viral disease. In: Owen, J.B. and Axford, R.F.E. (eds) *Breeding for Disease Resistance in Farm Animals*. CAB International, Wallingford, UK, pp. 315–327.

Kitching, R.P., Hammond, J.M. and Taylor, W.P. (1987) A single vaccine for the control of capripox infection in sheep and goats. *Research in Veterinary Science* 42, 53–60.

Kramer, J.W. (1966) Incidence of trypanosomes in the West African Dwarf sheep and goat in Nsukka, eastern Nigeria. *Bulletin of Epizootic Diseases of Africa* 14, 423–428.

Kuil, H. and Folkers, C.(1966) Een zeer anaemische geit. *Tijdschr.Diergeneesk* 91, 1247.

Kumar, A., Singh, S.V., Rana, R., Vaid, R.K., Misri, J. *et al.* (2001) PPR outbreak in goats: Epidemiological and therapeutic studies. *Indian Journal of Animal Sciences* 71(9), 815–818.

Kumar, A., Barua, S., Rana, R. and Vihan, V.S. (2004) Epidemiological behavior of foot and mouth disease virus (type O) in goats. *Veterinary Practitioners* 5, 134.

Lundeheim, N. (1979) Genetic analysis of respiratory diseases in pigs. *Acta agriculturae Scandinavica* 20, 209–215.

Mallick, K.P., Dwivedi, S.K. and Malhotra, M.N. (1979) Anaplasmosis in goats: report of clinical cases. *Indian Veterinary Journal* 56(8), 693–694.

Mandonnet, N., Aumont, G., Fleury, J., Arquet, R., Varo, H. *et al.* (2001) Assessment of genetic variability of resistance to gastrointestinal nematode parasites in Creole goats in the humid tropics. *Journal of Animal Science* 79, 1706–1712.

Matheron, G., Barre, N., Camus, E. and Gogue, J. (1987) Genetic resistance of Guadeloupe native goats to heart water. *Onderstepoort Journal of Veterinary Research* 54, 337–340.

Matthews, J.G. (1999) *Diseases of Goat*, 2nd edn. Blackwell Science Ltd, Hoboken, New Jersey, USA.

Menzano, A., Rambozzi, L. and Rossi, L. (2007) A severe episode of wildlife-derived scabies in domestic goats in Italy. *Small Ruminant Research* 70, 154–158.

Molina, A., Delgado, J.V., Madueno, R., Morera, L. and Llanes, D. (1991) Paratubercuosi incidence in different goat breeds; Canary Island breed, a case of genetic resistance. In: Owen, J.B. and Axford, R.F.E. (ed.) *Breeding for Diseases Resistance in Farm Animals*. Proceedings of an International symposium, Bangor, Wales, Sept. 1990. CAB International, Wallingford, UK, pp. 479–480.

Morris, C.A., Wheeler, M., Hosking, B.C., Watson, T.G., Hurford, A.P., Foote, B.J. and Foote, J.F. (1997) Genetic parameters for milk yield and faecal nematode egg count in Saanen does. *New Zealand Journal of Agricultural Research* 40, 523–528.

Muller, M and Brem, G. (1991) Disease resistance in farm animals. *Experientia* 47, 923–934.

Okoh, A.E.J. (1981) An epizootic of anthrax in goats and sheep in Danbatta Nigeria. *Bulletin of Animal Health and Production in Africa* 29, 355–359.

Osman, T.M., Ali, A.M., Hussein, M.O., El Ghali, A. and Salih, D.A. (2017) Investigation on *Theileria lestoquardi* infection among sheep and goats in Nyala, South Darfur State, Sudan. *Insights in Veterinary Sciences* 1, 17–23.

Pawaiya, R.V.S. and Shivasharanappa, N. (2014) Post-mortem techniques of goat and sheep. In: Kumar, A., *et al.* (eds) *Training Manual on Advances in Goat Production and Health*. CIRG Publications, Makhdum, Uttar Pradesh, India, p. 140.

Payne, L.N. (1985) Genetics of cell receptors for avian retroviruses. In: Hill, G.W., Manson, J.M. and Hewitt, D. (eds) *Poultry Genetics and Breeding*. British Poultry Science Series vol. 18. CAB International, Wallingford, UK.

Perry, B.D., McDermott, J.J., Randolph, T.F., Sones, K.R. and Thornton, P.K. (2002) *Investing in Animal Health Research to Alleviate Poverty*. International Livestock Research Institute (ILRI), Nairobi, Kenya.

Philipsson, J., Thafvelin, B. and Hedebro-Velander, I. (1980) Genetic difficulty in Swedish Holsteins at first and second calving. Studies on disease recordings in first lactation cows of Swedish dairy breeds. *Acta Agriculturæ Scandinavica Section A, Animal Science (Stockholm)* 30, 327–335.

Pralomkarn, W., Pandey, V.S., Ngampongsai, W., Choldumrongkul, S., Saithanoo, S. *et al.* (1997) Genetic resistance of three genotypes of goats to experimental infection with *Haemonchus contortus*. *Veterinary Parasitology* 68, 79–90.

Prince, A.M. (1958) Quantitative studies on Rous sarcoma virus on chorioallantoic membrane of chick embryo. *Journal of National Cancer Institute* 20, 147–157.

Purnell, R.E. (1981) Tick-borne diseases. *The British Veterinary Journal* 137, 221–240.

Radwan, A.I., Bekairi, S.I. and Mukayel, A.A. (1992) Treatment of Brucella melitensis infection in sheep and goats with oxytetracycline combined with streptomycin. *Revue Scientifique et Technique* 11, 845–885.

Robson, J. and Rickman, L.R. (1973) Blood incubation infectivity test results for *Trypanosoma brucei* subgroup isolates in the Lambwe Valley, South Nyanza, Kenya. *Tropical Animal Health Production* 5, 187–191.

Ross, J.D. (1981) Herd health program for Angora goats. In: Howard, J.L. (ed.) *Current Veterinary Therapy, Food Animal Practice*. WB Saunders Co., Philadelphia, Pennsylvania, USA.

Rout, P. K., Chauhan, K.K., Matika, O. and Bishop, S.C. (2011) Exploring the genetic resistance to natural gastrointestinal nematode infection in Indian goats. *Veterinary Parasitology* 180, 315–322.

RIRDC (2005) Proceedings of Australian Cashmere Workshop. Rural Industries Research and Development Corporation, Wagga Wagga, New South Wales, Australia.

Sapre, S.N. (1964) An outbreak of blue tongue in goats and sheep in Maharashtra state, India. *Veterinary Review* 15, 69–71.

Schat, K.A. and Davies, C.J. (2000) Viral diseases. In: Axford, R.F.E., Bishop, S.C., Nicholas, F.W. and Owen, J.B. (eds) *Breeding for Disease Resistance in Farm Animals* 2nd edn. CAB International, Wallingford, UK, pp. 271–300.

Schmidt, H. (1941) Vitamin A deficiencies in ruminants. *American Journal of Veterinary Research* 2, 373–389.

Scott, D.W. (2007) *Color Atlas of Farm Animal Dermatology*. Blackwell Publishing, Ames, Iowa.

Sharma, D.K. and Chauhan, P.P.S. (2006) Coenurosis status in Afro-Asian region: A review. *Small Ruminant Research* 64(3), 197–202.

Sharma, H.N. and Tyagi, R.P.S. (1975) Diagnosis and surgical treatment of coenurosis in goat. *Indian Veterinary Journal* 52, 482–488.

Shanks, P.I. (1949) Enterotoxaemia in goats. *Veterinary Record* 61, 262–263.

Shiferaw, G. (2004) Anthrax in Wabessa village in the Dessie Zuria district of Ethiopia. *Revue scientifique et technique (International Office of Epizootics)* 23(3), 951–956.

Singh, D. and Swarnkar, C.P. (2012) Evaluation of targeted selective treatment strategy in sheep farm in

Rajasthan. *Indian Journal of Animal Sciences* 82(7), 679–686.
Singh, N., Kala, S.N., Vihan, V.S. and Singh, S.V. (1990) Genetic study on the susceptibility to Johne's disease in goats. *Indian Journal of Animal Science* 60, 1163–1165.
Singh, D.D, Pawaiya, R.V.S., Gururaj, K., Gangwar, N.K., Mishra, A.K., Singh, R., Andani, D. and Ashok, K. (2018) Detection of *Clostridium perfringens* toxinotypes, enteropathogenic *E. coli*, rota and corona viruses in the intestine of neonatal goat kids by molecular techniques. *Indian Journal of Animal Sciences* 88(6), 655–661.
Smith, M.C. and Sherman, D.M. (1994) *Goat Medicine*. Lea & Febiger Publications, Philadelphia, Pennsylvania, USA.
Smith, M.C. and Sherman, D.M. (2009) *Goat Medicine*, 2nd edn. Wiley and Blackwell, Ames, Iowa, USA.
Songer, J.G. (1996) Clostridial enteric diseases of domestic animals. *Clinical Microbiology Review* 9(2), 216–234.
Stewart, M.A., Miller, R.F. and Douglas, J.R. (1937) Resistance of sheep of different breeds to infestation by Ostertagia circumcincta. *Journal of Agricultural Research*. 55, 923–930
Sumberg, J.E. and Mack, S.D. (1985) Village production of West African dwarf goats and sheep in Nigeria. *Tropical Animal Health and Production* 17, 135–140.
Thiaucourt, F. and Bölkse, G. (1996) Contagious caprine pleuropneumonia and other pulmonary mycoplasmoses of sheep and goats. *Scientific and Technical Review of the Office International des Epizooties (Paris)* 15, 1397–1414.
Uzal, F.A., Plumb, J.J., Blackall, L.L., O'Boyle, D. and Kelly, W.R. (1996) Detection by polymerase chain reaction of *Clostridium perfringens* producing epsilon toxin in faeces and in gastrointestinal contents of goats. *Letters in Applied Microbiology* 23(1), 13–17.
Vagenas, D. (2002) *Genetic and Quantitative Aspects of Resistance to Gastrointestinal Nematode Parasites in Small Ruminants*. PhD Thesis, University of Edinburgh, Edinburgh, UK.
Van Tonder, E.M. (1975) Notes on some disease problems in Angora goats in South Africa. *Veterinary Medical Reviews* 1/2, 109–138.
Van der Westhuysen, J.M., Wentzel, D. and Grobler, M.C. (1988) *Angora Goats and Mohair in South Africa*, 3rd edn. NMB Printers, Port Elizabeth, South Africa.
Vatta, A.F., de Villiers, J.F., Gumede, S.A., Krecek, R.C., Mapeyi, N.P., Pearson, R.A., *et al.* (2007) Benefits of urea-molasses block supplementation and symptomatic and tactical anthelmintic treatments of communally grazed indigenous goats in the Bulwer area, KwaZulu-Natal Province, South Africa. *Journal of South African Veterinary Association* 78(2), pp. 81–89.
Verster, A. and Tustin, R.C. (1990) Treatment of cerebral coenuriosis in sheep with praziquantel. *Journal of South African Veterinary Association* 61(1), 24–26.
Waller, P.J. (1999) International approaches to the concept of integrated control of nematode parasites of livestock. *International Journal for Parasitology* 29, 155–164.
Warren, A.L., Uzal, F.A., Blackall, L.L. and Kelly, W.R. (1999) PCR detection of Clostridium perfringens type D in formalin fixed, paraffin-embedded tissues of goats and sheep. *Letters in Applied Microbiology* 29(1), 15–19.
Whitford, H.W. (1982) Anthrax in dairy goats. *Southwestern Veterinarian* 35, 15.
World Organization of Animal Health (2019) OIE-Listed diseases, infections and infestations in force in 2019. Available at: http://www.oie.int/en/animal-health-in-the-world/oie-listed-diseases-2019 (accessed 2 May 2019).
Woolaston, R.R., Elwin, R.L. and Barger, I.A. (1992) No adaptation of Haemonchus contortus to genetically resistance sheep. *International Journal for Parasitology* 22, 377–380.
Wyk, J.A. and Bath, G.F. (2002) The FAMACHA system for managing haemonchosis in sheep and goats by clinically identifying individual animals for treatment, *Veterinary Research* 33, 509–529.
Braun Jr, W. (2007) Periparturient infections and structural abnormalities. In: Youngquist, R.S. and Threlfall, W. *Current Therapy in Large Animal Theriogenology*, 2nd edn. Elsevier Inc., Philadelphia, Pennsylvania, USA.
Yousif, Y.A., Dimitri, R.A., Dwivedi, S.K. and Ahmed, N.J. (1983) Anaemia due to anaplasmosis in Iraqi goats: I. Clinical and haematological features under field conditions. *Indian Veterinary Journal* 60, 576–578.

6 Herd Management

6.1 Housing Systems and Principles

Environmental factors such as heat, cold, humidity, solar radiation and wind speed have negative impacts on the productivity of goats. Goats can adjust to various environmental conditions; however, they should be protected from sun, rain and wind. Goats are distributed in diverse macroclimates in the tropics and have exhibited a remarkable ability to adapt to varying climatic conditions. Housing systems vary according to the production systems. The housing environment collectively covers factors such as physical, psychological and biological welfare and health of the animal. Housing requirements will be different for specific production systems, such as extensive, intensive and subsistence or pastoral agricultural systems. The environmental conditions prevailing in cold climates favour intensive goat production systems, which may require a high standard of housing with good insulation, slatted or straw-bedded floors and efficient ventilation. Similarly, goat houses in hot, dry environments require big open yards with shade, which reduces the radiant heat load on the animals. Goat houses in the hot, humid tropics are characterized by slatted floors raised above the ground and space boarding on the side walls to promote good ventilation, ease of cleaning and better parasitic control. Houses with well-drained and clean floors and yards are very important in humid climates. In semi-intensive and intensive production systems, decisions concerning the location of the building, the internal layout of the various compartments, the feeding, watering and milking facilities to be used, the feed storage space required, and the roughage and manure-handling system to be adopted are all important aspects of goat housing.

6.1.1 Housing design

Housing design should be based on the climatic conditions and production system. Climatic data can be used as a guide for the type of housing, insulation and ventilation. The housing design will depend on several factors, for example, the availability and cost of building material in the region. Housing design is an interaction between engineers, building material suppliers, livestock specialists and regulatory agencies. The housing design should be workable, affordable and spacious. A detailed cost analysis is required before starting the construction for housing and shelter.

Housing for goats depends on factors such as climatic requirements, animal comfort and welfare (Whates and Charles, 1994). All these factors directly affect the health and performance of the animal. Animal welfare is now a regulatory issue and tropical countries have to adopt these norms. Livestock production is becoming more consumer driven and has to address food safety and animal welfare issues. It is also necessary to define how the animals are raised and details of the production environment. The produce such as meat and milk should not contain any drug residues, insecticides and microbial load. Therefore, the production environment will play a vital role in the profitability and income of farmers in the future.

A comfortable house for goats should provide facilities for drinking, eating and expressing the animal's natural behaviour, such as jumping, running and fighting. The material used in goat housing should not cause injury to the animals. Goats must exhibit natural behaviour in confinement, which should be encouraged by the farmer. The welfare of an animal is a state of physical and mental health. Goats are friendly and readily approach human beings and are known as wandering animals. The type of housing structure used depends on type of goats being housed (i.e. meat, dairy or fibre) as it is necessary to modify the farm structures to fit requirements. Dairy goats will need an area for milking and milk storage, and fibre-producing goats will require an area for shearing, fibre collection and storage. If the farm is being used for demonstration and training purposes, all necessary facilities for biosecurity measures and safe visitor passage should be developed.

Climatic factors

Goats perform better in arid and semi-arid areas and the majority of goats are found in these regions of the world. Goats seem to be less tolerant to hot, humid conditions as compared to other climatic conditions. Tropical goat production areas with hot, humid climates need to develop specific management criteria for optimum productivity. The rate of metabolism depends on body size, feed intake and muscular activity, and therefore it is necessary to consider the influence of metabolic heat production in the body. Animals lose heat through radiation or non-evaporative heat transfer from the body or by evaporative heat transfer and respiration. Non-evaporative heat loss occurs at low temperatures and depends on coat colour and skin thickness. Evaporative heat loss occurs at high temperatures.

The climatic factors that affect the animal's performance include temperature, wind direction, humidity, concentration of CO_2 and NH_3. Animals become more productive around the thermoneutral zone, when metabolic heat production and energy expenditure are low. Animals use energy to regulate body temperature when it is above or below the critical temperature. Optimal productivity is achieved when metabolic heat production and energy expenditure are low, for example, dairy goats perform better in temperatures below the lower critical temperature. The limit of thermoneutrality for goats is considered to be environments with an air temperature in the range of 13–27°C, relative humidity of 60–70%, a wind velocity of 5–8 km/h, and a medium level of solar radiation (Trodahl *et al.*, 1981; Mennella and Borghi, 1983).

Ventilation

Ventilation is an important factor as it removes ammonia and CO_2 from an animal house. Ventilation is required to maintain temperature and remove gases, dust or microbial growth. The ammonia level should be below 25 ppm and CO_2 should be less than 0.3% in the animal house. Lactating and growing animals in humid climatic conditions require higher rates of ventilation. The ventilation rate will depend on the ambient temperature as they are directly proportional to each other.

Choice of site for housing

Wind direction should be taken into account when selecting the location for the animal house. There should be effective soil drainage at the selected site as poor drainage leads to flooding and foot and parasite problems. The slope of the site is also very important as it will affect the water run-off during heavy rain. There should be enough space for future expansion. The proposed site should also have access to road, water, utilities and services.

Space requirements

The size and natural behaviour of goats must be taken into consideration when finalizing the space requirements. This includes their natural tendency to wander and their bipedal stance, along with resting, feeding and socializing. Confinement can result in seasonal parasitic and coccidiosis problems in goat sheds. As a general rule, the space requirement will be around 1.4 m^2 (15 ft^2) for adult goats in the tropics. The space requirement is more for concrete floors as compared to elevated slatted floors. An additional area of 2.3 m^2 (25 ft^2) of well-drained and fenced area is required for exercise. It is important to provide adequate space for goats at feeding troughs. The incidence of displacement and aggressive behaviour is increased when there is less space at feeding troughs (Boe *et al.*, 2008). Low-ranked goats spend more time waiting to access the food troughs and are unable to eat sufficient feed, which will result in low growth rates.

The floor space for different age groups should be slightly more in hot weather conditions as compared to cool weather (Table 6.1). In tropical climates, huddling is disadvantageous to the health and productivity

Table 6.1. Optimum floor space requirements for different categories of goats.

Category/age of goats	Floor space (m^2/goat)	
	Covered space	Open space
0–3 months	0.20–0.25	0.40–0.50
3–6 months	0.5–0.75	1.00–1.50
6–12 months	0.75–1.00	1.50–2.00
Adult	1.50	3.00
Bucks, pregnant and lactating does	1.50–2.00	3.00–4.00

Source: Misra (2008).

of the animals. The knowledge of air space requirement per animal is important, because the cleanliness of ambient air with regard to airborne waste from animals is directly proportional to the actual volume of air into which that waste is dispelled. Similarly, the concentration of pollutants is directly proportional to stocking density.

6.1.2 Flooring

The floor plan and construction depends on the climate, production system and behaviour of the goats. The surface upon which an animal lies should not be a source of thermal or physical discomfort, injury and infectious diseases. The ideal floor needs to be hygienic, dry and reasonably temperature resistant. The floors should not be slippery – the ideal floor material is wood, plastic or metal. Wooden floors should be made of hard woods and should not have any cracks. A clay and sand floor is low maintenance and easy to clean. Concrete floors may be hard on the feet but they are easy to clean and disinfect. They can be bedded with sand, straw, sawdust or other absorbent materials. Dry straw can be used as ideal bedding for weaners and growers during the cool period, but a thin layer of straw is likely to be more suitable during warm or hot weather conditions. Perforated or slatted floors are mostly drier and more hygienic than solid floors with limited bedding, and are thus more suitable for humid and high rainfall areas (Devendra, 1964). On the basis of time spent by the animal in various activities on different types of floors, it was observed that the highest preference was for slatted wooden floors followed by slatted plastic (Constantinon, 1987). Slatted or raised floors have low labour requirements and provide a clean and healthy environment with low parasitic loads. Raised floors have a management advantage due to superior rain run-off and air movement and reduced parasitic load and contamination from faeces and urine. Mastitis cases are more frequent in dairy goats housed in concrete-floored sheds. A deep-layer litter system on the floor increases the chances of paratuberculosis and brucellosis.

6.1.3 Roof and wall construction

The shape of the roof will be determined by the levels of solar radiation and rainfall. Roofing material in the shed is an integral component that determines the cost of construction and the micro-environment within the shed. A wide range of roofing materials with different thermal characteristics are available for diverse climates. Thatched roofs are common in the tropics for both smallholder and organized goat farms. White washing (with lime and white cement in equal parts) or painting the roof surface white can reduce the temperature by several degrees (Bond *et al*., 1954). Building materials should have a black surface to reduce the reflection of radiation from the hot ground and surrounding objects, and the exterior walls should be painted in dull colours. Red mud plastic roofing sheets, polycarbonate sheets and polyvinyl chloride sheets have good insulation values and are easily available. Thin thatch panels may cover galvanized iron sheets to make them more effective.

The cost of improved thatch is about 70–80% more than ordinary thatch, but it may be better in the long term because it is fire and water retardant and it has longer durability. It has been suggested that it would be disadvantageous to use highly reflective materials as roofing because of the reflection of short- and long-wave radiation back to the housed stock. A roof with a dark colour on the internal surface is known to reduce reflection, which is a desirable feature. The outer surface of roofing materials will absorb less heat if it has a low absorptivity for short-wave radiation and a high emissivity. The inside coating of the shed will reduce the radiation load on the housed goats if it has a high short-wave absorbance and a low emissivity. The radiation load on goats in a house may, therefore, be reduced by use of appropriate materials for roofing.

6.2 Housing Facilities

6.2.1 Housing pattern

Different types of goat houses are constructed in different agro-climatic conditions by using locally available plant material and other resources. Principally, a sheltered area and a big open yard are primary requirements for comfortable rearing in housing. In semi-intensive and intensive production systems, decisions concerning the location of the building, the internal layout of the various compartments (e.g. feeding, watering and milking facilities), the feed storage space required, and the roughage- and manure-handling systems to be adopted are all important factors affecting the success of goat production. In dry arid regions, housing is constructed using bricks, cement and iron bars to provide sufficient protection from heat and cold winds (Fig. 6.1). Some houses are constructed with bamboo, and have an open space for ventilation during hot periods (Fig. 6.2). In coastal areas with more rainy days, goats are forced to graze on wet surfaces. In such climatic zones, the goats are reared on raised wooden slatted floors (Fig. 6.3). Again, the housing and resting place in an island region is provided with protection from rain and good ventilation (Fig. 6.4). An example of open housing in the cold desert of Ladakh is shown in Fig. 6.5. A simple thatched housing system in a dry region is shown in Fig. 6.6.

6.2.2 Housing facilities for different requirements

Goats thrive in outdoor environments and depend on browsing and grazing. Therefore, total confinement of goats causes thriftiness and leads to a general decline in health. Goats that are regularly rotated have fewer health problems with respect to internal parasites. Wet conditions also make the goats vulnerable to illness. Goat housing needs good ventilation for avoiding moisture accumulation, especially during the winter season. A dairy operation should include provisions for milking and milk storage. For meat goats, a simple wooden structure with an open access works well in most cases. Fibre goats require an area for shearing and fibre collection, grading, sorting and storage. There should be a separate facility for hoof care as fibre-producing goats require hoof trimming more often than other breeds. Wet conditions can also damage the quality of the fibre/hair. Mohair goat breeds require more maintenance and the utmost care. There should be a clean and dry storage facility for storing fibre to ensure good quality for better market prices.

There should be a separate clean and soft-bedded pen area for pregnant animals to avoid any traumatic injury from nearby goats, which may lead to risk of abortion. Goats in the advanced stages of pregnancy should be transferred to a kidding pen in a timely manner for their safety. The bedding should be replaced daily if possible.

Fig. 6.1. An example of a house constructed using bricks, cement and iron bars with sufficient protection from the heat and cold winds in dry arid regions.

Fig. 6.2. Open-space bamboo house for ventilation during hot periods.

Similarly, lactating goats should be housed in a well-ventilated area with a dry floor as wet floor conditions are the predisposing factors for mastitis and skin infections. As broken fence wiring can tear or lacerate the udder, careful and continuous monitoring of the fencing is important. There should be a facility for suckling of kids. Buck housing must be separated from the milking herd and in the opposite direction to get free access to an exercise lot. The space requirement should be adequate based on body frame, behaviour and temperament.

6.3 Working Facilities

Working facilities are helpful when performing routine management tasks such as handling, weighing, drenching, vaccinating, dipping, hoof trimming and loading. The required facilities need to be developed, otherwise, these jobs often get delayed or overlooked.

- **Breeding pens:** The breeding of does should be carried out in a controlled environment. The size of the pens or pastures will depend on the number of does breeding at that time. The pens should be sufficient to allow free movement of 10–20 does.
- **Buck pens:** This refers to the areas that house bucks or breeding sires and should allow adequate space for exercise.
- **Milking pen or parlour:** The most popular design is parallel-side openings as it allows individual animal handling for milking. It is therefore possible to identify any problems in an individual doe and recording the milk yield is easy. The milking parlour should have concrete flooring for easy cleaning. The milking platform should be 22.5–45 cm higher than the floor to permit easy milking. The platform should allow 45 cm in width and 105cm in length for each animal to be tied. Milk must be cooled immediately after milking and stored at a temperature of 4–8°C until processed or consumed. Cold water is more efficient than cold air for cooling milk. A refrigerator/cooler or a bulk tank cooler is needed for keeping large amounts of milk (Belanger, 1990).

Fig. 6.3. Examples of goat housing with raised slatted floors made of wood or plastic in high rainfall areas.

- **Milk stand:** Mobile milk stands are required for milking the goats during exhibitions, fairs and competition.
- **Milk house:** The milk house should have a hot water heater, refrigerator, a table for work to take place, a rack for drying and storing utensils, and a cooling tank.
- **Kidding complex:** The kidding complex should have a separate facility to protect the kids from rain, sun and dust. The kidding pen should have a parturition kit containing autoclaved scissors, blades and povidone iodine for navel cord cutting. Each individual kidding pen should have adequate bedding and feeding facilities. A suckling and milking facility should be attached to the kidding shed and it should be provided with minerals and water.
- **Weaning pens:** These pens are designed especially for weaned kids as the feeding requirement is completely different from other goats. The weaning kids should be provided higher digestible protein with high fibre content. Shelter, feed, water and minerals should be provided adequately in a weaning pen.
- **Post-mortem house:** May be placed next to an isolation ward. It should have a concrete floor with a post-mortem table.
- **Burial pit:** The burial pit should be located in an area away from the sheds, where infected carcasses will be buried.

Fig. 6.4. Example of a house for protection from rain with good ventilation in an island region.

Fig. 6.5. Examples of open housing in a cold desert region of Ladakh, India.

Fig. 6.6. An example of a simple thatched house in a dry region.

- **Dipping bath tank**: The dipping tank is required to dip the goats regularly to control external parasites. It should be located away from the goat shed and the routine migratory route.
- **Quarantine area:** A separate isolated area should be kept as a quarantine shed to keep the flock free from infections from new arrivals. Quarantine pens should be outdoors with proper shelter, handling and feeding facilities. The first pen is for general treatment and deworming or vaccination, if any. The goats may then be kept in the next pen for between 21–60 days for observation of any sign of diseases. They should be provided with good quality water, green fodder and dry fodder, and grains should be avoided for the first few weeks to avoid overfeeding. Goats in stressful conditions are more prone to coccidiosis, so an approved coccidiostat may be used in the feed. Proper biosecurity measures should be followed during handling of the animals. The handling person should not be allowed to work with the general flock during the quarantine period.
- **Sick animal ward:** This area should have at least one or two pens with adequate facilities of water, feed and minerals. The shed should be located in the buffer zone.

6.4 Utility Facilities

6.4.1 Designing the appliances/equipment

Equipment includes all types of feeders, waterers, corrals and other working facilities such as milking, fibre shearing, milk storage, weighing and manure disposal. The equipment must be easy to use and be fit for purpose. Feeders, waterers, corrals, portable plastic partitions, milking pales, milk storage, mineral feeders, water dispensers and cleaning equipment are all essential for goat farming. The equipment should be disinfected on a regular basis to prevent the spread of infectious organisms.

6.4.2 Feeding facilities

Villagers maintaining small flocks tend to use discarded metallic or earthenware utensils for feeding concentrate and dry roughage (straw) to their animals (Fig. 6.7), whereas in organized farms conventional

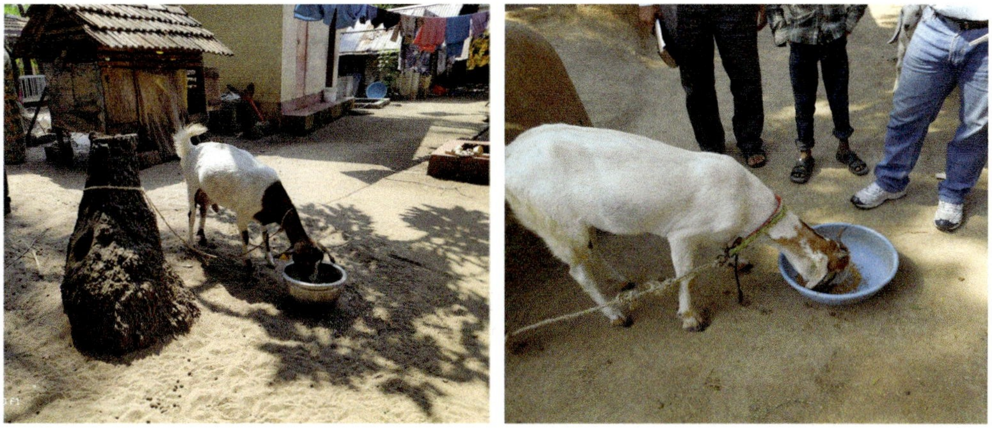

Fig. 6.7. Feeding equipment in village conditions.

troughs and hayracks are generally used. The animals tend to contaminate the feed by voiding their faeces and urine into or nearby the feeders and then standing inside or over these feeders. Farms in the organized sector spend about 60–70% of their running expenditure on feed materials. To minimize this cost, it is important to reduce the wastage and contamination of feed and to increase the efficiency of feeders. However, these devices are not suitable for all types of feed material, such as dry fodder, green fodder and concentrate.

Protocol 2 at the end of this chapter gives details of some of the devices developed by ICAR-Central Institute for Research on Goats (based in Mathura, India) that are suitable for use by goat farmers. Goat feeders should be suitable for feeding concentrates and dry and green fodders simultaneously, thereby reducing labour costs and feeding time in semi-intensive and intensive goat production systems. The height of the feeder trough should be designed to match the shoulder height of goats, which takes into consideration their feeding/grazing habits. The use of such appliances has reduced feed wastage from 35–40% (in conventional ground level feeding method) to 5–10%. The chances of contamination of feed with animal excreta is also reduced. A rectangular feeder can accommodate 10–12 goats whereas a hexagonal feeder can feed 12–15 goats. The adoption of technology reduces the cost of feed by 25–30%, which is estimated to be about 30 Indian rupees (US$0.43) per day for ten goats. Use of this technology under intensive feeding conditions also reduces the pressure on natural resources and mortality and morbidity in goats, facilitating sustainable goat rearing with better financial returns.

6.4.3 Watering facilities

An adequate quantity of clean water of reasonable temperature should be available at all times. The water requirements depends on the level of milk production and the ambient temperature. Adult goats drink about 400–700 ml water/day. Environmental temperature, lactation level, amount of body fat, age, water content of forage, exercise, and salt and mineral content of rations all influence the total water intake. The mean dry matter:free-water intake ratio for goats should be 1:1.2. Contaminated water may be one of the main sources of pathogens, for example, coccidia that affects the health of young animals in particular. Water tanks must be cleaned-out regularly. Water tanks with float valves may be suitable for large-scale intensive goat production. Some watering devices have been shown in Protocol 2 at the end of this chapter.

6.5 Flock Management

Generally, good flock management practices need to be followed for better health, production and optimum utilization of resources. Smallholders and pastoralist production systems need to follow best-management practices to ensure an efficient production system. A standardized management protocol should be followed by large flocks maintained in intensive and semi-intensive production systems.

It is important to define the objectives of the farm, which will be different for specific conditions. Research institutes or universities may develop farms for teaching and demonstration purposes or as a germplasm resource centre for producing high-quality foundation stock. Farmers may start a typical profit-making farm to produce high-quality foundation stock. Before starting the farm, it is necessary to assess the availability of resources. It is important to get an idea of the agro-climatic conditions such as the temperature, rainfall, wind direction, feed and fodder availability, water quality, land type, forage availability, and so on. The details about foundation stock and its availability need to be known, as it is very difficult to get good quality bucks and does from markets. Owners do not sell good quality females. Female goats at market tend to be repeat breeders, infertile, repeat aborters or poor yielders. The availability of good quality adult bucks at markets is also limited, so it is not wise to depend on markets. The newly procured goats should be put into quarantine as they may carry pathogens, disease vectors and internal parasites. The management of quarantine sheds should be carried out by skilled animal husbandry workers.

Effective management is necessary for optimum utilization of all the resources by a large flock. A successful farm operation has different manpower needs, such as skilled animal managers and workers (for kid management, milking of does, handling bucks, etc.), supervisors, veterinarians and security personnel. It is also important to assess different resources such as veterinary services, equipment services, feed and fodder availability, marketing of produce, storage, electricity and road access. Similarly, large farm housing and management depend mainly on the weather conditions throughout the year. In hot and arid conditions, the housing should not be exposed to the sun during daytime if possible; in a cold climate the housing should face the sun during the daytime. Better ventilation space is required for high rainfall, arid and cold regions. And in stall-fed goats, respiratory ailments are a major problem so a proper ventilation system needs to be developed. Ventilation design should be decided as per the requirements and climatic conditions, as should the flooring. In high rainfall areas slatted floors (wooden or plastic or local material) should be used, but an earth floor is desirable in arid and semi-arid areas. Sandy soil can be cleaned easily and intermittent soil replacement can be carried out. Housing space should include both a covered and open space. This should include resting, feeding and exercise spaces. The farm layout should include a goat shed, kidding shed, milking parlour, kid boxes, isolation ward, post-mortem space, feed and fodder storage, office space, feed-lot shed, water troughs, dipping tank, manure pit, burial pit, disposal pit, handling area, record room, etc. Subsequently, general equipment such as feeders (feed troughs and racks), a waterer (water container), a kidding pen, a weighing machine and a milk recording machine should be procured as per requirements.

The daily management activities include daily dairy maintenance, duty allotment, counting of goats in each shed, grazing time, feed distribution, identification of diseased animals, oestrous observation, mating, kidding, shed cleaning, and so on. It is also important to decide on the procurement strategy of dry fodder from the market and storage in proper conditions. Good quality fodder will be available during specific seasons.

The activity chart for each day and season should be finalized. General management activities include water container cleaning, bedding arrangement for kids, neonatal kid care, housing cover for winter, cleaning activities, soil replacement, lime application, equipment repair, building maintenance, and so on. Health management planning for vaccination, dipping, deworming and screening for brucellosis should be scheduled. Breeding management such as semen evaluation, a mating plan, culling and transportation should be taken care of. It is necessary to keep a record and develop a flock-specific health plan for effective management and optimum productivity. Record keeping in a daily diary should include information on mating, kidding, feeding, body weight, milk yield, reproduction, feed and fodder register, management and cleaning register, vaccination, treatment and sample collection, and so on (see Protocol 2 in Chapter 2 and Chapter 5).

6.5.1 Preferred management practices in a flock

The majority of goat populations are found in the developing countries of Asia and Africa. Good animal husbandry, nutrition, vaccination, parasite control and effective biosecurity programmes are required for a productive goat enterprise. The outcome of successful goat husbandry is a healthy and productive flock. Farm operations are routine or periodic in nature and have a direct effect on the goats' productivity.

Identification marks

Identification is required in order to maintain proper records on goat farms, appropriate feeding of animals, better management and animal insurance. Identification marks are either temporary or permanent. Colour marking or hanging small plastic/metal plates around the goat's neck are forms of temporary marking. Colour marking is done using silver nitrate in kids of white colour skin on either side of the body. The kid number along with the dam number is written on small iron discs hanging around the kid's neck; these are useful until weaning to trace the kids during suckling or treatment. The permanent methods are tattooing and tagging. For tattooing, a number is tattooed on the inner surface of the ear (preferably the left ear). For tagging, long-lasting metal/plastic tags (either self-piercing or non-piercing types) are fixed to ears. Tincture iodine should be applied to the piercing wound to prevent infection. Radio frequency identity technology is also used by farmers for better data recording, and to manage basic livestock health monitoring.

Castration

Castration is one of the management activities practised to prevent indiscriminate breeding. Castration is the process by which the testes, epididymides and a portion of each spermatic cord are excised from a male kid. There are three methods of castration: (i) Burdizzo's or emasculatome method; (ii) rubber band or elastrator method; and (iii) incision or knife method. Antiseptic solution should be applied on the site before and after castration to avoid infection, and tetanus antitoxin injection is recommended to avoid any complications.

Umbilical cord treatment

This is an important but minor activity required immediately after a kid's birth. The umbilical cord should be cut one inch away from the kid's abdomen using sterilized scissors or blades and should be cauterized by dipping the cut end in tincture iodine solution or povidone iodine. The application of these antiseptic solutions should continue until the remnant cord has dried and fallen off.

Colostrum feeding

Colostrum is the first udder secretions or milk after kidding. It is high in nutrients and antibodies. Colostrum should be fed at one-eighth to one-tenth of the kid's body weight as soon as possible, preferably within half an hour of kidding. This will mean that the antibodies are absorbed well through the kid's intestine to the circulation for better immune regulation during the early stages of life (see Chapter 4, Section 4.3.2 for more details).

Hoof trimming

Overgrown hooves make walking difficult in sand and damp or muddy soil. Flocks should be checked for hoof growth on a regular basis, and trimming is usually required two or three times a year. Hoof trimming is necessary for the well-being of animals and keeping them fit for normal production. Hoof trimming can be done in a standing position/lateral recumbency using a special trimming tool or sharp knife. After the animal is properly restrained, observe the colour and the smell of the hoof. A rotten smell is usually indicative of foot infection and appropriate treatment should be given. Clean out any mud, manure and small stones from between the walls of the hoof. Dry conditions can make hooves difficult to trim. The best time to trim hooves is immediately after rain or heavy dew, because the hoof wall will be much softer and easier to trim. The inside wall of the foot can also be trimmed, and heel regions can be trimmed when needed. The sole may be slightly trimmed until a slightly pink colour appears. The overall goal of the trimming process should be to make the foot match the angle of the coronary band. After trimming, the bottom of the foot and the coronary band should be parallel with each other.

Brushing/grooming of goats

Goats are groomed in a standing position on a clean floor, starting from the neck towards the hindquarter in the direction of the hair using a brush or curry comb. Before uniform brushing, sticky hair and dry faeces should be cut off using scissors. Grooming of goats makes the hair coat clean, glossy and pliable and helps in close inspection for observation of abnormalities on the body. It also makes the goat more docile and stimulates circulation of blood and lymphs in the body.

Clipping

Longer hair helps the animal to keep their body warmth in the winter season and shorter hair helps the goats to stay cooler in warmer seasons and allows sunlight to reach their body or skin, which prevents lice infection. Clipping is mainly done using hand or electric clippers to maintain the cleanliness of the coat hair. Goat breeds with long hair on their hindquarters should be hand clipped before the breeding season and during the rainy season. Similarly, clipping is often needed in hairy breeds (e.g. Marwari goats) and fibre-producing breeds at regular intervals.

Weighing

Weighing is routinely carried out at regular intervals to monitor the general health and body condition of the animals. In a small flock, a sling-type scale can be used whereby a cotton/leather rope is slung under the belly (one strand behind the elbows and other just before the stifles) and the other end is attached to a spring-balance scale. The animal is lifted off the ground in the sling and the weight can be read on the spring-balance scale. For larger flocks, a platform-weighing scale with a digital balance is more practical. (See Protocol 1 at the end of Chapter 2.)

Bedding

Bedding is mostly used in kidding pens for kids up to weaning age, especially in extreme winter to protect kids from chills and cold. Apart from this, bedding is also needed for preventing kids from licking the soil. Locally available materials such as dried waste weeds, any waste grasses, sawdust, and so on, can be used at a depth of two to four inches. The bedding materials should be replaced daily or rotated to keep them dry until the materials are sloughed and foul. Lime should be used while replacing the bedding materials to keep the floor dry.

Disinfection of goat housing and equipment

General hygiene and sanitation programmes need to be followed to maintain a better health environment. The shed floor (concrete/brick) should be disinfected with phenyl solution or any recommended antiseptic solution at monthly intervals. The upper layer of the soil should be replaced with fresh soil along with a dusting of lime. Walls need to be whitewashed at regular intervals. Feeders, waterers and milking equipment should be scrubbed regularly with 4% washing soda (sodium bicarbonate) in hot water. Quaternary ammonium compounds such as cetrimide can be used to disinfect utensils, udders, milkers' hands and towels for wiping udders.

Drenching

Mass medication is generally required for administering drugs such as anthelmintics and cocccidiostats. During drenching, the head should be held horizontally and not tilted up, to reduce the chances of the animal developing aspiration pneumonia. The drenching gun should be inserted at the commissure of the lips and the nostrils held closed while the medication is quickly dispensed. The volume of the drug should not be more than 5–10 ml. A stomach tube, passed via the mouth after proper restraining or with the help of suitable speculum, may also be used.

Wound management

Goats are prone to skin abrasions during grazing on thorny bushes and accidental contact with sharp wiry fences, which may lead to septic or maggot-infested wounds. Sometimes, the goat be a victim of an animal or bird attack that inflicts skin damage and leads to wounding. A wound is an injury that typically causes a break in the integrity of the skin, and develops to sores, abscesses, warts or inflamed skin lesions. The abrasions may be cuts, tears, burns, breaks or other damage to living tissue. Wounds are categorized as clean, contaminated, infected or maggot (myiasis). The flies responsible for the maggot wounds prefer a warm and humid environment. Therefore, myiasis mostly occurs in the summer months in temperate zones, but may occur all the year round in the tropics. The wound not only causes clinical illness, but decreases the hide value and productivity of animals. Hence, timely care is required to prevent worsening to any extent.

The major aim of wound management is to restore function, repair skin integrity and minimize infection. Lacerations are cuts or tears in the skin. Uncomplicated, simple lacerations are usually managed by complete closure; however, this may not be possible if the wound is dirty or infected. In deeper cuts, damage to muscles, tendons and other tissue must be treated before a wound can be closed.

For fresh wounds, apply a bandage impregnated with antiseptic solution (e.g. povidone iodine solution) to prevent further contamination. The dressing with tincture benzoin should be done to check for bleeding; heamostatic drugs may also be given.

In extensive bleeding, surgery is required to ligate the ruptured artery or vein. Sutures, staples or surgical glue can be used to close the wound. Sometimes, layers of closure with administration of antibacterial therapy is required to treat infections. In septic wounds, dead tissue and foreign material should be removed from the wound as it reduces bacterial contamination. It is suggested that a tissue sample is collected for culture and drug sensitivity for judicious selection of antibacterial therapy. The wound should be cleaned with antiseptic solution (lavage) to wash away both visible and microscopic debris. This reduces the risk of infection. A syringe is used to spray a solution onto or into the wound to clean it. Antibiotics are used to treat bacterial infections. Medications for pain relief are also usually given. For maggot-infested wounds, the maggots should be removed and the wound should then be covered with a maggot-cidal dressing, followed by an antiseptic dressing with antibacterial therapy. The use of fly-repellent cream should continue until recovery.

Administration of injections

The most appropriate site for intramuscular injection is in the neck muscle. Longissimus muscles over the back in the lumbar region may also be used. The volume of drug administered in one site should not be more than 5 ml and the needle size should be 2- to 3-cm long and not larger than 18-gauge unless the medication is highly viscous. Shorter needles should be used for young kids. The thigh muscle is not a preferred site for intramuscular injection in adults and particularly in young kids because the muscle mass is less and often leads to sciatic nerve damage and peroneal paralysis. Subcutaneous injections are commonly given in the neck in the same region described for intramuscular injections, or on the chest wall about 5 cm behind the point of the elbow. The jugular vein is commonly used for intravenous injection by using 2- to 3-cm long needles of 18 or 20 gauge. Intradermal injections are given using 26-gauge, 1-cm long tuberculin needles. Intraperitoneal injections are given to neonates to treat hypoglycaemia with glucose solutions or navel infections with aqueous-based antibiotics by inserting an 18- or 20-gauge needle perpendicular to the skin about 1 cm to the left of the navel, no deeper than 1 cm (Smith and Sherman, 2009). An intramammary injection can be administered with the help of a teat catheter or a specially designed teat injection after cleaning and swabbing the area with alcohol.

6.5.2 Parturition and weaning management

Parturition

The body condition score should be 3.0–3.5 during pregnancy. The period 6–8 weeks prior to kidding is very important, and quality feed should be supplied during this period. Clean water and mineral mixtures should be available to the doe at all times. Screening for brucellosis and faecal egg count should be carried out. A dose of selenium or supplementation should be provided. Kidding should be carried out in a clean environment or kidding pen, which should be bedded with straw, soft grass or other absorbent material. The first signs of impending kidding are udder enlargement, swelling of the vulva, restlessness, loss of appetite, frequent urination and mucous discharge from the vulva. There are three stages of parturition. The first stage is the uterine contraction and cervical dilation and usually lasts for 2–6 hours. The second stage is the abdominal contraction along with uterine contraction and will last from 30 minutes to 2 hours. The doe starts the second stage when her waters break; if the doe is straining for more than 1 hour after her waters have broken, then assistance is required for kidding. The third stage is the expulsion of placenta, which takes about 5–8 hours.

Kid management after birth

New-born kids should be kept on a clean and soft-bedded floor at a suitable environmental temperature. It is essential to check that the new-born kid is breathing with ease immediately after birth. Remove all the mucus from the nose and around the mouth for good breathing. Timely and sufficient colostrum feeding is of utmost importance; ensure feeding within 2 hours of parturition. If colostrum is not available, the kid can be fed by another mother that has recently given birth. Alternatively, frozen colostrum may also be provided. Suckling should be allowed after washing the udder with antiseptic water to prevent oral-borne infections such as colibacillosis. Then treat the navel with iodine to prevent disease-causing organisms from reaching the umbilical cord. This also helps in rapid drying and eventual breaking away of the cord from the navel. The digestive system of a kid is not fully developed at birth. Fibrous forage, leaves of legume trees, grass and hay should be introduced at 2–3 weeks of age which will help to develop the rumen and promote faster growth of kids (see Chapter 4, Section 4.3.2).

Weaning management option

The ideal weaning age is between 12 and 16 weeks. The weaning period may be a little earlier during a drought or unavailability of biomass due to low rainfall, in which case a suitable creep-feed ration is essential. Kids are completely moved from a milk diet to feed and fodder after weaning. Post-weaning can be very stressful on the kids, resulting in outbreaks of coccidiosis and them becoming susceptible to other diseases. Moreover, all vaccinations are also done around 3 to 4 months of age, which may add to the stress. The necessary arrangements for a pasture, holding pens, feeding and housing facilities should be checked.

The pens need good shelter that is well-bedded before weaning. A suitably sized water source that allows easy access for the kids is also important. Some water troughs are too high for young kids to use. The drinking trough height should not be more than 30 cm for weaned kids. Clean shelters should be provided before moving kids after weaning in order to reduce exposure of the kids to diseases. Kids need to be familiar with the feeding system. The addition of creeper rations, green fodder and dry fodder should be done gradually to acclimatize with the feeding schedule. Kids should be given adequate roughage before weaning to promote normal rumen development and proper rumen flora. Castrated and breeding kids should be kept in separate pens to ensure better care and nutritional supplementation.

Coccidiosis and pneumonia are the dominant problems in the post-weaning age group, particularly under conditions of confinement housing. Transition to a diet of solid feed encourages more ingestion of coccidial oocysts while feeding, and, coupled with the progressive decline of maternal antibodies and the stresses of weaning, appear to predispose kids to respiratory infections after 8 weeks of age (Smith and Sherman, 2009). Coccidia commonly reside in the intestine and multiply rapidly resulting in intestinal damage, which can cause decreased performance, reduced resistance to other diseases and even death. Coccidia are most often controlled through the use of coccidiostats in feed additives. Treatment for coccidiosis may be started at least 2 weeks before weaning and continue for 2 weeks after. In pasture-grazing kids, *Haemonchus contortus* infections may occur; therefore broad-spectrum anthelmintics should be administered after monitoring the parasitic load in faecal samples.

6.5.3 Management during natural disasters

Any country can experience a natural or man-made disaster (e.g. droughts, floods, earthquakes and cyclones) that can affect the animal population in different ways, leading to moderate to heavy economic losses. Man-made disasters may not harm the animals directly but may threaten their environment, for example, bomb explosions, pollution, industrial accidents or destruction of the natural habitat. Therefore, it warrants preparedness, mitigation or rehabilitation in a suitable manner (Ganguli *et al.*, 1993).

Goats are natural browsers and prefer open space, so confinement to a particular area is stressful for them. In confinement, goats are very sensitive and need more attention than cattle and sheep. They can also develop overeating behaviours. Goats will be kept in confinement during heavy floods, cyclones or any other type of natural disaster. Feeding and space management during natural disasters are major constraints. Moreover, providing safe and sufficient water is another problem during heavy rain and floods. A possible solution is to disinfect water and provide feed with supplement.

In drought conditions, additional drinking water should be supplied through tube wells, clean water tanks or harvested rainwater in tanks or large ponds. Nutrition support to animals in the form of dry fodder reserves, urea molasses licks, complete pellet feed and conventional feed and fodder should be provided. During regular drought spells, planting alternative drought resistant fodder crops should be promoted along with the use of seed/fodder reserves.

Earthquakes can cause damage to buildings, infrastructures, bridges, dams, roads and railways. Beside the scarcity of feed and fodder, the contamination of water through seepage of drainage water causes great discomfort to people and animals. In such crises, goats should be given the safest place for shelter so that the animals can survive for 2–3 days without any assistance. The goats should also be vaccinated against tetanus and the most prevailing infectious diseases.

Floods are one of the most common natural disasters causing extensive damage to property, livestock, crops and human life. Animals are natural swimmers and therefore have a chance to escape drowning if they are not tied or caged. In the event of a flood, the source of drinking water becomes contaminated.

The immediate actions include locating high ground for shelter, a clean or disinfected drinking water supply and arrangements for dry fodder. All injured animals should be promptly treated.

Many animals are likely to die in a disaster situation, and disposal of dead animals becomes a problem. One should not throw dead animal carcasses into rivers or streams. The carcass should be burned or buried as per the standard protocol. There is always a fear of disease spread in the herd. The risk can be avoided through mass vaccination programmes for diseases such as foot-and-mouth disease, haemorrhagic septicaemia, anthrax, PPR, *Escherichia coli*, and so on. Some precautions on pest control along with proper hygiene and sanitation may be followed. An awareness programme should be prepared for farmers to face any unwarranted disaster. Special teams should undertake preparedness and relief exercises to test their efficacy and readiness.

6.5.4 Handling climate stress and management

The management during climatic variations such as extreme heat or extreme cold or humid conditions is essential to maintain productivity. Climate change could influence the animal in three different ways:

- heat-related diseases and stress;
- extreme weather events;
- emergence or re-emergence of infectious diseases, especially vector-borne diseases critically dependent on environmental and climatic conditions.

Indigenous goat breeds in their own habitats have the ability to survive dry periods of feed scarcity and water shortage for considerable time periods. Goats reared in low input–output systems are less vulnerable to climate change, while those reared under intensive farming systems on high output are not well adapted to heat stress making them highly vulnerable to climatic variability. In long-term strategies, promoting local indigenous livestock species that are resilient to climate variability should not only be conserved for future use, but should also be utilized as a potential tool to mitigate against climate change effects.

6.6 Integrated Goat Farming

Integrated goat farming is a viable income-generating sector, especially for marginal farmers.

6.6.1 Agriculture and goat farming

Integrating goat farming with cropping systems like agro-forestry models such as agrisilviculture, hortipasture, boundary plantations, wind belts and silvipasture models with trees and grasses can be established in wastelands to supplement farm income. The plant most commonly used for goat fodder in specific regions should be planted for continuous fodder supply, saving on the cost of feed and fodder (see Chapter 4, Section 4.7).

6.6.2 Vermicomposting and goat farming

Vermicomposting involves introducing earthworms and microorganisms into a pit containing waste biomass and goat faecal balls. Vermicomposting has shown improvement of soil quality, prevention of run-off, preservation of groundwater quality, reduction in greenhouse gas emissions and conservation of biodiversity. The vermicomposting pit can be prepared by the availability of goat pellet and waste and the guidelines should be followed to prepare the manure.

6.6.3 Aquaculture and goat farming

The basic principle of a goat–fish integrated farming system is the full use of livestock waste by converting it into valuable fish protein. The production of manure from a goat is around 0.5–1.0 tonnes per year. It is a good source of organic matter and contains 2.7% nitrogen (N), 1.78% phosphorus (P_2O_5) and 2.88% potassium (K_2O) (Table 6.2). This is reported to be several times richer in NPK content than manure from cows and other animals (Gaddie and Douglas, 1975). Goat urine is equally important as it contains both nitrogen and potash. Moreover, these droppings are coated with mucus and are

Table 6.2. Nutrient contents (%) of animal faeces.

Source	Organic content	Nitrogen	P_2O_5	K_2O	Protein
Goat	60	2.7	1.78	2.88	17.31
Cow	30	0.7	0.30	0.65	4.38
Horse	60	0.7	0.34	0.53	4.38
Pig	30	1.0	0.75	0.85	6.25
Rabbit	50	2.0	1.33	1.20	12.50
Sheep	60	2.0	0.54	1.54	12.54
Chicken	50	1.6	1.25	1.25	10.00

Source: Gaddie and Douglas (1975).

semi-dry. Deterioration of water quality is low when compared to other types of manure. On average, 60–80 goats are reared to fertilize 1 hectare of pond area. Manuring helps in the formation of plankton bloom and also acts as a feed material for fish. Feed usually accounts for 40–60% of the total cost of fish farming and this can be saved by utilization of goat manure for fish production (Nanda and Naik, 2001).

6.6.4 Integrated goat farming for goat products

This model ensures introducing products to the market at consumer level in desired quality and quantity within particular time frames. In this supply chain a farmer or a smallholder sells their live goat to a slaughterhouse owner or any authorizing animal purchasing agency. The meat processing industry prepares standardized meat products, value-added meat products and functional meat products and sells them on to the market. Goat milk and milk products have a potential worldwide market. Goat milk collected at farm level should adopt the standard milk collection procedures and maintain hygienic protocols. It should be transported from door to door, maintaining the cold chain until ready for the milk processing factories. The factories convert raw milk into value-added products that retailers sell in attractive and informative packages for human consumption. Farmers raising transgenic goats for therapeutic milk protein production must follow the standard supply chain of milk to the processing unit. Other auxiliary goat products such as skin, fibre, vermicomposting and blood may also be sold to specialized processing units. The farmers working in this chain can earn money for profitable goat farming. A model for planning integrated goat farming is shown in Fig. 6.8.

6.7 Protocols

Protocol 1: Goat house with improved thatch roof material

The process of making an improved thatch roof involves four main steps: (i) making thatch panels; (ii) preparation of non-erodible mud plaster; (iii) fire-retardant treatment and (iv) preparation of stabilized cow dung slurry.

Making thatch panels

1. Prepare two identical frames of half split bamboos of 5 cm diameter by placing them across each other about 15–20 cm apart.
2. These are tied to each other with the help of 20 standard wire gauge (SWG) galvanized iron wire or strong rope. The length and breadth of the panels is decided as per the size of the structure.
3. Spread thatching material on one of the frames. The thatching material should be about 10 cm thick. After this the other frame is placed over the thatching material.
4. Both the frames are tied together with the help of 20 SWG galvanized iron wire and by putting body weight over the joints and applying pressure. The thickness of the thatch material should be about 5 cm after compression.
5. The thatch panel should then be fixed properly on the roof structure of the shed.

Preparation of non-erodible mud plaster

1. Preparation of the mud mortar: Pond-based soil containing at least 25% clay and wheat straw (preferably one year old) lopped into 2-cm lengths are used for preparing the mud mortar. Add straw at the rate of 64 kg per cubic metre of soil and allow decomposing in the presence of water; knead it three to four times a day for about 7 days in summer or 14 days in winter. The mud mortar should not be allowed to dry during this process or during the construction work.
2. Preparation of bitumen cut back: 80/100 grade bitumen is mixed with kerosene oil in the ratio of 5:1. Bitumen is heated till melting and then poured into kerosene oil (never pour kerosene into bitumen); keep stirring while pouring bitumen. The mixture is called bitumen cut back.
3. Preparation of non-erodible mud: Pour the bitumen cut back (64 kg of this mixture in 1 cubic metre of mud mortar) onto the previously prepared mud mortar. Overturn with spade and paddle so that it is thoroughly mixed. Ensure that no black spots of unmixed cut back are left in the mud.

Fire-retardant treatment

1. First, a layer of mud mortar (without bitumen cut back) is applied to the previously prepared thatch roof to give the roof a smooth, even finish. Allow the plaster to dry and crack for 1–2 days. Fill the cracks with the same mud mortar.

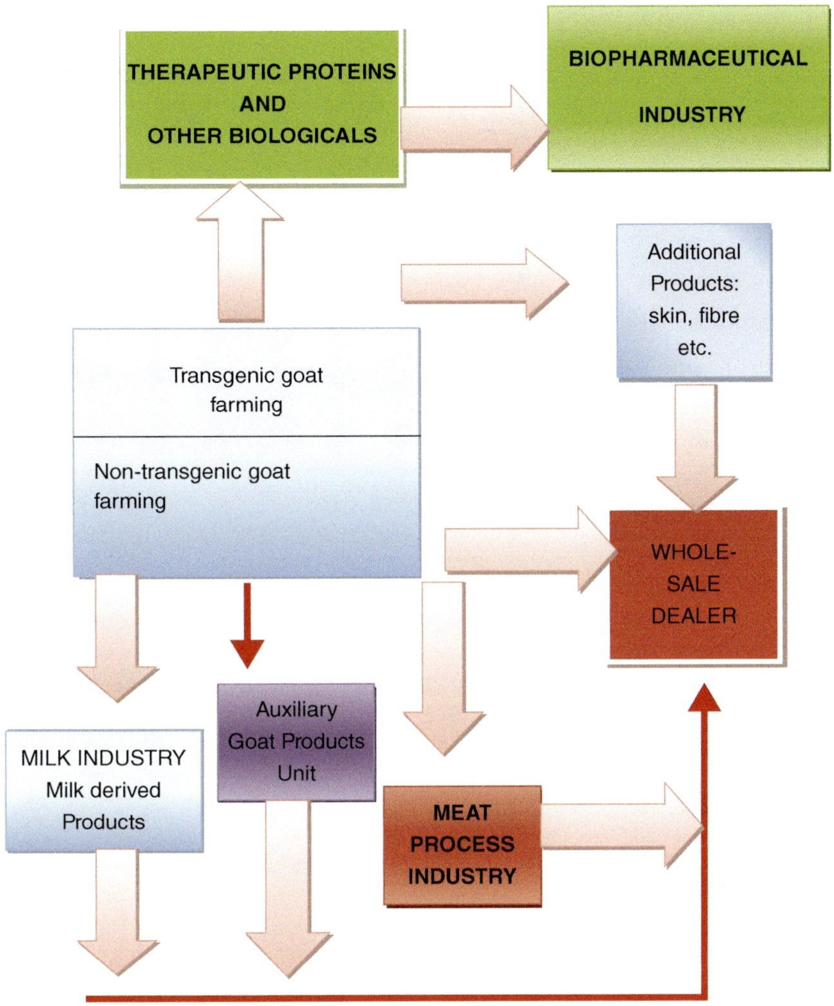

Fig. 6.8. Integrated goat farming for the production of milk, meat, therapeutic proteins and other additional commercial products.

2. The roof is then plastered with non-erodible mud mortar (12 mm) and left to dry for a few days. All the cracks are filled with non-erodible mud plaster.
3. The same treatment can be given to the ceiling of the structure.

Table 6.3. Material required for the roof area (each m^2) of a goat house.

Dry soil	0.014 m^3
Wheat straw	900 g
Bitumen (80/100 grade)	750 g
Kerosene	150 ml

Preparation of stabilized mud–cow dung slurry

1. Cow dung and soil is mixed thoroughly in equal quantities. Sufficient water is added to form a uniform thin paste.
2. As described above, prepare and mix bitumen cut back into the mud–cow dung slurry (material requirement is given below).
3. Two coats of this slurry are applied on the roof to give a smooth finish.

Table 6.4. Material required for mud–cow dung slurry (two coats) for preparation.

Soil	0.001 m^3
Cow dung	0.001 m^3
Bitumen	210 g
Kerosene	46 ml

Protocol 2: Feed and watering devices

Rectangular feeder

These feeders consist of a four-legged angle iron stand, a galvanized iron (GI) rectangular-shaped feeding trough with a triangular pyramid in the middle and a vertical and sloping hayrack mounted on the feeding trough (Fig. 6.9). The upper portion of the rack is covered with a GI sheet and the lower portion consists of iron bars. These feeders have provision for individual or simultaneous feeding of various types of feed materials. The feeders can accommodate about 12 goats. The feeding height should be kept at the average shoulder point height of the goats. These feeders can be locally fabricated and save space and labour. There is less wastage of feed material and no contamination through faeces and urine (Misra, 2008).

Fig. 6.9. Rectangular feeder for adult animals.

Hexagonal feeder

This feeder consists of a six-legged stand onto which a hexagonal bottom trough with a six-faced central cone is fitted (Fig. 6.10). A six-faced cylindrical hayrack is mounted on the feeding trough. The lower portion of this rack has vertical slanting outward iron bars, and the top portion is made of GI sheet. This feeder also has provision for individual or simultaneous feeding of various types of feed materials. The feeder can accommodate about 15 animals. The animals can eat from all the sides. The feeding height should be kept at the average shoulder point height of the animals. These feeders can be locally fabricated and save space and labour. There is less wastage and no contamination through faeces and urine (Misra, 2008).

Fig. 6.10. A hexagonal feeder for adult animals.

Hanging-type circular kid feeder

The hanging-type circular kid feeder was developed for kids up to 3 months of age. It consists of a jointless aluminium pan (used as a feeding trough) and a conical storage bin (made of GI sheet) (Fig. 6.11). Feed material is kept in the storage bin fixed above the aluminium-feeding trough, from which it flows into the feeding trough. This feeder may accommodate about ten kids. Animals can take feed from all sides (Misra, 2008).

Fig. 6.11. A hanging-type circular kid feeder.

Kid feeder with hay rack and watering device

This feeder consists of a feeding trough, hayrack, watering device and adjustable stand (Fig. 6.12). It is suitable for the feeding of all types of feed materials. Watering devices are attached to the width sides of the hayrack. It can accommodate 8–10 kids. It is suitable for kids from the day of birth to 12 months of age. There is no contamination of feed. The rate of feed consumption and wastage of feed was found to be satisfactory. For kids up to 3 months of age, a lower feeding trough height (25% less than average shoulder point height) may be more comfortable and will give better performance. After

Fig. 6.12. Kid feeder with watering device.

removal of the stand and watering devices, this feeder may work as a hanging-type feeder for kids up to 3 months of age (Misra, 2008).

Kid feeder with watering device

The performance of this feeder (Fig. 6.13) was found to be better than all the other kid feeders. The visual observations indicated that if the width of the feeding trough is increased, it might become more comfortable for kids above 6 months of age. An increase in the width also changes the dimensions of various components of the feeders like the hay rack, stand, etc. The stand is a fixed height, and a lift cock is added to the water container to reduce contamination and wastage. This feeder can accommodate about ten kids of 6 to 12 months of age. After removing the stand and watering devices, this feeder can be very successfully used as a hanging-type feeder for kids below 3 months of age (Misra, 2008).

Circular watering device

This equipment consists of an aluminium pan from which water may be taken by kids (Fig. 6.14). A water storage container is fitted over this pan from where water percolates to the pan to maintain a fixed water level in the pan. The whole device is fitted to

Fig. 6.13. Kid feeder with watering device.

Fig. 6.14. Circular watering device for kids.

an adjustable stand made of GI pipes. Animals can drink water from all sides. Ten kids can be accommodated by this device (Misra, 2008).

Barrel-type kid watering device

With this waterer kids can drink water from two sides (Fig. 6.15). A barrel is made of GI sheet and fitted over a water trough also made of GI sheet. The barrel is filled with water. The whole device is fitted on an adjustable stand made of GI pipe. It can accommodate about ten animals up to 12 months of age (Misra, 2008).

Fig. 6.15. Barrel-type watering device for kids.

6.8 References

Belanger, J. (1990) *Raising Milk Goats the Modern Way*. Storey, North Adams, Massachusetts.

Boe, K.E., Andersen, I.L. and Jorgensen, G.H.M. (2008) Effect of reduced feeding space for dairy goats on feed intake and social interactions. American Society of Agricultural and Biological Engineers, St. Joseph, Michigan, USA.

Bond, T.E., Kelly, C.F. and Ittner, N.R. (1954) Radiation studies of painted shade materials. *Agriculture. Engineering* 35, 89.

Constantinon, A. (1987) Goat housing for different environments and production systems. *Proceedings of the IV International Conference on Goats Brazil*, 241–268.

Devendra, C. (1964) The housing of goats in Malaysia. *Malayan Agriculture and Horticulture Association Magazine* 21, 52.

Gaddie, E.R. and Douglas, D.E. (1975) *Earth Worm for Ecology and Profit: Scientific Earthworm Farming*, Vol 1. Bookworm Publishing Company, Ontario, California.

Ganguli, S.K., Urmil, A.C. and Somiya, P.A. (1993) Natural disasters: an overview in Indian context. *Indian Journal of Community Medicine* 18(3), 110–113.

Mennella, V.G. and Borghi, P. (1983) *Building Standards and Equipment for the Intensive Keeping of Dairy Goats*. International Commission for the Southeast Atlantic Fisheries, Madrid, Spain.

Misra, R.P. (2008) Role of housing and management to improve productive and reproductive efficiency of goats. In: Kharche, *et al*. (eds) *Advanced Production and Reproduction in Goats*. CIRG Publications, India, pp. 35–43.

Nanda, P.K. and Naik, P.K. (2001) *Integrated Goat and Fish Farming*. Available at: https://www.researchgate.net/publication/258725647 (accessed 23 May 2019).

Smith, M.C. and Sherman, D.M. (2009) *Goat Medicine*, 2nd edn. Wiley and Blackwell Publications, Hoboken, New Jersey.

Trodahl, S., Skjevdal, T. and Steine, T.A. (1981) Goats in cold and temperate climates. In: Gall, C. (ed.) *Goat Production*. Academic Press, London, UK.

Whates, C.M. and Charles, D.R. (1994) *Livestock Housing*. CAB International, Wallingford, UK.

7 Production System Characterization and Development

7.1 Introduction

Livestock is the major source of livelihood security as well as employment in tropical countries. As the result of expanding economies and higher population growth, there is a higher demand for livestock products across Asia and Africa. It is predicted that this demand will continue over the next 30 to 50 years (FAO, 2005) – there will be an increase in consumption of meat and milk products, and a rapid rise in demand for livestock products all over the world. Therefore, the livestock sector has to expand to meet the growing requirements and adapt to changes in the environment.

The nature of livestock production in the tropics varies depending on the local environment and culture. The importance of livestock production goes beyond its food production function. Livestock is used as draught power, organic manure, dung as fuel and other by-products such as leather, bone and horns. The rapidly growing livestock market, the way the livestock product is retailed and prospects for global trade will create income-generating opportunities. Therefore, the current situation presents a unique opportunity to reduce poverty through livestock production and marketing. Simultaneously, the present situation is really complex with respect to the following factors: (i) food production is showing almost no growth and there may be a reduction in grain production for human consumption; (ii) the possibility of cultivating biofuel is adding supplementary problems; (iii) grazing areas are diminishing due to an increase in cropping areas to accommodate the growing human population; and (iv) climate change will bring harsher environments and heat stress. Therefore, the present situation demands an analysis of adaptability to formulate strategies to enhance adaptability, productivity and marketing for sustainable livestock production.

The climate in tropical countries varies from tropical monsoon to temperate dry. Livestock production is the world's largest user of land, either through grazing or through consumption of food grains and fodder. The future of the livestock production environment will be governed by market demand and will move towards more intensive and integrated processing. Livestock rearing is one of the most important economic activities in rural areas and contributes significantly to the national economy. Livestock contributes 30% of the total protein requirement worldwide. The livestock market provides an opportunity for global trade and is a potential income-generating opportunity for livestock farmers. Livestock products are consumed locally and 70% of the produce is sold informally in tropical countries. Therefore, smallholder production systems should be the target area for enhancing sustainable livestock productivity. To frame the area- and ecological-specific strategies for best performance outcomes it is necessary to characterize the production environment in various ecological regions for livestock production. Smallholder livestock production systems can be transformed by:

- agribusiness/innovation;
- more value-added products, resulting in increased return;
- sustainable, equitable and healthy management; and
- carbon/environmental footprint management.

7.2 Production Systems

7.2.1 Goat production systems in the tropics

Livestock production and management practices vary around the world. Therefore, different agro-ecological zones are defined based on climate, landforms, soil, land cover and land use. The agro-ecological zones are broadly classified into five categories including tropics, sub-tropics, temperate, boreal and arctic. The Food and Agriculture Organization of the United Nations (FAO) has defined agro-ecological zones (AEZs) in order to assess agricultural resources (FAO, 2017).

Goats in the tropics are kept under varying production environments, and production objectives vary according to environmental conditions. Goats serve as a source for food security and provide cash to meet day-to-day expenses. Therefore, it is essential to improve production performance through improved husbandry. Moreover, production systems need to be characterized in order to identify problems with respect to specific regions. The output goals of different goat production systems are presented in Table 7.1.

7.2.2 Goat production systems in Africa

Pastoral systems: arid and semi-arid

This production system is characterized as a low rainfall region, ranging between 200 mm and 700 mm per annum. Goats are kept as large flocks and may or may not be mixed with other species. The production system faces high kid mortality rates, particularly weaning mortality; variable feed and fodder availability; larger kidding intervals; and the occurrence of epidemic diseases such as *peste des petits ruminants* (PPR) and contagious caprine pleuropneumonia. Moreover, water availability is limited and parasite infection remains a problem in both kids and adults. The main opportunities include the possibility of selection within breeds/ groups, and improving neonatal kid mortality and pregnancy care can also enhance performance (Peacock, 1996).

Agro-pastoral system: semi-arid

Livestock keepers grow crops during the rainy season and move into livestock during the dry period. The Fulani ethnic group in West Africa are typical agro-pastoralists. There is seasonal variation in the feed availability, and the available forage tends to have high fibre and low protein levels, resulting in low productivity rates. During the dry season, crop residues are available to goats for forage (Peacock, 1996).

Mixed farming: humid

Mixed farming systems in Africa are characterized by increased rainfall and limited land availability. The major crops are maize, beans, rice, sweet potatoes, cassava and taro. Similarly, tree crops such as cocoa, rubber, plantain and fruit trees are available for goat forage. Goats are kept in small numbers/ groups that are mostly kept in confinement or tethered. Goat breeds are small and prolific. This type of farming system is found in Nigeria, Benin, Togo, Ghana and Cameroon. Internal parasite

Table 7.1. Output goals of tropical goat production.

Ecological zone	Production system	Output	Region
Arid	Transhumance (horizontal)	Milk, meat, hair	North Africa, Near East, India
	Sedentary Ranging	Meat, milk, fibre	Latin America
	Ranching	Fibre (meat)	USA, southern Africa, Turkey
Semi-arid	Semi-sedentary pastoralism	Meat	East Africa
	Agro pastoralism	Meat	West and southern Africa
Sub-humid	Mixed farming	Meat	Africa, Latin America, South Asia
Humid	Lowland mixed Farming	Meat	Southeast Asia
	Upland perennial Cropping	Meat	Southeast Asia, West Africa
Highland	Transhumance (vertical)	Fibre, meat, milk	Northern India, Turkey
	Lowland mixed Farming	Fibre, meat	Lesotho
	Upland perennial Cropping	Meat	Ethiopia

Source: based on Peters and Horst (1981).

infestations and occurrence of PPR are the major health problems (Peacock, 1996).

Mixed farming: sub-humid

This region covers West and Central Africa with an annual rainfall that varies between 1000 mm and 1500 mm. Sorghum and maize are the main crops. Goats may be kept in transhumant (seasonally moved) pastoral flocks during the dry season and return to grazing lands during the rainy season. Pastoral flocks are large in size and production performance is affected by seasonal fluctuations in feed and fodder availability (Peacock, 1996).

Mixed farming: highland

This production system is found above 4000 m in Ethiopia where the crops grown in highland farms include maize, wheat, barley, oats and potatoes. Goats graze on step-hill sides and feed on crop residues. The flock size is small in this region. The quality of feed is low and goats graze on crop residues to thrive (Peacock, 1996).

7.2.3 Goat production systems in Asia

Mixed farming: humid (irrigated)

This is a region that has intensive year-round crop cultivation. Rice and other crops are cultivated in regions with higher human populations. Goats are usually kept by the roadside. This farming system is found in Indonesia, Malaysia, India and the Philippines. Goats graze in rice bunds or at the roadside or in common property resource areas. There is a low intake of fodder due to the high moisture content of forage and the high fibre content of crop residues. The flock size is small and maintained as backyard animals. Internal parasite infestations throughout the year affect the health and production of goats (Peacock, 1996).

Mixed farming: humid/sub-humid (rain-fed crops)

This is the region mainly found in South and South-East Asia. Rain-fed crops include rice, maize, vegetables and other legume crops. Goats are the major animal component in rain-fed agriculture. The goats have poor growth, low production rates and high pre-weaning mortality rates (Peacock, 1996).

Extensive systems

Extensive goat rearing is found in the arid and semi-arid areas of Pakistan, India, Bangladesh and Nepal. Seasonal pastoralism is also observed in the region. In the high-altitude Himalayan region, pastoralism is found throughout the year (Peacock, 1996).

7.2.4 Goat production systems in Central and South America

Extensive system: semi-arid

Goats are mainly kept in extensive grazing systems in the semi-arid area of Central and South America. Small flocks (up to ten) are reared by marginal farmers or labourers; flock sizes of 20–50 are generally herded by family groups. The goats are kept for meat and skin by-products. Goats are known for their drought-tolerant ability and remain productive (Peacock, 1996).

Mixed farming: sub-humid

Goats are kept mainly in intensive production systems in the sub-humid zones of Central America, northern regions of South America and the West Indies. The natural vegetation includes tree leaves, legumes and crop by-products such as banana stems. Goats are a significant contribution to the cash income of farmers in the region (Peacock, 1996).

7.3 Pastoralist Production Systems in India and Africa

Pastoralism is mainly spread in the rangelands of Asia (36%) and Africa (30%). The rangelands have wide variations in climatic conditions, from severe cold to hot (harsh climates), with limited resources of water and pasturelands (Dong *et al.*, 2011; Reid *et al.*, 2014). Pastoral cultures rely heavily on livestock rearing which plays a very important role in sustaining their livelihood and acts as an insurance against disaster. It is a highly innovative production system for livestock rearing as there is very little scope for agricultural expansion. The pastoralist mostly depends on mobile animal grazing, which allows the animals to utilize a wide variety of forage vegetation types that are widely dispersed. Traditional pastoral systems preserve natural ecosystems through extensive ranching and rotational grazing and by using a variety of livestock. This increases seasonal grazing and the carrying capacity of the land. As a result of well-skirted livestock movement, the herds stay healthy and produce a reliable supply of milk and meat that meets the demands of pastoral households.

The different types of pastoral production systems in the tropics, particularly in Africa and Asia, have similar management principles and common problems in their pastoral production systems. The contribution of pastoralist to fragile ecologies is manifold. They increase the carrying capacity of land, fertilize the high-altitude grazing area and forest area with manure, and maintain biodiversity in the region. The different pastoralist systems have some common norms for selecting their animals, for example Maasai farmers in Africa and Gaddi farmers in India do not prefer twinning in does, etc. The pastoral system is facing a major problem as the present generations are not interested in following the traditional methods. They are more interested in moving to urban areas for better education and employment opportunities. At present, Maasai farmers are hiring labour for herding and managing their flocks, and the Changpa nomads in Ladakh region (India) and the Gaddis nomad in Himachal Pradesh (India) are facing similar problems and therefore need urgent attention to manage fragile ecological zones.

7.3.1 Maasai pastoralists

In Africa, pastoralism makes up 66% of the continent's lands for pastoral production as it is considered to be the most efficient use of rangelands, for example, the arid and semi-arid lands (ASALs) of Kenya (Idris, 2011). The pastoralists continuously face environmental stresses as they constantly navigate in the hot and dry climate of the ASALs in search of pasture lands and water sources (Pavanello, 2009). Maasai (also spelt Massai) are nomadic pastoralists of East Africa who range along the Great Rift Valley of Kenya and Tanzania, the Samburu of Kenya, and the semi-pastoral Arusha and Baraguyu (or Kwafi) of Tanzania. In Kenya, Maasai pastoralists inhabit ASALs (e.g. Laikipia county) that are geographically located in the north, east and south of the country; they account for 36% of the total population, which is around 14 million Maasai people (GOK, 2014). The pastoral Maasai of East Africa keep cattle, goats and sheep, and move their herds throughout the year to optimize the utilization of rangeland resources for maximum meat and milk production. Elders direct warriors to conduct ecological skirting, which includes identifying and classifying plants and accurately assessing the water-holding capacity of distant pastures. Then they draw up movement itineraries on the basis of the warriors' reports. Maasai pastoralists are familiar with numerous plants in their rangelands and pastures. They can also describe each plant's seasonality, nutrition value, toxicity and medicinal properties for the different animals they keep. Maasai can distinguish between those plants that are good for increasing milk and those that fatten livestock and improve their condition.

The household among the Maasai pastoralists is the basic unit of production, distribution, packaging and consumption in the pastoral value chain. Household heads are considered to be the sole owner of the pastoral livestock and take overall decisions regarding livestock activities. They are also responsible for herding the livestock and have the power to define access to pasture and water for livestock. The household heads are also involved in the sale of livestock as they make the final decision on which livestock should be sold and when it should be sold. It has been observed that white and black goats were valued as a source of blessings and are thus regarded as a source of wealth. The household heads also decide on when livestock should be slaughtered for household consumption as well as the continuous extraction of other food components from the livestock, namely, blood and milk (Ameso *et al.*, 2018).

7.3.2 The Gaddis pastoralists

Migratory pastoralism is common in the high-hill mountainous region of Himachal Pradesh (India). The migratory nomads are called the 'Gaddis'. They derived this name from their native land 'Gadheran' that lies on both sides of the Dhauladhar ranges. The Gaddis belong to different castes of *Rajputs*, *Khatris*, *Rana* and *Thakurs*. The nomads keep both goats and sheep in their flocks. The majority of Gaddi goats are reared under the transhumance (migratory) system of management in large-sized flocks. The migratory route of each herd is fixed, guided by elders who also follow predefined night stay areas on the migratory route. The nomads also have fixed sites for kidding and kid rearing in their migratory routes. The activities of Gaddis pastoralist are described in Fig. 7.1.

Only very small fractions (around 5%) of Gaddi goats are reared under semi-intensive systems in

Fig. 7.1. Himalayan migratory system: the Gaddi pastoralists at Himachal Pradesh, India. (a, b) Flock movement; (c) Gaddis community person; (d) a horse used for carrying kids and logistics; (e) a dog used for guarding the flock; and (f) making a rope using coarse fibres.

small stationary flocks, where flock sizes range between 5 and 20 animals. Due to various factors such as ease of handling, shifting occupations, shrinkage of forest land and changing forest policies, the farmers prefer medium-sized flocks. One of the distinctive features that differentiate Gaddis from other nomads is having appreciable land holdings and a permanent house in the mid hills. Elderly people and women live in these houses while males accompany the animals during migration periods. In some families, rotational substitution of young people is followed. The migratory flock size varies between 100 and 350 animals. Each flock has three or four horses for carrying loads during migration and two or three dogs (of distinct Gaddi breed) for guarding and navigation. In general, the flocks consist of equal numbers of goats and sheep. Exclusive goat flocks are also owned by some shepherds, which was previously not the common practice.

These goats are entirely dependent upon grazing. During winter, the flocks graze in the valleys, while in summer they move to high-altitude pastures. The main reason for their migration to the lower hills during winter is the absence of grazing facilities due to heavy snowfall at high altitudes. The shepherds migrate from the mid hills to plains in November and return back to the same region in April. From April onwards they migrate to high altitudes (2600–2700 m above mean sea level) and remain there until May. They also move to higher hills (3000–3600 m above mean sea level) of the Lahaul and Spiti, Churah, Satrundi and Rohtang areas from June to August. During winter (November to March), they migrate around Paonta Sahib, Sataun and Sarahan areas of the Sirmour district and Arki in Solan district. The migration to these places is accomplished in large groups.

7.3.3 Changpa pastoralists

Pastoralism is the only source for Changpa nomads as the possibilities for agriculture farming in Changthang (Ladakh region) are quite limited. Changthangi goats, sometimes called 'Changra' goats, are primarily used for pashmina production and secondarily used for meat production. Pashmina goats are of great importance for vitalizing the economy in the poverty-stricken region of Changthang in the Leh districts of Ladakh, India. The Changthangi goats are reared by the tribal Changpa nomads of the Changthang region of Greater Ladakh who remain in migration in the vast barren land of Changthang for good pasture throughout the year. The Changpas (nomads of Changthang) of Ladakh region of Jammu and Kashmir keep goats, sheep and yak in their flocks. The Changthang region of Ladakh is classed in the cold arid zone and the climate type is cold desert. The climate is dry with scanty rainfall and very low thermal and hydric indexes. The temperature ranges from −40°C in peak winter (Jan) to +35°C in summer. The average humidity varies between 13–90% and the highest humidity corresponds to the lowest temperature. The snowfall in the area is very scanty and when it occurs it remains throughout the winter due to the very low winter temperatures. Wind velocity in the area is very high. The animals graze the grasses in the natural pasture area from morning to evening. Each village has their own allotted pasture areas where the farmers migrate throughout the year. They follow the typical rotational grazing established by their ancestors. They migrate according to the availability of pasture grasses in different seasons and have a reserved pastureland for winter. A member of the family, mostly females, takes the flock for grazing early in the morning (by 7 a.m. in summer and 9 a.m. in winter). They return from grazing by 6 p.m. during summer and 4 p.m. during winter. Kids are allowed to suckle their mothers up to 2–3 months of age, and thereafter the kids are released for grazing outside grasses (Fig. 7.2).

In the Changthang area, the goats are kept in open paddocks (Lass) made up of either stones, bricks or sometimes made by dried bushes and their roots (see Fig. 6.5 in Chapter 6). There are no housing facilities even during the cold harsh winter. The floor of paddocks has a 7–10 cm thick compact dried manure layer. The housing space for these animals varies during summer and winter as it is necessary to accommodate large numbers of goats in compact spaces. Animals of all age ranges are kept in the same paddock and the goats are cohoused with the sheep. No special housing is provided during parturition, and kids are kept in separate open paddocks while the dams are allowed out for grazing 2–3 days after parturition. In some areas, the goats are kept in the open in all the seasons with a watch dog (Changthangi dog) guarding the animals. The kids require protection from cold, hence they are kept in small burrows covered with stones and cloths during night hours. This is done to avoid direct wind exposure to kids, as the winds blow at very high speeds. The kids are brought out of the burrows only for suckling. Sick kids are also kept in these holes. During rainfall, snowfall and disease conditions, animals are kept in the tents where the Changpas themselves live. Since all the flocks are migratory, there is no proper housing facility for these goats; the main cause of kid mortality is

Fig. 7.2. Dry mountain migratory system: the Changpa pastoralist of Leh and Laddakh, India. (a) Goat herd; (b) horses carrying logistics; (c) a dog kept for guarding the flock and house; (d) a tent inhabited by the pastoralists; (e, f) the grazing landscape.

hypothermia. The goats are provided supplements during the peak winter season in the form of barley grain, wheat grain, local peas and dry alfalfa grasses. The health-related problems reported were contagious ecthyma, gid, liver fluke and pediculosis.

7.3.4 Gola pastoralists

Ganjam goats are medium-sized animals mostly raised by farmers of the 'Golla' community (Fig. 7.3). Ganjam goats are also known as 'Golla' goats, 'Lanka' goats or 'Dalua' goats. The farmers of the Chhatrapur, Rambha and Khallikote areas of the Ganjam district, Odisha (India), belong to the Golla community and rear their goats under the extensive or range system. Golla farmers do not have any sheds to keep their goats in. The native tract of Ganjam goats falls in east and the south-eastern coastal plain region and the north-eastern Ghats. The agro-climatic condition is hot and humid with a mean annual rainfall of 1577 mm and a relative humidity between 80% and 90%. The average summer temperature is around 39°C and the average winter temperature is around

Fig. 7.3. Costal migratory system: Gola pastoralist in the Ganjam district, Odisha, India. (a) Ganjam goat; (b) portable kid house; (c) preparing own food; (d) herd of goats; (e) suckling kid; (f) movement during migration.

11.5°C. The Ganjam goats of Gola farmers are entirely raised on natural vegetation. The farmers do not provide any supplementary feed to the animals whether they are migrating in the jungles or in the villages. Kids are allowed to suckle the does for 2–3 months. The farmers provide tree lopping for grazing. This community base their livelihood in vulnerable coastal areas that face climate risks such as cyclones, heavy rainfall and coastal erosion.

7.4 Characterization of Production Systems

Global livestock production systems have been classified based on the value of production and also based on types of rearing. The FAO classified livestock production systems mainly as (i) where less than 10% of the total value of production comes from non-livestock farming practices and (ii) a mixed farming system where more than 10% of the total value of production is from non-livestock farming activities (Sere and Steinfeld, 1996). The grassland-based production system and landless livestock production system have been adopted in developed countries, including Central and South America. Grazing and mixed farming systems are concentrated in temperate and tropical highland zones, humid and sub-humid tropical and sub-tropical zones and arid and semi-arid tropical and sub-tropical zones of the world. A landless livestock production system is otherwise known as an intensive production system, which is characterized by less than 10% of animal feed produced on the farm and the animal density being more than 10 livestock units per hectare. A grassland-based production system (extensive system) is characterized by more than 10% of animal feed being produced on the farm and an animal density of less than 10 livestock units per hectare. Mixed farming systems represent livestock and crop production systems, and includes rain-fed as well as irrigated mixed farming systems. Globally, mixed farming systems provide more than 54% of total meat and 92% of total milk production followed by landless systems and grazing systems (Solaiman, 2010).

In Asia, mixed farming systems dominate the livestock production system in both arid/semi-arid and humid/sub-humid climatic zones. Mixed irrigated systems support the production system and make crop production more feasible (FAO, 1996). The mixed rain-fed farming system is also widespread in the region, and is especially difficult for livestock due to high temperatures and high humidity. The integrated crop/animal systems include rice/wheat/goat/sheep in rain-fed and irrigated mixed farming systems in Asia (Devendra *et al.*, 2005). Goats are kept in all types of climatic zones and the population density is less in mixed rain-fed systems. Goats also have an important role in the livelihood of grassland production systems and are mainly used for meat and milk production (Robbins, 1992). The type of goats or livestock varies in different production systems in different geographical regions. Information on the productivity of goats in different production systems is not available. Moreover, goats are selective browsers, therefore information on grazing of other ruminants may not be directly applicable to goats. It is important to establish a sustainable goat production system. The mapping of goat production systems can be carried out by analysing the seasonal feed calendar, rainfall patterns, cropping calendar, feeding methods, disease calendar, flock size and distribution by household. Similarly, it is also necessary to know the ratio of lactating goats to new-born kids, the number of breeding males to females, non-productive breeding females, reproductive rate, and so on. Moreover, it is essential to rank the problems of a village or an area with respect to goat production (see Protocol 2 at the end of this chapter).

7.5 Assessing Goat Production Systems

Goats are found worldwide and reared in a wide variety of production environments. The local breeds are well adapted to local ecological conditions. Goats are reared in different production systems in the tropics. It is essential to characterize the production system, in the particular agro-ecological region, and devise appropriate methods to maximize the gain for farmers/producers. Generally, farmers keep goats for meat, fibre and milk production. The farmer's income can be enhanced by increasing the growth rate or per day milk yield of the goats. In order to optimize the management system it is important to reduce the input cost in a sustainable manner to increase profitability. Similarly, the system requires the farmer to reduce the risks associated with goat production in pastoralist systems. Moreover, intensive production systems require more efficiency by optimum resource management and value addition. Therefore, it is necessary to assess the production system, and the desired interventions should be implemented for effective management and better output. Farmers are the key players in the system, and the evaluation of the production environment should be carried out based on the farmers' perceptions. Mapping of production systems requires effective interactions with all the stakeholders, including farmers. It is necessary to analyse the social and ecological conditions, the weather patterns along with the irrigation systems and associated factors for better output results. The production system needs to be dissected from agricultural, livestock and other angles to decide the effective intervention for profitability. Similarly, the disease conditions of the humans, livestock and plants need to be analysed to develop a preventive health programme for a sustainable life. Soil health analysis is also an added parameter for sustainable growth in all the agro-ecological regions. The factors affecting human and livestock reproductive health and other aspects need to be analysed. For ruminant production, it is necessary to analyse the rainfall pattern, cropping pattern, feed and fodder availability, feeding method, feed calendar, reproductive rate, the breeding males and females, kid mortality and the health of animals.

7.5.1 Methodology to assess production systems

Rapid rural appraisal (RRA) or participatory rural appraisal (PRA) are techniques that are used to identify the problems in agriculture. PRA is one of the techniques that is employed to assess and analyse the socio-economic and agro-ecological conditions. PRA is defined as a method of learning about the rural landscape and conditions from active participation of rural people. PRA techniques help in rapport building with villagers and play a major role in not only initiating a process of dialogue with the rural poor, but also allowing them to participate in planning and decision making.

The main aim of PRA techniques is to make rural people participate in their own investigation and analyse the situation related to their socio-economic and agricultural development. Villagers are supported to carry out the task and researchers need to start the process and then sit back and observe. In PRA, knowledge is gathered in a more participatory manner where interviews, investigations, transects, and mapping and diagramming are carried out by the villagers themselves (Gomez, 1985; Ruttan, 1987). Collection of information for PRA requires inquisitive observation by the interviewer and in-depth analysis of the situation.

The following PRA tools and techniques were used for the collection of information.

Techniques

- **Direct observation:** A method to observe the objects, events, processes, relationships and record systematically.
- **Key informants:** Those persons who will provide detailed information about the village.
- **Group interview:** A group consisting of 15 people is interviewed to get community level information.
- **Structured interview:** All the stakeholders are interviewed based on the selected/set questions about the major problems of the villagers.
- **Semi-structured interview:** An interaction between farmers, development agencies and scientists based on predetermined and new questions as per situation analysis.
- **Secondary data review:** Information collected from secondary sources such as government departments, research institutes or universities in order to cross check the data about the villages.
- **Participatory village mapping:** One important aspect of PRA technique is participatory village mapping. This mapping technique is done mainly by villagers to cover different aspects of the village such as social issues, village resources, agro-ecological conditions and different technologies adopted by the villagers, mobility of the villagers to different areas for different needs and sources of motivation.
- **Family interview:** A personal discussion with an individual family of the village about socio-economic and farming conditions.

Tools

- **Social map:** Describes the value system, leadership pattern, different social norms of different organizations and social interactions. The social map should be drawn first and can be modified based on inputs from other villagers.
- **Resource map:** Describes different natural and man-made resources present in the village. The natural resources include different types of land, water, forest and waste land. Man-made resources include transport, communication, irrigation types, marketing facilities, processing facilities, banks, schools, post office and community centres. The problems and opportunities can be described and discussed on the basis of a resource map.
- **Agro-ecological map:** Describes the soil type, topography of the land, types of vegetation, rainfall pattern and distribution, cropping calendar and feed resource availability in a village.
- **Technology map:** All the technologies that have been adopted, discontinued or over-adopted are listed on this map.
- **Mobility map:** Depicts the movement of villagers to different places for education, medical facilities, fetching of drinking water, marketing facilities or work. Mobility of the villagers depends on several factors in the village.
- **Farm household map:** Describes the structure of a typical household in the village.
- **Transect:** A transect walk is taken up through the village streets and fields in order to get an in-depth knowledge regarding different aspects of land use, AEZs, soil conditions, land-use patterns, irrigated crops, livestock resources, fisheries and microenvironments.
- **Timeline:** Describes the events that took place in the village chronologically. This includes introduction of radio, television, hospital/medical centres, new crops or varieties, use of fertilizers and artificial insemination. It describes natural

disasters such as outbreaks of drought, flood, epidemics, famines and other social events.

- **Time trend:** Depicts the quantitative changes that took place in the village over the last 10 or 15 years. For example, the changes in area under cultivation, yield, irrigated area, livestock population, fish production, forestry, fodder crops, horticultural crops.
- **Seasonal analysis/seasonal calendar:** Helps in identifying months of greatest difficulty or vulnerability and other variations that have an impact on village lives. It provides ideas about the cropping calendar, climate variation, crop pests and diseases, the food calendar, fodder availability, and human and livestock disease occurrence.
- **Venn diagram:** Depicts the key institutions and individuals in a community and their relationships and importance in decision making. Each institution or individual is represented by a circle in the Venn diagram. The size of the circle represents the degree of importance or scope.
- **Wealth ranking:** Ranking of households in a community based on certain criteria such as land holding, crop production or jobs.
- **Livelihood analysis:** This analysis is in the form of diagrams used to interpret the behaviour, decisions and cropping strategies of households with different socio-economic characteristics.
- **Daily routine analysis:** Depicts the average daily life of villagers. The activities they perform in a day from early morning until they go to bed at night are described in the daily routine analysis.
- **Impact diagram:** Shows the impact of certain adopted technologies by the farmers. Impact diagrams show the technology adoption behaviour of the farmers.
- **Matrix scoring:** A form of analysis used for determining technology decision behaviour of farmers. In this analysis, a matrix is prepared by placing the reasons for adoption of a certain technology in one column and other technologies in a separate column. Key informants are asked to give reasons for a particular technology and to rank them according to the importance as they feel appropriate. The higher the number, the greater the importance placed on the technology under review.
- **Flow diagram:** Shows the flow of information in and out of the village.

7.6 Problem Identification in Agro-ecological Regions

One of the major objectives of agricultural research is to understand the problems that farmers experience and to propose adequate measures or technologies to control these issues. Agricultural scientists work on this strategy and enhance productivity in different parts of the world. It is important to identify, discuss field problems and provide suitable technologies to overcome such problems through an interdisciplinary or multidisciplinary approach. In tropical countries, most farmers are poor and literacy rates are low. It is a challenge to identify problems in farmers' fields. Finding solutions that fit with their practical needs will bring a radical change in the villagers' life by which economic conditions can possibly be improved. By increasing the adoption rate of agricultural technologies, problem identification at the farmers' level is required to formulate needs-based programmes for research and extension systems (Gomez, 1985; Sabarathnam, 1988). The methodology for problem identification for agricultural research is described in Protocol 1 at the end of this chapter.

7.7 Indicators for Sustainability in Goat Production Systems

Goats are the best animals for arid, dry-land and rain-fed agro-climatic zones, providing a livelihood to millions of households in tropical countries. Goat keeping has been a traditional activity and a primary source of livelihood for many smallholders. Goats are distributed in all types of extreme climates and are well adapted to specific environments. They have comparatively low susceptibility to prevailing diseases and can be easily raised with locally available feed and fodder resources or under intensive and semi-intensive grazing systems. Goats are the animal champion for social justice – they provide a potential pathway from poverty to prosperity for resource-poor farmers globally. Modern scientific inputs in goat rearing have tremendous potential for social transformation in many disadvantaged regions. Goat production systems fulfil all the indicators for sustainability as the system is ecologically stable, economically viable and energy efficient.

Goats as a species have become important due to concern over sustainable development and for the upliftment of the rural mass in terms of economic and nutritional stability. Goats have a wider genetic base as they are distributed over varied geographical and climatic conditions with differential production capabilities. Due to their wider genetic variability, goats can be manipulated to develop suitably adapted animals in relation to different tropical regions. They can be reared alongside other livestock without any serious competition due to their preferential grazing habits: they spend more time in browsing whereas sheep and cattle graze grass. Goats are most efficient and economical for small holders in all types of production systems. These animals have faster rates of growth and an early reproductive age under harsh climatic conditions.

7.7.1 Goats as a resource

The tropical region is highly affected by climate change, and goats in the region play a significant role in providing livelihoods to farmers. Goats have wide adaptability and genotypes are observed in all extreme environmental conditions. Therefore, the genotypes can be utilized for better use during varied climatic conditions. Goats are of special economic importance in arid and semi-arid areas and in dry-land agriculture, where other livestock farming is not economic. The productivity of different goat breeds in different production systems gives an insight into their importance (see Protocol 4 at the end of Chapter 2).

7.7.2 Ecological stability

The ecological stability of the farming system should be the foremost criterion for sustainable development. Goats can thrive on many food stuffs that are not utilized by any other species of livestock. They are able to survive in hot, humid and water-logged conditions more effectively than other ruminants. Goats integrate well in mixed agriculture as they consume leafy wastes, clear weeds and provide manure. They can graze in rough terrain and desert areas for longer periods without water and utilize the roughage not used by any other animals. Goats also show a considerable degree of resistance to foot-and-mouth disease (FMD), mange and internal parasites.

Goats generally have a long snout and upright tail. The mouth is unusual in having a mobile upper lip and grasping tongue, which permits the animal to nibble tiny leaves from thorny trees. However, goats have also been labelled as an enemy of ecology with regard to their grazing habits, but there are several scientific works to disprove this suggestion so it is no longer valid. Studies at different institutes indicate that small ruminants do not play a major role in ecological destruction and their role in ecological degradation has been exaggerated (RTF, 1987). In contrast, some experts opine that they improve the grazing land by dispersing seeds, and goats have been widely recognized for controlling many weeds such as *Nasetta trichaforma, Rubus Fracticosus* and *Rosu rubiginosa*. In countries like France, Italy, Australia and Syria, there is no official bias against goats. There is no correlation between goat density and deforestation. Higher concentrations of goats are found in semi-arid and rain-fed regions in the tropics and these areas are likely to be targeted for future agricultural exploitation.

7.7.3 Viable economy

Goat farming systems are generally capable of fulfilling most of the basic needs and providing work for all local inhabitants with the guarantee of a satisfactory living standard. Goats are generally an adjunct to cropping systems where they make use of natural vegetation and crop residues supplemented by tree leaves. Goat farming is the most profitable of all the livestock in different farming conditions. As it is a low- investment proposition for farmers, it is a very assured method of income for marginal and smallholder farmers who survive in part-time employment as agricultural labour. The small size of goats is especially significant as it relates directly to economic, managerial and biological advantages; this is especially true in marginal and rangeland areas where rainfall and available feeds are sparse.

It has been reported that a goat keeper with a small flock could earn about INR 250 (US$4.17) per month, whereas the incomes of goat keepers with medium and large flocks are INR 475 (US$7.92) and INR 800 (US$13.33) per month, respectively (Singh and Ram, 1987). The economics of owning one buffalo versus 5 goats in Rajasthan (India) was studied and it was observed that the profit per annum was INR 1945 (US$32.42) from the goats and INR 1755 (US$29.25) from the buffalo (Pasha, 1990). Similarly, a study of drought-prone total gross income of smallholder and marginal farmers was accounted from small ruminants (Pasha, 1990).

The average gross income per household was highest in goat farming as compared to other livestock and crop farming systems. In arid regions, goats were more economical than sheep and cattle. Goats in deserts that are kept under range conditions are 40–160% more economical than sheep, and that an indigenous goat may provide INR 250 (US$4.17) per year to small farmers besides 2 quintals of manure and the benefits of clearing obnoxious weeds and thorny bushes (Ghosh and Khan, 1980).

7.7.4 Energy efficiency

The energy efficiency of a system is measured as input–output energy ratios. Therefore, it is necessary to understand the ecosystem and the farmers' indigenous practices and knowledge to design highly energy efficient systems. Inputs such as support energy, land and labour are not that costly for a goat production system. As far as the grazing habit is concerned, goats can thrive on vegetation that is not suitable for other animal species. As a ruminant it can use roughages and non-conventional feeds for its energy utilization and growth. Land mainly targeted for goat grazing is rangeland and wasteland. Another major input is labour, which is lowest in goat farming in tropical countries (Sastry, 1995). Goat rearing is mainly handled by women and children in village conditions. The output of the system is milk, meat, fibre, manure and breeding stock. Meat demand is high, so the question of overproduction of animals does not arise in the present situation. The excreta of goat urine and dung have been proved to be rich sources of nitrogen for soil fertility.

The energy ratio is defined as the relationship between energy output of a system and energy inputs that are required for the operation of a system. The energy ratio is mathematically expressed as:

$$E_R = E_o / E_t$$

where E_0 is energy output and E_t is energy input.

The energy ratio is otherwise known as the energy efficiency of a system.

Energy input

Energy intensity is estimated as the ratio of the energy inputs per mass of product. Energy inputs are expressed as direct or indirect energy. In a goat dairy farming system, the direct energy inputs are feed and fodder, grazing land, breeding bucks and electrical energy used for the management system of a goat farm. In a goat farm, the electricity is used for milk pasteurization, maintaining the temperature of the milk storage tank, water heating, lighting, ventilation, and so on.

The indirect energy is embedded in the products used by goats. Feed is the dominant energy input used by the goat, whether as conserved forage or grazed grass. Forage can be fed to goats because they can convert forage cellulose into digestive nutrients through microbial fermentation. Other major energy inputs include fertilizers, farm machinery and fuel.

Energy output

Energy output for goat production systems comprises of food and non-food (manure) items. Food energy content is calculated on the basis of product fat and protein content. A large portion of non-food energy is in the form of manure.

In general, the animal system is a poor converter of energy as basically it is a second source of energy. Solar energy is the primary source of energy. By means of autotrophs, solar energy is converted to green energy through photosynthesis. Grazing animals are dependent on green biomass as foliage. The major share of energy intake is spent on keeping up body metabolism and a small portion is used to produce milk and meat (Fig. 7.4)

For goat production, fossil energy is used for farm system management such as transport, storage and processing of feed. The energy efficiency of goat farming varies with different geographical locations, climatic conditions and socio-economic structures.

7.8 Goat Development: Options and Strategies

Different genetic improvement projects have been discussed in Chapter 2 (see Chapter 2, Section 2.2). This section presents an analysis of the different goat development projects in Africa and India

7.8.1 Development projects in India

The Indo-Swiss Goat Development and Fodder Production Project (ISGP)

This was started in 1981 in Rajasthan, with the objective of improving goat production through

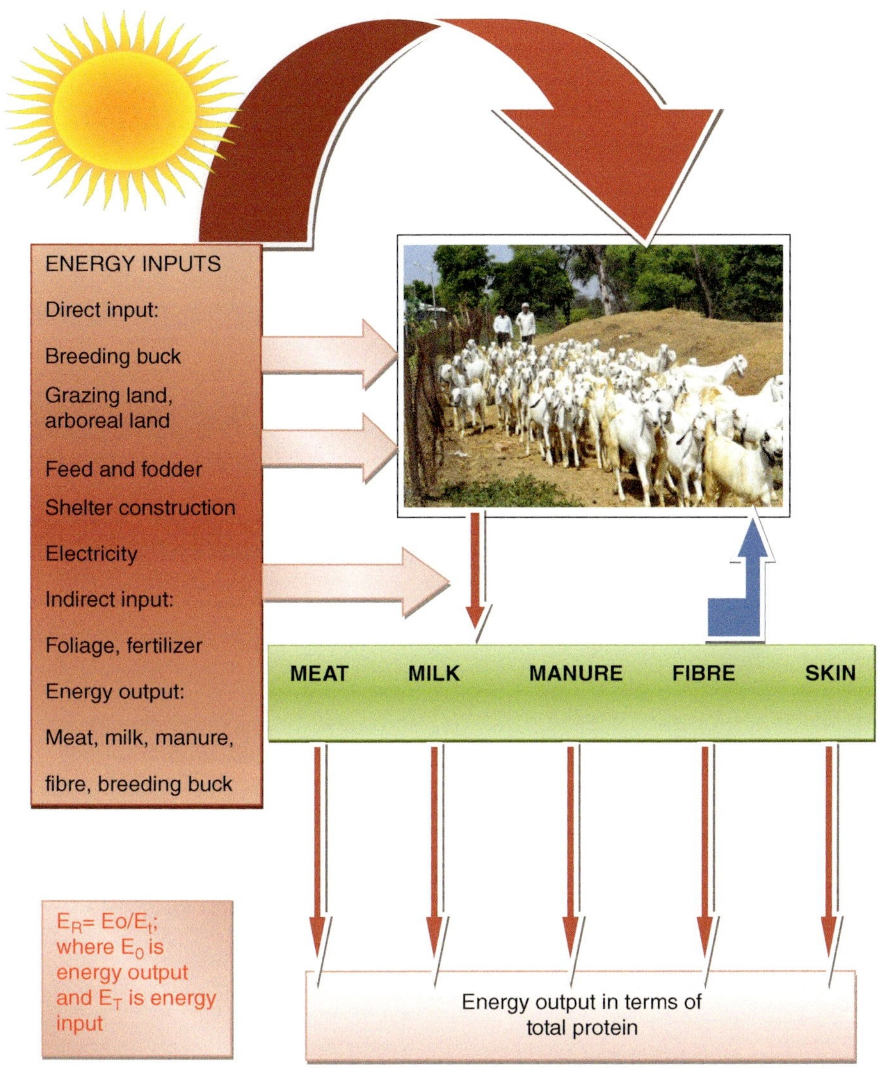

Fig. 7.4. Schematic representation of the energy efficiency model in goat production.

genetic improvement and increasing fodder production. It was a collaborative project between the governments of India and Switzerland. The project used two exotic breeds (Alpine and Toggenburg), but the crossbred offspring did not perform as expected; the pure-bred Sirohi performance was considered better with improved nutrition (Kropf *et al.*, 1992). Subsequently, it was decided to improve the performance of the Sirohi breed through selection with improved management of feed and fodder in field conditions. Sirohi goat owners are particularly known for taking good care of their animals and providing the goats with concentrate feed. The performance of Sirohi goats was excellent in smallholder production systems. These findings confirmed the importance of the Sirohi as a well-adapted and well-performing, dual-purpose breed for semi-arid conditions

The Food and Agriculture Organization of the United Nations (FAO)

A project was implemented by the department of animal husbandry and veterinary services of the government of Sikkim in collaboration with the FAO. Sikkim is a state in the north-east region of India. The programme included an RRA exercise and a characterizing production system with a large number of women participants.

The Malabari Goat Breed Improvement Programme

This project was implemented by 11 non-governmental organizations (NGOs) in the Kerala, Tamil Nadu region. Malabari breeding bucks were selected on the basis of phenotypic superiority for size and appearance and given to a cluster of goat-keeper families in villages (1 buck per 25 families). The project established community organization, strengthened existing ones and empowered women livestock keepers. It is a high rainfall area, but there is still a fodder and feed shortage in summer. Therefore, tree fodder and other forage needs to be developed and made available during the summer season. An integrated programme for improvement in health, nutrition and breeding is also needed.

Some other development projects in India have been review by Nimbkar and Ghalasi, 2012.

7.8.2 Developmental projects in Africa

FARM-Africa is promoting learning and good practice and design of goat projects across the region. FARM-Africa has learned a great deal from the successes and failures of its own grassroots projects. Some projects will be discussed for learning from past experience.

Developmental projects in Kenya

USAID supported the Small Ruminant Collaborative Research Support Program (SR-CRSP) during 1980–1992 that was implemented by the Kenyan government. The project output was a new breed called the 'Kenya Dual-Purpose Goat' (KDPG) for smallholder systems in East Africa. This synthetic breed was developed by following a four-way cross between two local breeds (Small East African and Galla) and two European breeds (Toggenburg and Anglo-Nubian) at a government breeding station in Naivasha (Fitzhugh, 1982). The ending of donor support effectively terminated the programme, and KDPGs are now rare in the region.

Another project was supported by the British Overseas Development Administration to develop goat improvement in northern Kenya. The project was carried out at the Marimanti Breeding Station in Tharaka-Nithi District between 1983 and 1989. Galla goats from northern Kenya were procured and selected for growth and mothering ability. As the funding period ended, the managers were unable to run the farms and left to find work elsewhere. The goats unfortunately developed beznoites, a disease that is hard to control, forcing many goats to be culled. Subsequently, the remaining goats were sold or stolen and the buildings are now used as a district headquarters.

Another goat development project was initiated in Burundi and Kenya during 1980–1990 with the support of the German government. The Burundi project at Ngozi pioneered the use of the buck station as a cost-effective means of crossbreeding in smallholder systems (Rey and Jacob, 1991). The project imported German Alpine goats for crossbreeding with local goats. A large number of crossbreed goats were produced but the farmers were not able to replace their bucks after the end of the project. This led to the gradual dilution of Alpine blood levels and reduced performance (Rey, 1992). This project also introduced the disease caprine arthritis encephalitis into Burundi. This complicated the management of the breeding unit and reduced its output. The project was discontinued due to the failure to supply breeding stock. The projects could not impart skills to farmers to handle the improved goats and could not provide reliable healthcare systems and delivery of limited veterinary services.

FARM-Africa initiated a Dairy Goat and Animal Healthcare Project in the Meru and Tharaka-Nithi districts of Kenya in 1994. The farmers received the support they needed and managed the project in a cost-effective manner. Breed improvement was through crossbreeding local goats owned by group members at buck stations keeping pure Toggenburg buck. The first crossbreed or F1 would be crossed again with a pure Toggenburg to produce a 75% Toggenburg goat, which was called the Meru Goat, and stabilized at that blood level. Replacement bucks were bred at a small number of breeding units consisting of four females and one Toggenburg buck, managed by a farmer nominated by their group. Training was provided to farmers on improved goat management, group dynamics, breed improvement and goat health. Buck keepers were trained how to manage the Toggenburg bucks, use them effectively, record their performance and promote their use in their community.

Development projects in Ethiopia

FARM-Africa's first goat project was implemented in partnership with the Ethiopian government and two academic institutions, Alemaya University of Agriculture and Awassa College of Agriculture (now part of Debub University), and several NGOs. The project specifically targeted women-headed households in the dry, drought-prone, highlands of east and southern Ethiopia over 1200 km apart. The project focused on the poorest households and provided two local goats on credit to each woman. The women had to repay the credit received in-kind by returning a weaned kid to the group for lending to another poor woman. A training package was developed, and women were trained in improved goat management and fodder growing. This was the first time women owners were trained in Ethiopia. Crossbreeding was planned to take place at two breeding stations at Alemaya University of Agriculture and Awassa College using the Anglo-Nubian breed.

The project generated many benefits to over 5000 households. The project helped the poorest of the poor to gain self-respect and some level of household food security. Women were able to sell goats during periods of drought and avoid receiving food aid (Zewdu, 1999). The households benefited from the increased milk production and better nutritional care for children. Crossbred male goats were in great demand and commanded high prices, particularly during festivals and holidays. The female community animal health workers earned income and their status grew in their communities as they provided treatment to sick animals. The fodder component was particularly successful in Hararghe where forage strips helped prevent soil erosion and the manure from goats supported vegetable production. Each of the 120 women's groups set up their own savings and credit association or ekub, which they used to invest in small businesses and petty trading activities. The mutual support women derived from group membership empowered them to develop skills and confidence. Lessons learned from this project formed the basis of FARM-Africa's Women's Enterprise Development Project. The project mainly focussed on the breed improvement component in breeding stations at two academic institutions and little interest was put into community development. Managing the breeding units was extremely difficult and production of crossbreeds for distribution to farmers was very slow and inefficient.

Development projects in Tanzania

FARM-Africa had a small goat component as part of its broad-based agricultural programme, the Babati Rural Development Project from 1991–2006. The goat programme targeted women in the poorest households and provided pure Toggenburg goats on credit to groups of four women. Farmers were provided training in all aspects of goat production and they developed good forage plots. The Toggenburg Breeders' Association (TOBRA) was set up in 1999 to manage the breeding stock and organized a popular goat show that became a regular event. TOBRA was the main supplier of pure Toggenburg goats in Tanzania. FARM-Africa also initiated the Tanzanian Goat Network (TAGONET), which continues to link goat practitioners together. This system of credit was very slow and did not make the best use of the valuable Toggenburg goats.

Development projects in Uganda

FARM-Africa's project in Mbale in eastern Uganda was modelled on the Meru project design. The project was implemented in eastern Uganda in the districts of Mbale, Sironko, Manafwa, Bududa and Kapchorwa in partnership with the local governments in the respective districts. The project supported about 1500 farmers organized into 39 farmer groups in nine sub-counties. The initial breeding stock was imported from FARM-Africa Kenya and Tanzania. The enthusiastic Elgon Dairy Goat Breeders' Association is currently working closely with the project in implementation and monitoring of the activities. The project has already started to generate significant benefits at both the household and community level (Alokit-Olaunah *et al.*, 2007). The main benefits to farmers were income from the sale of castrates, goat manure and goat milk. The farmers were consuming goat milk and the manure was used by them to improve the productivity of their coffee, banana and vegetable crops. The project faces an overwhelming demand for Toggenburg goats and crossbreeds, and several district councils are adopting the model in order to support more farmers' groups and are incorporating this into their district budgets. The government extension staff work closely with the project team and are part of the animal health delivery referral system. Many requests for assistance from the project have been referred to the Director, Animal Resources at the Ministry of Agriculture.

7.9 Planning and Implementing a Goat Enterprise

7.9.1 Entrepreneurship in goat farming

Goat farming is a demand-led revolution for entrepreneurship in the tropics. Entrepreneurship is the key driver for economic development. It is necessary to plan and implement entrepreneurial programmes on livestock production for employment. Thus, entrepreneurship development in agriculture and livestock production is the best potential alternative to find employment avenues for the rural population. Entrepreneurship in livestock production is a demand-led revolution as it increases job opportunities, provides nutritional security, encourages a better use of existing natural resources and manages the production system in a smarter way. There are vast opportunities for entrepreneurship in livestock. World food demand is expected to increase up to 20% by 2050 and meat demand by 76%. Therefore, the demand itself attracts various entrepreneurs to work in this area.

7.9.2 SWOT analysis

Strengths

Goats are found more in ecologically fragile arid and semi-arid areas and mostly with poor people. Strong consumer preferences for quality animal foods especially from goats are likely to change the prevailing goat production scenario in Asia and Africa during the coming years.

- Goats are comparatively tolerant to different environmental stresses and have higher feed utilization efficiency on forest-based resources/rangelands. They have varied genotypes adaptable to a wide range of agro-ecological conditions.
- They sustain themselves by a top grazing habit and can be raised with locally available feed and fodder resources or under intensive or semi-intensive grazing systems.
- Goats fit well into existing farming systems with no competition from traditional large livestock.
- They have an efficient production system due to short gestation and early maturity.
- Goats have better rearing potential under constraints of limited land and forage area. They are comparatively less susceptible to prevailing diseases.
- They are economically more viable and profitable, i.e. low investment, high and consistent returns, flexible for farmers in capital investment (space and feed, management), saleable at any point in time and a steady source of income (sell one group, get another ready).
- Entrepreneurs realize commercial viability due to the remunerative price of goat produce.
- Changing consumers' preferences for better quality meat.
- Goat milk has medicinal value and is a source of nutrition for children and pregnant women. Goat cheese and butter have export potential.

Weaknessess

Goats are associated with the poorest people and distributed over the most disadvantaged places in the world. Due to low socio economic status and lack of awareness, resource-poor goat keepers often fail to adopt or implement modern methods of goat husbandry. Poor production of local goats, poor nutrition and inadequate health care are limiting factors for optimum productivity. The major weaknesses of goat production systems are:

- Goat rearing is not well accepted by all classes of people; mostly confined to disadvantaged classes and smallholders.
- Lack of value-added and demand-based goat production systems and unscientific slaughter practices.
- Lack of definite breeding, grazing, marketing and goat slaughter policies.
- Shrinking grazing land and inadequate feeding practices.
- Scarce and inadequate health programmes and affordable disease preventive measures for goats. Kid mortality up to 1 month of age is very high in different production environments.
- Limited availability of superior germplasms for improving native/local goats.
- Unorganized and middleman/buyer-oriented marketing.
- Limited credit and insurance schemes for goat farming.
- Misconception that goats are an ecological threat is hindering goat farming in forest-rich states.
- There is no platform for all players (breeders, farmers, various input suppliers, butchers, middlemen, marketing, and professionals) to interact and find solutions to their problems for increasing and sustaining productivity.

Opportunities

Various opportunities are present in goat farming management and production systems. The major opportunities have been identified as:

- Designing programmes and strategies to improve welfare in goat farming management and better scope for using artificial intelligence in this field.
- Future therapeutic protein production by biopharma industry for human use.
- Gain support from people below the poverty line.
- Role to play in the cottage industry.
- Goat milk is being recognized as an important health food for the future and can be produced for different health requirements.
- Goat milk powder and cheese have better significance in human nutrition and excellent quality cheese has export potential.
- Tailor-made foods from goat milk to better fit human needs.
- Lean meat from goats is preferred by consumers.
- Valued as the best bio-organic manure producer.

Threats (and challenges)

Activity-based and/or commodity-based groups never focus on improving the backward and forward linkages required by producers to extract maximum profit from their enterprises. Some of the challenges and threats associated with goat farming in the present scenario include:

- Unavailability of superior bucks and their indiscriminate slaughter at an early age.
- Loss of germplasm due to a change of strategy in agricultural production patterns and intensity.
- Lack of supply chain systems.
- Poor conditions of farms.
- Endemic and emerging diseases are causing high mortality rates. Lack of proper diagnostic tools and immunoprophylactic measures has led to high mortality rates.
- Indiscriminate breeding with available breeds and crossbreeds.
- The pasturelands where the goats are traditionally reared is diminishing.
- Due to frequent natural disasters such as droughts and floods, farmers are compelled to dispose of good animals including kids, as they do not have the resources to face such eventualities.

7.10 Project Formulation and Implementation

7.10.1 Writing a project proposal

In goat science, a project can be focussed on different aspects such as goat improvement, goat biology, energy efficiency, better solutions for a known problem, environmental factors affecting goat production, and so on. The formulation of a research project or writing a proposal requires a specific approach and is part of the role of scientists or researchers. The formulation of societal development projects, technological ideas or improved methodologies in goat improvement programmes is the responsibility of developmental agencies, NGOs and farmers. The basics of project writing should be known to all the development workers and should be prepared according to the requirement of funding agencies. The funding agencies support the programme and so it is important to fulfil their requirements and to follow their project submission format. The basic structure of a project proposal is as follows:

1. Title of the project.
2. List of participants.
3. Project summary.
4. Description of work: Background: physical context, socio-economic analysis, current situation; problem description: knowledge gap, past work done; objectives; methodology; activities to fulfil the objectives.
5. Project duration: Phasing of activities and targets; timeline; activity milestones, targets and deliverables.
6. Inputs/resources: Feed, goat, equipment, veterinary medicine, training needs and extension material.
7. Budget: Project cost; human resource requirements, consumables.
8. Expected output: Impact of the proposal and innovation in proposal and societal impact; sustainability of the project and social change.

7.10.2 Planning a goat business

Starting a goat farm enterprise

Goat farming is a suitable venture in various environmental conditions. Before starting a business operation, it is essential to know about the agro-climatic region, local goat genotype, and feed and fodder availability. The individual must be clear about the objectives and careful decisions should be taken before starting the programme. It is also important to analyse the marketing channels to determine the long-term sustainability of the business venture. A business plan along with a feasibility study is required to start goat farming. Goat farming can be established as an enterprise, where an enterprise budget analysis is required to determine the profitability and risk management. A market situation analysis is also required for long-term sustainability of the programme. Detailed and accurate record keeping will support the manager to take effective decisions about farm management and optimize productivity and profitability. Record keeping will guide the management and identify the strengths and weaknesses of the management system.

Record keeping

An accurate record-keeping system is required for successful decision making and managing the operation efficiently. It will provide insight for better management of a flock, optimal use of resources and an idea of future expansion. There are different types of records, for example, management records and business records.

MANAGEMENT RECORDS This includes records on animal identification, the mating and kidding registers, feeding register, reproduction record, health record, vaccination and deworming (See Protocol 2 in Chapter 2 for more details). These records help the farmer to take appropriate management decisions. It is essential to analyse the system efficiency, feed conversion efficiency and identify the best individuals in a particular environment. The records are helpful to take appropriate measures to enhance production efficiency.

BUSINESS RECORDS Business records are based on general farm income and expenses. These include payroll, veterinary services, equipment, building and facilities, utilities, sales records, purchase records, insurance, income tax, and so on.

Goat enterprise budget analysis

A budget analysis is essential to start any new venture. First, it is essential to determine the herd composition, which includes the number of does, bucks and other available animals. Second, the general assumption with respect to breed characteristics such as kidding rate, survival rate (adult mortality and kid mortality) and replacement rate (doe replacement, buck replacement; the standard is around 20%) needs to be established. The major heads under the budget analysis are discussed for raising goats for meat, milk and fibre purposes; however, the cost varies according to the genotype, the agro-climatic region, type of management and product quality. The end user can consider the major heads as a guide, allocating the costs according to prevailing rates for preparing the budget analysis report. The following parameters should be assumed before planning the goat enterprise: kidding rate, weaning rate, mortality rate, conception rate, feed price, and the price of meat, milk and fibre. Feed cost is around 60% of total variable costs; the second highest cost is labour which is around 28–39% of operational costs (Solaiman *et al.*, 2010).

Goat farming budget analysis is the major tool to examine both physical and financial targets within a particular time frame. The budgets will depend on market price, location and specific output. The details for budget analysis for meat, milk and fibre production are discussed in Section 7.10.3.

OPERATIONAL COSTS Operational costs of any enterprises may be variable or fixed costs. Variable costs include costs associated with feeding, housing, veterinary services, labour cost, marketing, equipment repair, and so on. Fixed costs include costs such as sheds, equipment depreciation, building repair, insurance, and taxes and interests. Interest on capital investments such as building purchase and equipment purchase is also considered to be a fixed cost.

REVENUE OR INCOME Revenue is defined as the number of items sold multiplied by the price. The revenue or income of a farm can be increased by targeting specific market requirements of the consumer during a specific period of the year. Milk production and sale during festivals provides good prices to farmers. Similarly, mature kids available for slaughter during the festival season fetch good prices. Profit can be calculated on a per doe basis and can be expressed as income above variable cost or fixed cost.

7.10.3 Budget analysis for different production environments

Budget analysis for meat-producing goats

OPERATIONAL COSTS

- **Fixed costs:** Building costs, equipment costs, insurance, income tax and interest (property tax, office expansion, bank interest).
- **Variable costs:** Feed costs (dry fodder, green fodder, concentrate feed grazing cost, feed for kids), healthcare costs (vaccinations, booster doses, deworming, other veterinary costs), management costs (bedding, marketing, supplies, equipment repair and labour).

GROSS RECEIPTS

- Market kids, sale of does, sale of bucks.

Budget analysis for milk-producing goats

OPERATIONAL COSTS

- **Fixed costs:** Milking units and other aspects such as building costs, equipment costs, insurance, income tax and interest (property tax, office expansion, bank interest).
- **Operation costs:** Feed costs (grass fodder, dry fodder [legumes], grazing, pelleted feed, minerals, kid feed, milk replacers), healthcare costs (vaccinations, booster doses, deworming and other veterinary costs), milk transport costs, bedding, power (electricity), vehicle repair, equipment repair, housing improvement, taxes and insurance, interest on capital.

GROSS RECEIPTS

- Milk, market kids, sale of does, sale of bucks.

Budget analysis for fibre-producing goats

OPERATION COSTS

- **Fixed costs:** Building costs, equipment costs, insurance, income tax and interest (property tax, office expansion, bank interest), dogs (working dogs, 1 dog for every 100 goats), ranch horse and truck, bucks, labour (approximately 70 hours/ week 10 h/day)
- **Operation costs:** Feed costs (grass fodder, dry fodder [legumes], grazing, pelleted feed, minerals, kid feed, milk replacers), healthcare costs (vaccinations, booster does, deworming, other veterinary costs, veterinary medicine), crop insurance, predator control, shearing costs, store costs, marketing costs, electricity, repair and maintenance, interest on capital

GROSS RECEIPTS

- Market kids, sale of does, sale of bucks, sale of Mohair or pashmina (adult, yearling, kids), sale of fibre.

Budget analysis for subsistence goat farming

Goats are raised in subsistence farming systems in tropical countries. In this system, social values and nutritional security are major factors and can outweigh economic value. Economic analysis of different subsistence farming systems such as pastoral systems, arid desert and high humid in smallholder rearing is required for improving the management systems and increasing farmers' incomes. The components of budget analysis need to be determined for each production environment. The first step will be to describe the enterprise with respect to a production system and general assumption about flock size, kidding rate and mortality rate, and so on. The next step will be to assess the total gains and losses.

It should include: (i) description: using an improved breeding buck for improved goat production; (ii) gains (comparative additional gains); and (iii) costs saved, which can be prepared as per Table 7.2.

Table 7.2. Partial enterprise budget for subsistence goat production.

Description	Using an improved breeding buck for improved goat production
Gains (comparative additional gains)	
Costs saved	
Extra revenue	Higher priced offspring
	Food security
	Women prestige
Losses (comparative additional losses)	
Extra costs	Payment for buck service
	Extra labour for animal care
	Extra tree fodder for bucks
	Improved shelter for bucks
Loss of revenue	Use fodder for fire/timber income from selling corn used for extra feeding
Extra profit	Total additional gains - Total additional losses

Source: Andreasen and Qwist-Hoffmann (1997).

7.11 Protocols

Protocol 1: Problem identification

It is important to understand the farmer's problems and to propose adequate measures or technologies to control those problems. This is one of the major objectives of agricultural research for enhancing productivity in different parts of the world. As progress and economic development depend on agriculture, it is necessary to identify and discuss field problems and provide suitable technologies to overcome such problems through interdisciplinary or multidisciplinary approaches.

Methodology

Interview ten key informants for problem identification and interview 30 farmers for problem identification.

Farmers rear different livestock species such as cattle, sheep and goats. Thus 30 farmers should be selected based on their skills and experience from different categories of farmers. The selected farmers are asked to indicate their technological needs and the problems that they face in specific areas of goat management, goat diseases, breed performance or marketing, and so on. Suppose the farmers identify five or ten problems, they will then be asked to rank these problems by using the key ball/snowball sampling technique.

Farmers identified problems in goat production management and breeding and the problems were quantified for the purpose of analysis on the basis of the number of farmers who had given a rank to each problem. A rank-based quotient (RBQ) was then calculated for each problem:

$$\text{RBQ} = \frac{f_i(n+1-i)}{N \times n} \times 100$$

where F = frequency of farmers for i^{th} rank, N = the total number of farmers, n = the number of ranks and i = rank.

The farmers were also asked to estimate the loss or extent of damage due to each problem, which was then converted into a percentage. For example:

$$\%\,\text{yield loss in milk due to a problem} = 100 - \frac{\text{milk yield in problematic goat}}{\text{milk yield of healthy goat}} \times 100$$

The correlation for RBQ and the average loss between the values of the key informants and the farmers were then estimated using Spearman's rank correlation:

$$r_c = \frac{1 - \text{di}2 \times 6}{N(n+1-1)}$$

If the correlation is significant, then take the average of RBQ and average loss values due to the problem for the two sets of data, otherwise use only the farmer's data for further analysis. The magnitude of the problem is calculated by using the formula:

Village magnitude volume = RBQ × average yield loss percentage × number of animals

Correlation for RBQ between the values of key informants and farmers is calculated using Spearman's rank correlation:

r_c = 0.98 (significant at 5% and 1% level)

Correlation for average loss between the values of key informants and farmers:

r_c = 0.90 (significant at 5% and 1%)

Correlation coefficient is significant for both the parameters, so the average of RBQ and average loss was taken for calculating the village magnitude value.

Based on the problem identification and rankings, the problems will be classified for research and extension programmes, after which the perception report will be prepared.

Classification of problems

The problems should be ranked into priority needs for research and extension.

Perception report

An example of a perception report is given below, showing the research problem identification with justification.

No availability of good milking cows The farmers do not have good genetic potential in their milking cows in their region, and depend on markets in neighbouring regions. It is necessary to state a research project for improving the genetic potential of cows in local areas	Extension and research

Continued

Repeat breeding problems	Research
Loss in productivity of cows in subsequent generations	Research
No availability of good sheep breeds	Extension and research
Improvement of production of good breeds and good management	Extension and research
Crossbred animals are more prone to production diseases like mastitis	Research
No availability of good quality semen and artificial insemination problems	Extension
Deshi cows are not producing well	Research
Improvement of buffalo productivity and management	Extension and research

Protocol 2: Production system characterizations and analysis

PART A: Format for characterization of goat production systems

1. NAME OR ADDRESS OF GOAT UNIT/STATION

____

____

2. FARMERS' CLASSIFICATION

Landless farmer / female-headed farm families / tribal farm families / others

____

3. FARMERS' CATEGORIES

Size of holdings	No. of farmers	%
Small		
Medium		
Large		
Landless		

4. DESCRIPTION OF FARMING SITUATIONS

Farming situation	Area (hectare)
Cultivable land/net sown area (ha)	
Total area (ha)	
Cropping intensity (%)	
Topography	
Plain, hilly, coastal, others	

5. CROPPING SYSTEM

(i) Cropping System 1----, Cropping System 2----, Cropping system 3----

Major crops/ horticulture crops/ agro-forestry		Area (ha)	Varieties grown	Productivity	Major problems
i)					
ii)					

6. LIVESTOCK

Major livestock category	**Number**	**Productivity**	**Major problems**

7. MAJOR PROBLEMS IDENTIFIED IN THE VILLAGE

Problem category		**Problems**
(i)	General	
(ii)	Agriculture	

8. ORGANIZATIONS WORKING IN THE VILLAGE

Name and address of the organization	**Brief activity and impact**

9. GEOGRAPHICAL SITUATION

Name of agro-climate zone	**Altitude, latitude, rainfall (cm/day), temperature range, relative humidity (%)**

PART B: Background information

1. ABOUT THE FAMILY:

Centre:	Cluster :	District:			
Village:	Date of survey :				
Name of registered farmer:					
Members in family	Male:	Female:		Total :	
Education level of family head Tick (√) where applicable	Illiterate	Primary		Secondary school	
	College	Technical			
Main occupation Tick (√) where applicable	Agriculture	Animal Husbandry	Service	Labour	Other
Economic status Tick (√) where applicable	Above poverty line	Below poverty line		Other	

2. LAND TYPE

Land type	Total land (acres)	Land under cultivation (acres)
Irrigated, non-irrigated, wasteland		

3. OWN ASSETS

Type	Area (sq. feet)	Kachcha / Pucca	How old (no. of years)	Remarks
House				
Shed				
Store				
Back yard, etc.				

4. SOURCES OF IRRIGATION

Type	In use since when	Area under irrigation (acres)	
		Seasonal	All year round
Well, bore well, lift irrigation, canal irrigation, others			

5. RESOURCE MAPPING
(Tick √ where applicable)

Agricultural implements	Oil engine	Tractor	Tractor trolley	Sprayer pump	Chaff cutter	Bullock cart, etc.
Vehicles	Cycle	Motorcycle		Four wheeler (jeep, truck, car)		
Information	TV	Radio	Newspaper	Internet	Kisan helpline	Mobile based

6. GETTING INFORMATION REGARDING TECHNOLOGY/INPUT (AGRICULTURE/LIVESTOCK)

Agriculture department	
Village level work	
Veterinary department	
NGO/developmental agency	
Any other	

7. LIVESTOCK INVENTORY
Information about animals (write down in table as per age group).

Type of Animals	Below 1 year		1 to 3 years		Above 3 years					
					Male			Female		
	Male	Female	Male	Female	For Breeding	Breeding & Agri. purpose	Only for Agri.	In milk	Dry	Pregnant Animal

PART C: Information about goat production system

1. METHODS OF BREEDING

Natural service by bucks:

- Fee charged for service by bucks or free?
- In case of repeat service, fees charged or free?
- Which points are considered for selection of bucks to be used for natural service? (Breed conservation / Cross breeding / Production / Physical Appearance)

Which aspects are considered at the time of breeding of animal?

- Age of animal
- Animal's physical development

Which aspects are considered at the time of breeding of goats?

- When animal comes on heat first time post-partum
- Fixed time after parturition
- Bred after kidding when milk production is reduced

Problems faced in breeding of animal

Repeat breeding	Yes / No
Difficult birth (Dystokia)	Yes / No
Retention of placenta	Yes / No
Abortion	Yes / No
Any Other (Specify)	Yes / No

2. FEEDING SYSTEM

Is colostrum fed?	Yes / No			
If yes, within how many hours of kidding	2 hours	2 to 5 hours	After 6 hours	After 12 hours
How is colostrum fed?	By bottle		By suckling	
Is navel dressing done?	Yes / No		Antiseptic solution used:	
Navel cord cut by:	Using new blades		By any other means	
Feeding of milk	By suckling		Separately after pouring in bucket	
If allowed to suckle, how many teats?	One teat		Two teats	
When is kid given milk?	After milking		Before milking	
If fed after pouring into bucket, how much quantity?	____ litres			
How many days is kid allowed to suckle? (circle the number of days)	30, 40, 60, 75, 90 days			

3. FEEDING OF ADULT ANIMALS

1. Grazing hours:				
2. Provided at home:	Yes / No			
3. Name the item and feeding procedure:				
Green fodder	Dry fodder	Leaf creepers	Fodder trees	Plants
Dry fodder availability	Regularly	Occasionally		Not available
Green fodder availability	Regularly	Occasionally		Not available
4. If green fodder is less which type of animals are given it and at what rate?				
Concentrate availability	Regularly	Occasionally		Not available
5. Concentrate is given to which type of animals?				
Milking animals:				
Pregnant animals:				
2 months before kidding:				
6. Are special feeds given to animals after kidding?				
7. Is mineral mixture fed?	Yes / No			
If yes, what quantity and how it is fed:				
8. How is fodder fed?				
After chaffing by chaff cutter:				
By chaffing by spade :				
Without chaffing:				
9. If sent for grazing?	On own land	Pastureland		In Jungle
If yes – how many hours sent for grazing?				
10. Are feeds offered at night after grazing?		Yes / No		
If yes, in which season	Monsoon	Winter	Summer	Round the year
11. Is concentrate feed given?	Yes / No			
12. Method of feeding:				
By pellet	By grinding the grains	Simple grains		Any other
At the time of milking	After milking		Before milking	
13. How is concentrate feed offered?				
In a pot/bucket	Manger	Floor	Designated feeder	
Particulars	Milking	Dry	Pregnant	Young Growing

Concentrate (kg)				
How many days?				
14. Have pregnant animals been given additional concentrates in the last two months?				Yes **/** No
Type of feed:		Balanced concentrate		Homemade
If homemade, components used in homemade concentrate mix			Yes / No and quantity (kg)	
Cake				
Rice bran				
Wheat bran				
Mineral Mixture				
Oil cakes				
15. Is Silage used in feeding?				Yes / No
Main fodder used in silage making				
16. Seasonal fodder availability				
Types of fodder		Summers	Winter	Rain
Name of available grasses				
Name of available tree lopping				
Name of available dry fodder				
Name of available green fodder				
17. Feed availability and prevalent rates				
Type of fodder	Green fodder	Rate per quintal	Dry fodder	Rate per quintal
Different grasses				
Lucerne				
Cow pea				
Berseem				
Oat				
Concentrate component				Rate per kg
Groundnut cake				
Mustard Cake				
Readymade concentrate				
Others				

4. ANIMAL MANAGEMENT SYSTEMS

Kid management:

- Provenance of kid diseases:
  (1) Diarrhoea (2) Pneumonia (3) Naval illness (4) Tetanus
- Whether preventive vaccination done:
  (1) Regularly (2) Occasionally (3) Not done
- Vaccination against which disease?
  FMD / HS / PPR / ET / POX
- Are medicines for ectoparasite used?
  (1) Regularly (2) Occasionally (3) Not given
- How many times is medicine used in a year?
- Are deworming medicines given:
  (1) Regularly (2) Occasionally (3) Not given
- At what age is deworming started? Months of age _______
- Is there an interval between two deworming months?
- Coccidiosis (stunted growth, diarrhoea in 2–6 months-old kid) observed or not and medicine given

Adult management

Goat shed type	Kachcha (thatch) / Pucca (cemented/concentrate) / Plartiz (plastic)		
When is preventive vaccination done?	Regularly	Occasionally	Not given
Vaccination against which disease?	FMD / HS / PPR / ET / Goat Pox	Repetition_______	
Mortality in last year?	Yes / No		
Mortality rate in kids (0–3 months)			
Mortality rate in adults			
Do animals get water for a whole day?	Yes / No		
How many times is animal offered water in a day?	Once / Twice / Thrice		
How much water is given to each animal?	___ Litters		
Source of water for animals	From own farm	From outside	
	Well/bore well	River, Rivulet	
	Supply from village water committee	Pond	

Buck management

- Specially fed with concentrate:
- Any special feed:
- Any other special case:

5. ANIMAL HEALTH SYSTEM

Main diseases	Yes	No	Mortality (adults stock)	Mortality (in young stock)
FMD				
HS				

ET/ PPR/ GOAT POX				
Parasitic diseases				
Poisoning				
Other illness				
1. Is there incidence of tick and ectoparasites?			Severe / Medium / No	
2. Use of anti ectoparasite medicines		Regularly	Occasionally	Never
3. Which medicine is used?	______________________			
4. Is any traditional medicine used for removal of ectoparasites?			______________	
5. Incidence of Mastitis	More		Less	None
6. Is treatment of mastitis undertaken?			Yes / No	
7. Are males castrated?			Yes / No	
8. Number of kid deaths				
Up to 15 days:			Up to 3 Months:	
Up to 1 month:			Up to 6 Months:	
Above 6 Months:				
9. Number of deaths			Summer:	
Winter:			Raining:	

6. REPRODUCTION MANAGEMENT SYSTEM

- At the time of parturition, what is done to aid the expulsion of retained placenta?
- How is placenta disposed of?
- What are the methods of milking?
- Uses of milk produced (% usage).

Price for treating animals

Type of treatment (for the herd)	Amount invested
Vaccination	
Expenses on medicines, doctor's visit fees, etc.	
Insurance fee if animal is insured	

Any ideas regarding anthelmintic resistance:

- How do know about worm load?
- Do you examine faecal samples?
- Any other method?

7. ANIMAL SALE/CULLING SYSTEM

- Is the owner taking technical advice? Regularly / occasionally / never
- Changes in flock size envisaged in the next 12 months:

Particulars	Yes	No
Want no change		
Want to increase flock size		
Want to decrease flock size		

- How is the sale/purchase of animals done?
- Who takes the decision for the sale of adults/young and male/female animals? What are the criteria for that?
- How is the sale/purchase of animals done?
  Through local market   Through agents   Others

Price of goat sale (as per age and breed):

Below 1 year		1 to 3 years	
Male	**Female**	**Male**	**Female**

Are other by-products (e.g. vermin-compost) used?

- How are goat dung and urine used?
  - For making dung cakes
  - For making organic manure
  - Other use
- How are the above items utilized?
  - Used in one's own field
  - Sold to other farmers

8. MISCELLANEOUS INFORMATION:

How many hours are spent on labour work?

Work done by	How many hours / which type of work
Owner	
Owner's spouse	
Children	
Other members of family	
Hired labourers	

Does the owner keep records?

Particulars	Yes	No
Milk production		
Feeds and feeding		
Financial matters		

Awareness about farmers on:

Particulars	Yes	No
Proper feeding systems		

Oestrous cycle and symptoms of heat		
Pregnancy diagnosis		
Diseases spread due to ectoparasites		
Preventive measures for ectoparasites		
Coccidiosis		

Reasons for keeping animals:

Economic details of milking animals:

- Rate of goat milk sold per litre
- Rate of cow milk sold per litre
- Amount of milk kept for home consumption in litres
- Purchase rate of cow milk in market per litre
- Purchase rate of buffalo milk in market per litre
- Labour charge rate for male labour
- Labour charge rate for female labour
- Hours for labour

Mortality

- Number of goats in the year
- Number of kids born
- Number of animals sold
- Number of adult animals died
- Number of kids died (0–3 months)

Income from animals

Particulars	Total amount
Sale of kids	
Sale of heifers	
Sale of bucks	
From sale of buffaloes	
From buffalo heifers	
From buffalo males/bulls	
From sale of sheep, if owned	
Sale of manure	
If above manure used in own field, amount saved due to smaller quantity of chemical fertilizers purchased	
Sale of empty gunny bags	
Sale of dead animals	
Other income	

Part D: Traditional medicine used for goat illness

Illness/diseases	**Local ingredient used and method of application in brief**	**Period since using**	**Experience (recovery score 1–5)**
Diarrhoea			
Pneumonia			
Ectoparasite			
Internal Parasite			
Wound Management			
Fever			

Part E: Economics of goat rearing

1. FEED AND FODDER QUANTITY CONSUMED AND COST: PER GOAT PER DAY (INR)

Feed	**Name of feed/fodder**	**Summer**	**Rainy**	**Winter**	**Average price (INR/kg)**
Adult doe (>1 year)					
Green fodder					
Dry fodder					
Concentrates					
Other if any					
Adult male (>1 year)					
Green fodder					
Dry fodder					
Concentrates					
Other if any					
Kids: 3 months to 1 year					
Milk (<3 months)					
Green fodder					
Dry fodder					
Concentrates					
Other if any					

2. LABOUR COSTS

- Number of hired labourers:
- Wages (month or year):

3. ACTIVITIES PERFORMED BY FAMILY LABOUR (NO. OF HOURS SPENT PER DAY)

Work performed	Men	Women	Kid
Feeding and watering of animals			
Bringing grass from field			
Grazing of animals			
Cleaning of animals			
Milking of animals			

4. VETERINARY CARE COST

- Vaccination
- Boosters
- Anthelmintic drugs
- Coccidiostat
- Preventive medication
- Other

5. MORTALITY

- Number of kids died
- Number of adults died
- Price received from butcher for dead animals

6. WHAT ARE THE CONSTRAINTS IN GOAT REARING IN YOUR AREA? GIVE THE RANK:

Constraints	Rank
Unavailability of feed and fodder	
Unavailability of veterinary care	
Unavailability of quality bucks	
Unavailability of goat market	
Lack of credit facility for goat farming	
If other please specify	

Part F: Innovation in goat production

Name of innovation:
Detail of innovation:
Category of innovation:

Goat health	Goat housing	Milk and meat processing	Goat nutrition	Goat management	Buck selection	Others

Use of innovation:
How you know about this innovation:
How long have you used this innovation?
Can it be commercialized? Yes No
Approximate cost of innovation:

7.12 References

Alokit-Olaunah, C., Misinde, N., Bisagaya, A. and Wabule, C. (2007) *Mbale Dairy Goat and Animal Healthcare Project*. Impact Assessment Report, March, 2007, Mbale, Uganda.

Ameso, E.A., Bukachi, S.A., Olungah, C.O., Haller, T., Wandibba, S. *et al.* (2018). Pastoral Resilience among the Maasai Pastoralists of Laikipia County, Kenya. *Land* 7, 78.

Andreasen, L. and Qwist-Hoffmann, H. (1997) Inter-Regional project for participatory upland conservation and development (Field Document 6/97) - NEPAL - A framework for a participatory economic evaluation of improved goat production by women groups in the Bhusunde Khola watershed. Food and Agriculture Organization of the United Nations, Gorkha. Available at: http://www.fao.org/docrep/x5676e/x5676e08.htm (accessed 27 May 2019).

Devendra, C., Morton, J.F. and Rischowsky, B. (2005) Chapter 3: Livestock systems In: Owen, E., Kitalyi, A., Jayasuria, N. and Smith, T. (eds) *Livestock and Wealth Creation: Improving the Husbandry of Animals Kept by Resource Poor People in Developing Countries*. Nottingham University Press, Nottingham, UK.

Dong, S., Wen, L., Liu, S., Zhang, X., Lassoie, P.J., *et al.* (2011) Vulnerability of worldwide pastoralism to global changes and interdisciplinary strategies for sustainable pastoralism. *Ecology and Society* 16, 10.

FAO (1996) *Forest Resources Assessment* 1990. *Survey of Tropical Forest Cover and Study of Change Processes Based on Multi-Data High Resolution Satellite Data*. FAO Forestry Paper 130, Food and Agriculture Organization of the United Nations, Rome, 170 pp.

FAO (2005) Geographical trends in livestock densities and nutrient balance in South, East and South-East Asia. By Gerber, P., Chilonda, P., Franceschini, G. and Menzi, H. Lead. Electronic Newsletter V3N1, March. Available at: http://pigtrop.cirad.fr/subjects/environment_and_natural_resources_protection/geographical_trends_in_livestock_densities_and_nutrient_balances (accessed 27 May 2019).

FAO (2017) *The State of Food and Agriculture*. Available at: http://www.fao.org/3/a-i7658e.pdf (accessed 27 May 2019).

Fitzhugh, H.A. (1982) Dual-purpose goat production systems for smallholder agriculturalists. Proceedings of the 7th SR-CRSP Kenya Workshop. Held in Nairobi, Kenya, 15 March 1982. pp. 10–11.

Ghosh, P.K. and Khan, M.S. (1980) *The goat in desert environment. Research Bulletin* No. 12. CAZRI, Jodhpur, India.

GOK (2014) *County Assembly of Laikipia: Strategic Plan, 2013–2018*; Draft 3; vol. 1. Government Printers, Nairobi, Kenya.

Gomez, A.A. (1985) A farming systems research approach to identify farmers' production problems. In: Cornea, M.M., Coulter, J.K. and Russell, J.F.A. (eds) *Research Extension Farmer*. The World Bank, Washington DC, USA, p. 61.

Idris, A. (2011) Taking the camel through the eye of a needle: Enhancing pastoral resilience through education policy in Kenya. *Resilience Interdisciplinary Perspective Science and Humanities* 2, 25–38.

Kropf, W., Narayan, A. and Prasad, R.A.A. (1992) Comparison of reproductive performance and milk production of Sirohi goats with Alpine and Toggenburg crosses. *International Conference on Goats* 5, New Delhi, India, pp. 565–575.

Nimbkar, C. and Ghalsasi, P. (2012) Small ruminant rearing: Breed Conservation and Genetic Improvement. Document on South Asia pro poor livestock policy programme. NDDB and FAO, New Delhi, India.

Pasha, S.A. (1990) *Sustainability and viability of small marginal farmers with reference to animal husbandry and common property resources*. Paper presented in National Seminar held October 8 to October 10 at NIRD, Hyderabad, India.

Pavanello, S. (2009) *Pastoralists' Vulnerability in the Horn of Africa: Exploring Political Marginalization*, Donors' Policies and Cross-Border Issues—Literature Review, Humanitarian Policy Group Report, ODI, London, UK.

Peacock, C (1996) *Improving Goat Production in the Tropics*. Farm-Africa and Oxfam, Oxford, UK.

Peters, K.J. and Horst, P. (1981) Development potential of goat breeding in the tropics and subtropics. *Animal Research and Development* 14, 54–71.

Reid, S.R., Fernandez-Gimenez, M.E. and Galvin, K.A. (2014) Dynamics and Resilience of Rangelands and Pastoral peoples around the globe. *Annual Review of Environment and Resources* 39, 217–242.

Rey, B. (1992) Contribution of cross-bred goats to milk production and social welfare in Burundi. In: Kategile, J.A. and Mubi, S. (eds) *Future of Livestock Industries in East and Southern Africa*. Proceedings of a workshop held at Kadoma Ranch Hotel, Zimbabwe, 20–23 July 1992. ILCA, Addis Ababa, Ethiopia.

Rey, B. and Jacob, U. (1991) *Farm structure and participation in the Ngozi goat cross-breeding project in Burundi*. In: Proceedings of the 9th SR-CRSP annual meeting, Nairobi, Kenya, 8–9 March 1990.

Robbins, P. (1992) Goats and grasses in western Rajasthan: interpreting change. In: *A Collection of Papers from Gujarat and Rajasthan*. ODI, London, p. 612. Available at: https://www.odi.org/sites/odi.org.uk/files/odi-assets/publications-opinion-files/5405.pdf (accessed 27 May 2019).

RTF (1987) *Government of India. Report of the Task Force to Evaluate the Impact of Sheep and Goat Rearing in Ecologically Fragile Zones*. Government of India, Ministry of Agriculture, Department of Agriculture and Cooperation, New Delhi, India.

Ruttan, V.W. (1987) *Agricultural Research Policy and Development*. Food and Agriculture Organization of the United Nations, Rome, Italy.

Sabarathnam, V.E. (1988) *Manual on Field Experience Training for ARS Scientists*. National Academy of Agricultural Research Management, Hyderabad, 30.

Sastry, N.S.R. (1995) *Livestock Sector of India: Regional Aspects*. International Book House, Lucknow.

Sere, C. and Steinfeld, S. (1996) *World livestock production systems: current status, issues and trends*. FAO Animal Production and Health Paper 127. Food and Agriculture Organization of the United Nations, Rome, Italy.

Singh, K. and Ram, K. (1987) Economic analysis of goat keeping in the goat breeding areas of Punjab. *Indian Journal of Animal Sciences* 57, 317–323.

Solaiman, S.G. (2010) *Feeds and Feeding Management in Goat Science and Production 2010*. Wiley Blackwell Publication, Hoboken, New Jersey, USA.

Solaiman, S.G., Kebede, E. and Aviki, E.M. (2010) *Business plan, production enterprise and marketing strategy*. In: Solaiman, S.G. (ed.) *Goat Science and production*. Wiley- Blackwell Publishers, Hoboken, New Jersey, USA, pp. 339–357.

Zewdu, A. (1999) Reducing drought difficulties through goats: The experiences of FARM-Africa's Dairy Goat Programme in Eastern Hararghe. In Proceedings of the Third Annual EAGODEN Workshop, Harar, Ethiopia, November 1999.

8 Goat Products Supply Chains

In the early 1950s producers used to be completely involved in the whole process, from feedstock procuring and product development and the production process, through to the distribution to wholesalers or retailers. This process was generally managed by one promoter to minimize the product cost with good profit. The promoters would not make partnerships with other businesses and used a self-managed logistic system. Keeping large inventories in the store to sustain mass production was considered to be necessary. At the start of the 1970s, as a result of heavy competition, the promoters started to realize that they needed to reduce their inventory quantity in order to reduce costs. In this connection, manufacturers started introducing material requirement planning (MRP) and manufacturing resource planning (MRPII) systems to minimize inventory holding. This could only be possible by the application of highly sophisticated software packages and related information technology. Manufacturers were able to track goods in the logistic process to understand inventory numbers. Manufacturing processes like just-in-time and total quality management became an integral and unavoidable method for getting real-time information on goods transition and manufacturing. The 1980s witnessed the complete systematic process of feedstock procuring methods, manufacturing process management and logistic aspects of product transfer to end users. Manufacturers started focussing on product quality, quality assurance and quality management, efficient logistic processes for transferring products from the site of manufacturing to end users before planning the development of an industry.

The goat product supply chain management process involves the movement of a variety of goat products from the site of production to the consumer (end user), taking into account both the quality and quantity of the products, including product safety and guarantees. The product movement process has to follow certain sequential steps that are regulated by different organizations that are specialized in the process assigned. In certain steps of the supply chain process the value of the products in movement stages gets increased due to quality control and upgrading. The processes involved in upgrading the original product are known as the value chain. The main difference between a value chain and a supply chain is that a supply chain is the process of all parties involved in fulfilling an end user (consumer) request, whereas a value chain is a set of interrelated activities that a company uses to create a competitive advantage.

There are generally five steps a small ruminant farmer or cooperative farmer can use to gain a competitive advantage over their competitors:

1. Inbound logistics: The overall process of managing how quality breed goats are acquired, nurtured and transported to their final destination as per customer requirements. In addition, the farmer has to find prospective distributors and suppliers, and determine their efficacy and accessibility. Logistics managers (actors) are known as logisticians. Warehousing and inventory control management is also an important aspect at the beginning stage of farmhouse development and establishment. This intermediate step is a collateralized debt obligation (CDO) transaction. The warehousing period typically lasts 3 months and it comes to an end upon closing of the CDO transaction. The CDO is a structured financial product that pools together cash flow-generating assets into discrete tranches that can be sold to investors.
2. Small ruminant farm operation processes involve value-creating activities that transform input into the production of quality goats fit for milking and auction at the market (local or state level) for meat production.
3. Outbound logistics system development to supply high-quality breeds to customers on a demand basis.

4. Marketing and sales to suitable buyers to purchase goats, mainly on a cash payment basis or through bank transitions.
5. Effective additional efficiency services should be provided to enhance the goat value, such as customer support.

8.1 Goat Milk Supply Chain Management

Generally, a niche product is defined as a product produced or developed for a particular segment of a larger community or larger industry or larger market. Goat milk is a niche product. Four continents, namely Asia (57.8%), Europe (13.7%), Africa (25.2%) and America (3.2%), are the main producers of goat milk.

In order to promote social inclusion and the empowerment of women, Gujarat Co-operative Milk Marketing Federation Ltd. (GCMMF) was formed in 1948, which today is jointly owned by 3.6 million milk producers in Gujarat (The Amul Story, 2005).

The Anand Milk Union Limited (or Amul) is an Indian dairy cooperative based at Anand in the state of Gujarat for collecting, manufacturing and selling milk to both retail and wholesale dealers all over India. It is an excellent example of a milk supply chain management system (Fig. 8.1) (Alexander, 1977). It is managed by the GCMMF.

Goat milk is as nutritious as cow milk, and it can be used in place of cow milk if it is scarce (Table 8.1). India, China, Bangladesh and Pakistan keep 46.7% of the world's goat livestock, and 55.60% of the world's goat milk is produced by India, Bangladesh, Sudan and Pakistan (Table 8.2). The production of goat milk by continent in 2013 is given in Table 8.3. The highest goat milk producing continent in the world is Asia (10.65 million tons), followed by Africa (4.65 million tons), Europe (2.53 million tons) and, lastly, the Americas (0.59 million tons).

No substantial difference is noticed between goat milk and cow milk, apart from goat milk being whiter than cow milk. So, butter and cheese made from goat milk are whiter, but may be coloured while processing. Goat milk is tasty, nutritious and wholesome.

Goat milk is naturally homogenized like cow milk, where the cream remains suspended in the milk instead of floating on the surface. The most value-added quality of goat milk is its non-allergic nature, whereas cow milk can result in allergic reactions, often in children. Due to the presence of smaller fat globules, the curd made from goat milk is softer than cow milk. Goats produce milk with the yellow-beta-carotene converted directly into vitamin A. Fat globules present in goat milk are one-fifth the size of cow milk, and easy to digest.

Diversified products like different cheeses, butter, yoghurt, fluid beverages, and so on are manufactured from goat milk (Table 8.4). Goat milk has therapeutic value, and the World Trade Organization has long since reduced the price of goat-derived products (Business Wire, 2018).

In the global market, goat milk production increased by 39.2% during the period 2000–2012, with an average increase per year of 3.27%. The global goat milk market is estimated to reach revenues of around $15 billion by 2024, growing at a compound annual growth rate of more than 7% from 2018 to 2024 (Technavio Research, 2018).

8.2 Goat Meat Supply Chain Management

The supply chain management process involves introducing a product into the market with the guarantee of discharging the product at the consumer level at the desired quality and quantity in a timely manner. In the case of goats and the product supply chain, farmers or smallholders sell their goats at a local auction where a livestock dealer purchases them to sell on to a distributor either directly or through a regional auction. Generally, a distributor is a meat packer (slaughterhouse owner) or wholesale dealer. The wholesale dealer sells the carcasses or meat cuts to retail dealers in hygienically packed units based on consumer demand.

Goat meat from adults is also called chevon, cabrito or capretto, or kid when it is from young animals. Chevon is a French-derived word while cabrito is Spanish and Portuguese in origin and specifically refers to a young, milk-fed goat. In some parts of Asia (India, Pakistan, Nepal, Sri Lanka and Bangladesh) mutton is used to describe both goat and sheep meat. Developed countries enjoy goat meat as exotic livestock, whereas in developing nations like South-East Asia and Africa goat is the major source of meat production (Dhanda *et al.*, 2003).

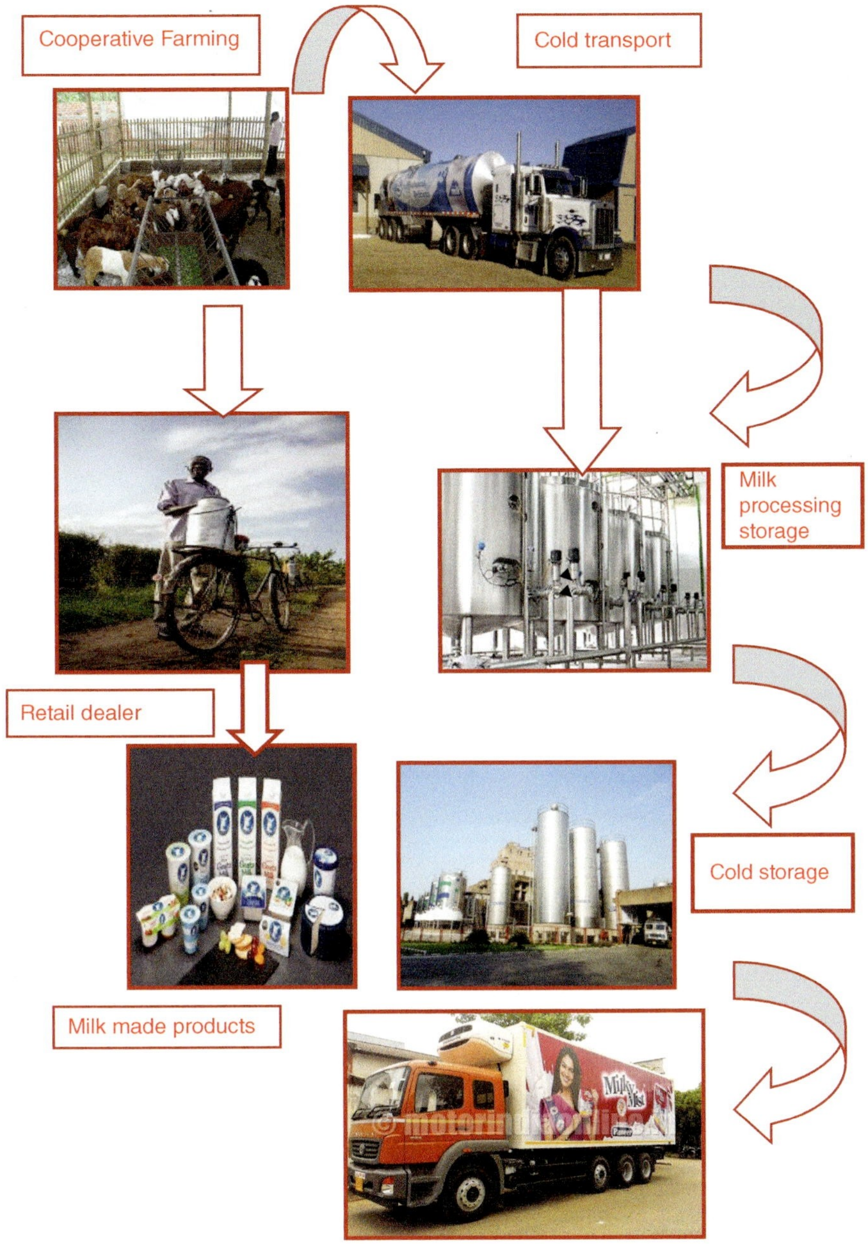

Fig. 8.1. Goat milk supply chain management.

The supply chain for goat meat is a process of events involved in the logistics systems for the movement of goat's meat from producers to consumers (Fig. 8.2). This operation process is handled and monitored by different players and needs optimization for better financial returns and benefits. Generally, the promoter breeds goats for slaughtering for a variety of purposes.

The brokers and traders buy goats directly from the farmers, and then pass the animals on to goat collectors, exchangers and transporters. Ultimately, the goats end up in an abattoir for meat production.

Table. 8.1. Physico-chemical characters of goat, cow and human milk.

Composition	Goat	Cow	Human
Protein (%)	3.0	3.0	1.1
Fat (%)	3.8	3.6	4.0
Calories (/100ml)	70	69	68
Vitamin A (IU/g fat)	39	21	32
Vitamin B thiamine (µg/100ml)	68	45	17
Riboflavin (µg/100ml)	210	159	26
Vitamin C (mg as citric acid/100ml)	2	2	3
Vitamin D (IU/g fat)	0.7	0.7	0.3
Calcium (%)	0.19	0.18	0.04
Iron (%)	0.07	0.06	0.2
Phosphorus (%)	0.27	0.23	0.06

Source: Yangilar (2013).

Table 8.2. The top goat milk producing countries in the world

Rank	Country	Production (metric tonnes)
1	India	5,000,000
2	Bangladesh	2,616,000
3	Sudan	1,532,000
4	Pakistan	801,000
5	Mali	720,000
6	France	580,694
7	Spain	471,999
8	Turkey	415,743
9	Somalia	400,000
10	Greece	340,000

Table 8.3. Major goat milk produced by continent in 2013.

Continent	Tons (million)
Asia	10,65
Africa	4,65
Europe	2,53
America	0.54

Source: FAOstat (2013)

Meat packing companies facilitate the slaughter, processing, packaging and distribution of animal meat for human consumption. These meat production methods also produce by-products that are sold as animal food and other products sold to other industries. Trains, container ships and airplane cargo holds are the usual shipping methods for most meat products around the world. New Zealand and Australian farmers from Oceania work in co-operatives to produce more than two-thirds of the world's exports of fresh and frozen ovine and caprine meats. The marketing channel describes the logistic system for goat meat from production (mostly from goat farming) to the place of consumption with the practices of cold storage transportation, handling and storage in highly hygienic facilities, ownership transfer, well-managed abattoirs, processing and packing, and distribution. These practices depend on the efficacy of the value chain and innovation.

Table 8.4. Diversified products from goats.

Types of goat products	By-products
Milk	Flavoured milk Skim milk (low fat)
Milk fermented products	Cheese Yoghurt Butter milk Kefir
Condensed milk	Instant milk powered Baby milk Lactose-free instant milk powered
Milk frozen products	Ice cream Frozen yoghurt
Biologics from genetically engineered goat milk	Therapeutic proteins
Goat meat	
Alternate goat products	Goat milk soap Goat milk paint Leather Goat hides Goat horn Goat candle Goat milk beer Mohair/cashmere Garden tea Colostrum milk/beestings

Presently, in most of the developing countries the main problem with small ruminant farmers is the variable, complex and inefficient marketing chain with different levels of participation. The main difficulty is access to the marketing chain, which is mainly due to inadequate access to market outlets, poor marketing management concepts and lack of training for small farmers. Efficient goat production and marketing is not possible without quality control and consistency in animal breeding, body weight and

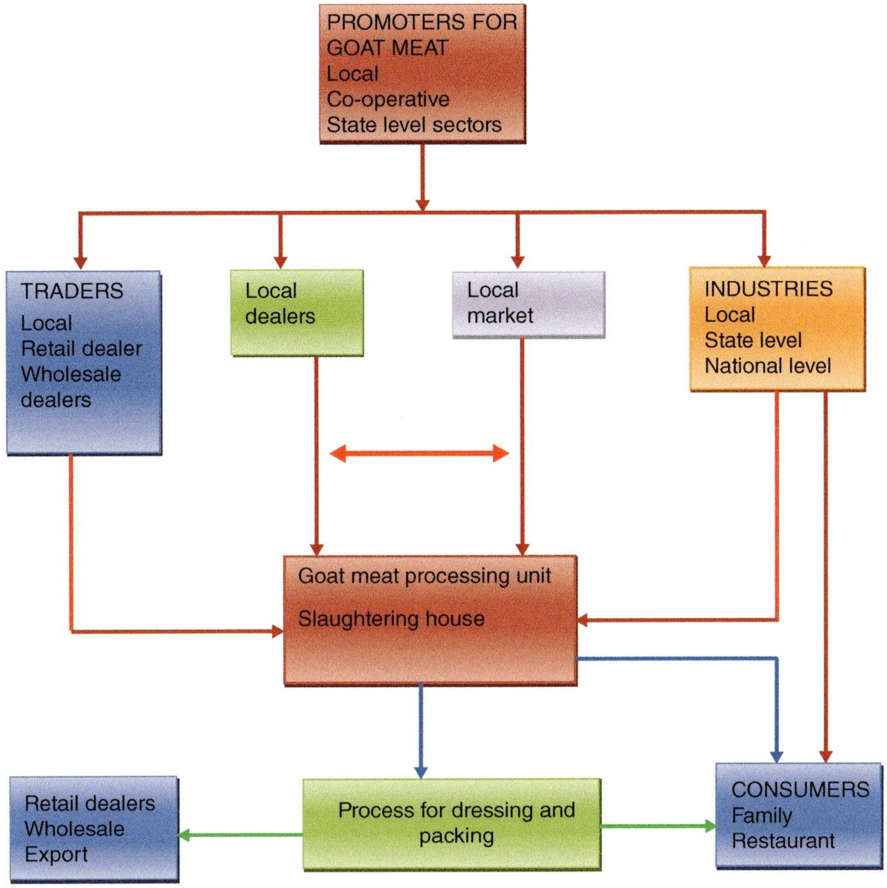

Fig. 8.2. Goat meat supply chain management.

health condition and proper processing at the abattoir. In addition, goat supply may not be properly managed on the basis of consumer demand across time. This is due to the wide variation in the price paid to producers versus that paid by consumers, which can be disheartening to the producer and could discourage them from improving the quality supply of goats in the market. The geographical location variability in goat production and availability of suitable facilities for goat slaughtering and processing are the major components of marketing cost.

The demand for goat meat in the global market can be improved by creating value-added quality to the goat meat, for example, by improving the breeding quality of goats and other environmental factors, particularly better nutrition and healthcare. By this practice the consistency in goat body weight over time can be well managed to attract the attention of buyers. The value of goat meat sales can be increased by fabrication of goat carcasses into cuts based on demand; however, this is not acceptable to customers in some areas. Deboning goat carcasses is labour intensive and expensive, but it can improve the quality of meat to make acceptable products for consumers.

8.3 Goat Fibre Supply Chain Management

Mohair and cashmere are the most common goat fibres in the market. Cashmere fibre is mainly produced by China, Russia, Mongolia and India. Producers sell to middle men or directly to processors (Fig. 8.3). In China, raw cashmere is available at central markets at prices set by the Chinese government's quality evaluation team. The price depends on demand in

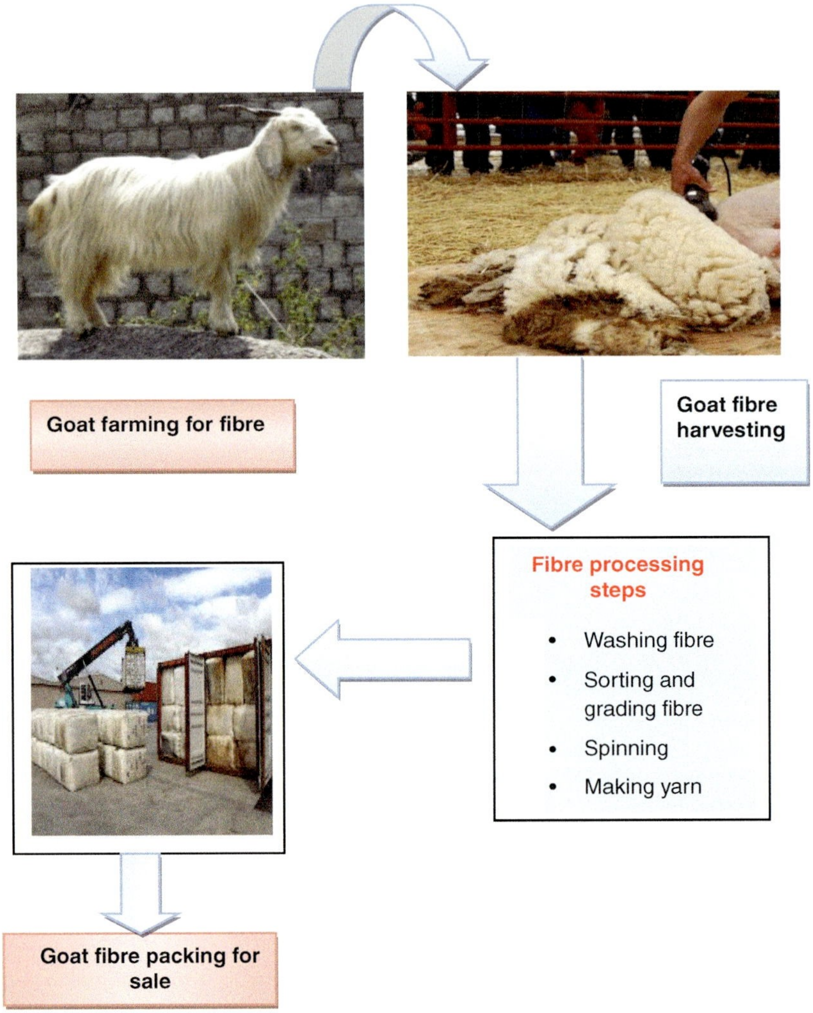

Fig. 8.3. Goat fibre supply chain management.

the world market and the requirements of different textile industries. Mohair is produced in Turkey, the US and South Africa. In South Africa, raw mohair is marketed through brokers at spot market auctions, while a private treaty method for the sale of mohair is common in the US. Mohair needs a broad market for the benefit of farmers.

8.4 Goat Skin Supply Chain Management

Globally, goat skin is produced for a variety of products like shoes, bags, gloves and other products that need a soft hide. Untanned goat skin is used for making containers for water, kefir, wine, and so on. Tanned leather from goat skin is used for making rugs and binding. Goat skin is used for traditional Spanish wine bags. Untanned goat skin is used for parchment or for drumheads or sounding boards of some musical instruments, e.g. *misnice* in medieval Europe, *bodhran* in Ireland, *esraj* in India, and for drum skin called *bedug* in Indonesia.

8.5 Goat Therapeutic Protein Supply Chain Management

Therapeutic protein production for human healthcare first started when *Escherichia coli* and

Saccharomyces cerevisiae were used as host cells for the production of simple polypeptides such as insulin and human growth hormone. But these microbe production systems have limitations in that the microbial model is not suitable for producing complex post-translational modifications or intricate folding requirements, such as coagulation factors and monoclonal antibodies. Industrialization of complex molecules, such as antithrombin and 1-antitrypsin, using recombinant techniques has not yet been reported by using microbial cells or mammalian cells as host systems. Currently, human plasma is the only alternate source for antithrombin and 1-antitrypsin, both quality and quantity-wise. In the late 1980s, reports of successful expression of therapeutic proteins in milk from genetically engineered animals such as sheep, goat, pigs and cattle created a major breakthrough in biopharmaceutical sciences (Simons *et al.*, 1987, Gordon *et al.*, 1987, Meade *et al.*, 1990)

Transgenic livestock animals like sheep, rabbits, goats and pigs are the best sources for producing recombinant therapeutic proteins (Hammer *et al.*, 1985) because the mammary gland is the best choice of expression of recombinant therapeutic proteins. Interestingly, past literature says that 100,000 kg of human serum albumin can be produced from 5400 cows, which meets the annual global demand for human serum albumin. Similarly, 5000 kg of antitrypsin (α-AT) production can be achieved by 4500 sheep; 100 kg of monoclonal antibodies (mAbs) can be produced from 100 goats; 75 kg of antithrombin III can be produced from 75 goats; and just two pigs produce 2 kg of human clotting factor IX (Bösze and Hiripi, 2012).

Thus, the progress of therapeutic proteins technology from microbes in bioreactors to expression in animal mammary glands has proven to be a better option for human protein expression (Fig. 8.4). Interestingly, it has been observed that goat milk contains several types of human proteins. (Table 8.5) (Archer *et al.*, 1994; Zhang *et al.*, 1997; Fan *et al.*, 2002, 2004; Sánchez *et al.*, 2004; Toledo *et al.*, 2006, 2008; Han *et al.*, 2007, 2009; Xiao *et al.*, 2009; Yang *et al.*, 2009).

Genetically modified goats can be used for the production of therapeutic recombinant proteins via the production of adequate milk at low cost with minimum maintenance. The state-of-the-art production of transgenic goats has diverted attention from therapeutic protein production by the pharmaceutical industry. A genetically engineered goat is thought to be a better alternative for the production of recombinant therapeutic proteins (Fig. 8.5).

Antithrombin-AT was the first human biological drug to be approved by the European Agency of the Evaluation of Medical Products (Schmidt, 2006) and the US Food and Drug Administration (FDA) (Lavine, 2009). Goats produce about sevenfold to eightfold more serum compared to rabbits. Besides this, goat serum contains about 20 mg/ml of total IgG, which is about two to three times more than rabbit serum. Thus, a goat model bioreactor is a more cost effective alternative for producing large amounts of antibody in a highly commercialized, competitive market. The most important factor is harvesting the antibody from a single animal rather than several different individuals.

The global therapeutic proteins market is worth about US$140 million and it is predicted that at the end of 2023 the market size will be US$217.23 million, registering a compound annual growth rate of 6.5% from 2017 to 2023 (ResearchAndMarkets.com, 2018). This is highly encouraging for researchers to pursue the commercialization of the goat model as a bioreactor to compete with other models presently using animal and plant cells and bacteria as host cells for producing recombinant therapeutic proteins.

In 2013, GTC Biotherapeutics changed its name to rEVO Biologics This company is the first to produce a therapeutic protein drug (recombinant human AT [ATryn]) from genetically modified goats that has been approved by the US FDA for use in the US market. The product was funded by two loans from its French partner company 'LFB Biotechnologies'. The drug prevents fatal blood clots in people with rare conditions such as haemophilia. The FDA also approved the goats used to make the drug, which was produced by a herd of 200 bioengineered goats living in carefully controlled conditions on a farm in central Massachusetts. The 'farm animals' are better alternatives to produce bioengineered drugs at lower cost and higher quality and quantity compared to the existing methods of extracting proteins from donated human blood or growing them in large steel vats of genetically engineered cells.

Other nations are also using transgenic goats for therapeutic proteins industrialization. For example, the biopharmaceutical industry in Iran produces human IX factor (Yekta *et al.*, 2013). Currently, promoters have shown interest in manufacturing therapeutic human monoclonal antibodies through goat mammary glands (Baruah *et al.*, 2006).

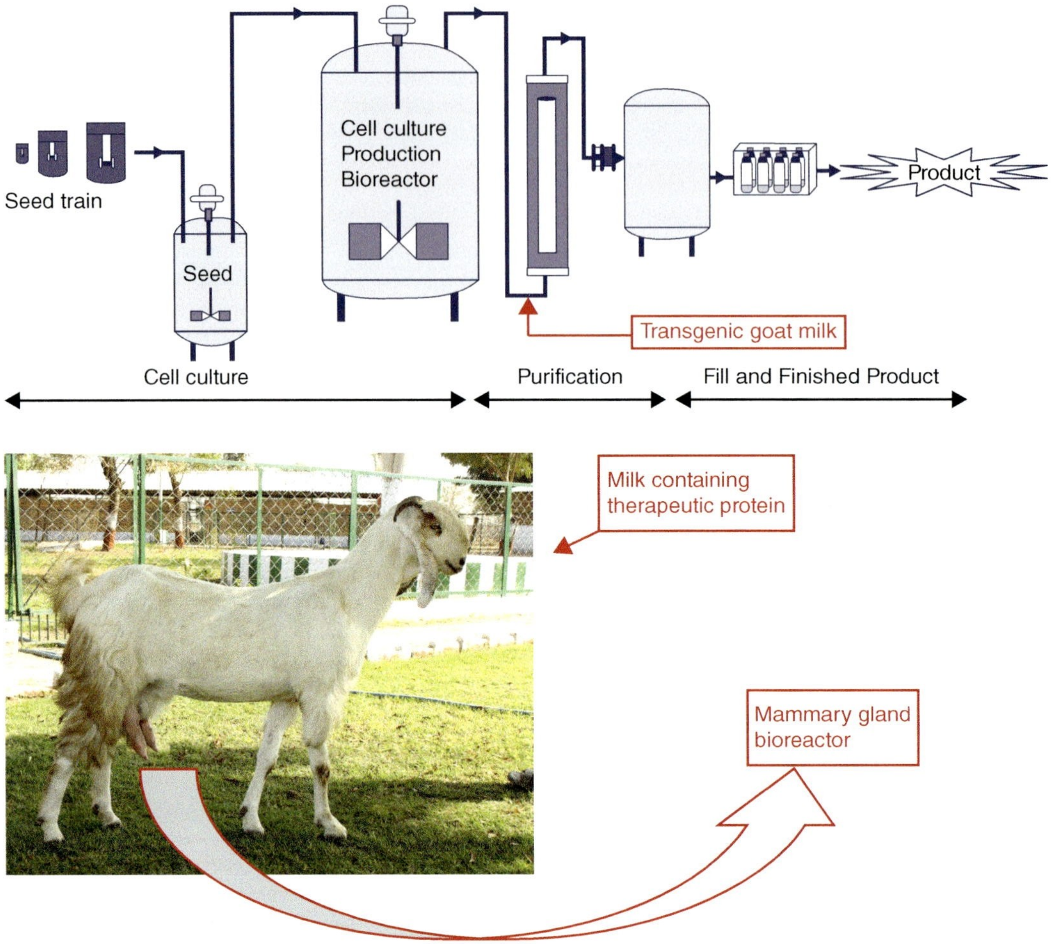

Fig. 8.4. Therapeutic protein production by an animal (mammary gland) bioreactor.

Table 8.5. Composition of different types of therapeutic human proteins in goat milk.

Gene (source)	Vector	Promoter	Expression (mg/ml)	References
Antithrombin III (human)	Adenovirus	CMV	2.8	Yang *et al.*, 2009
Growth hormone (human)	Adenovirus	CMV	2.4	Han *et al.*, 2009
Erythropoietin (human)	Adenovirus	CMV	2.0	Toledo *et al.*, 2006
Lactoferrin (human)	Adenovirus	CMV	2.0	Han *et al.*, 2007
E2-CSFV (swine)	Adenovirus	CMV	1.2	Toledo *et al.*, 2008
Growth hormone (human)	Adenovirus	CMV	0.3	Sánchez *et al.*, 2004
NGF-β (human)	Adenovirus	CMV	0.2	Xiao *et al.*, 2009
Lysostaphin (bacterium)	Adenovirus	RSV	0.02	Fan *et al.*, 2004
Lysostaphin (bacterium)	Adenovirus	RSV	0.001	Fan *et al.*, 2002
Growth hormone (human)	Retrovirus	MoMLV	6.0×10^{-5}	Archer *et al.*, 1994
Factor IX (human)	Retrovirus	β-actina	4.0×10^{-5}	Zhang *et al.*, 1997

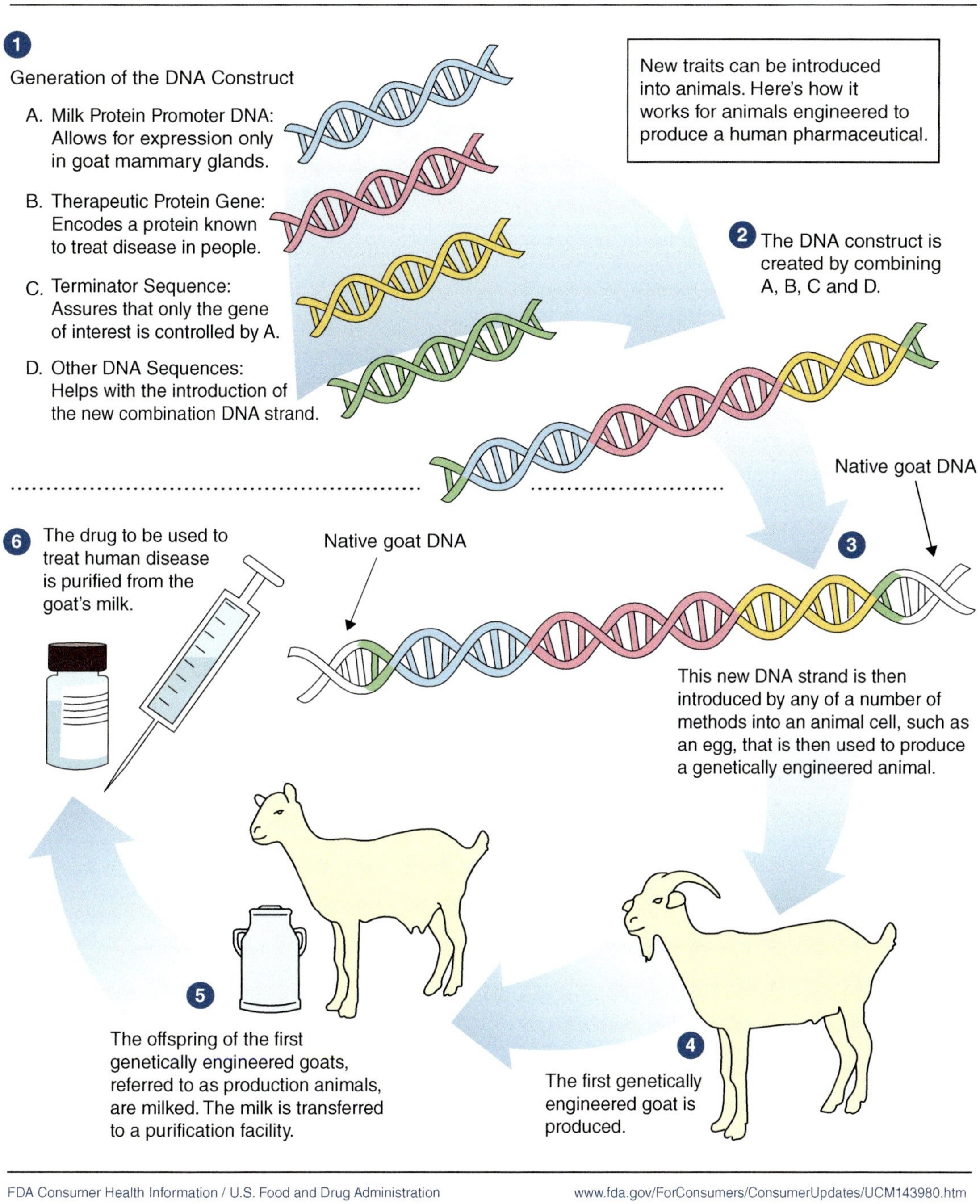

Fig. 8.5. Genetically engineered goats. Source: U.S. Food & Drug Administration, 2015.

8.6 Cold-Chain Logistics for Goat Products

At present, refrigerated and frozen logistics costs are about US$13.4 billion/year with an annual growth rate of 5–6%. It has also been assumed that this value may increase to 8–9% in the future, a trend that may last for some time. Some of the goat products need proper cooling. Milk is stored over 4°C. Proper cooling is critical for the maintenance of quality and for preventing any microbial infection. As compared to cow milk, goat milk is more sensitive to temperature and pH (Zadow *et al.*, 1983; Ram and Sindhu, 1991; Montilla and Calvo, 1997; Anema and Stanley, 1998.).

The heat sensitivity of goat milk depends on environmental factors and farming management practices. In goat milk, the transforming process from ultra-high temperature treatment (UHT, 136°C/6 s) to low temperature needs special care.

Presently, the problems related to goat milk marketing are due to inadequate commercial process technologies involved in the logistics process. Generally, the raw milk is thoroughly inspected before its transfer to the distribution system. Goat milk is an alternate feed source for dairy goat producers involved in buying and marketing lambs, veal calves or piglets. So it is necessary to have additional marketing skills. While marketing fluid goat milk, the dairy owner should have a proper strategy for obtaining additional income from the kid goat crop. Generally, kids are raised to different market weights and for different market seasons. The market value for goat milk production is mainly from six varieties of dairy goat breeds (Saanen, Nubian, Toggenburg, La Mancha, Oberhasli and Alpine). The lactation period for dairy goats averages 284 days, with peak production usually at 4–6 weeks after kidding. At present, powdered goat milk is gaining commercial importance due to its long shelf-life without any substantial changes in quality (Silanikove *et al.*, 2010; Fonseca *et al.*, 2013).

In addition, powdered goat milk is easy to distribute at low costs (Bylund, 2003; Tehrany and Sonneveld, 2010). The dehydration technology is helpful for ensuring consumers have a year-round marketable product. Spray drying is a common technology for making powered versions of any milk. Powdered caprine milk is used for a variety of products like ice cream, chocolate, yoghurt, infant formula and cheese (Kajal *et al.*, 2012). Due to the lack of cooperation and coordination among goat breeders and milk producers of cheese and other dairy goat products it has been difficult to establish large-scale production. The global goat milk market is expected to post a compound annual growth rate of over 6% during the period 2018–2022 (Business Wire, 2018).

The main reason for the increased demand for goat milk is the production of low lactose or lactose-free products, which are required for people with lactose intolerance. Goat milk is used as human infant milk formula for bottle feeding. This market fulfils the requirement of bottle-feeding or cup-feeding babies that are allergic to cow milk. The Dairy Goat Co-operative (DGC) ranks first in terms of revenue share in the global market for goat milk infant formula having 9.99% of market share in 2016, followed by Ausnutria Dairy with a market share of 6.75%.

Some of the key players in the goat milk infant formula market include DGC, Ausnutria Dairy, Baiyue Youlishi, YaTai-Precious, Red Star, Guan Shan, Milk Goat, Herds, Fine boon, Jinniu, Shengfei, ShengTang, Holle, FIT, Vitagermine, Xi Yang and the Baiyue Diary Group.

8.6.1 Facility growth

Current good manufacturing practice (cGMP) uses highly sophisticated processing units with integrated operation management and temperature-controlled warehouse facilities. Globally linked supply chain management is an integral part of third-party logistics services. The most important salient feature of third-party logistics services is based on customers' needs and market conditions, such as demands and delivery process requirements on the basis of product and materials. On the basis of special requirements, third-party logistics services can provide value-added services related to the production and procurement of goods, i.e. services that integrate parts of the supply chain, and when this integration occurs the provider is then called a third-party supply chain management provider or supply chain management service provider. The third-party logistics services target particular fractions within supply chain management such as warehousing, transportation or raw material provision (Research and Markets, 2019).

It has been observed that global third-party logistics companies reported a compound annual growth rate of 5% during the period 2014 to 2016. This market is estimated to be worth US$1054 billion by the end of 2024 (Allied Market Research, 2018). Presently, the cool chain supply management system has become sophisticated, with temperature controlling systems as per manufacturers' requirements. In addition, the provision of real-time tracking guarantees the life of the product as prescribed by the manufacturer. There is strong competition between bulk air-cargo container suppliers (e.g. Envirotainer and CSafe) and suppliers of single-use pallet shippers (e.g. Sonoco ThermoSafe and Cold Chain Technologies).

Some of the multinational groups (e.g. DGC, Ausnutria Dairy, Baiyue Youlishi, YaTai-Precious, Red Star, Guan Shan, MilkGoat, Herds, Fineboon, Jinniu, Shengfei, ShengTang, Holle, FIT, Vitagermine, Xi Yang and the Baiyue Diary Group) involved in goat milk processing use highly integrated technology with minimum expenditure of water and energy. The DGC plants have world-class facilities for manufacturing milk formula, and they can also monitor the entire process from milking to distribution. With the strong European market, DGC recorded a revenue of US$193 million in 2016. DGC specializes in developing, manufacturing and marketing its own brand of nutritional powdered goat milk for infants and children. The products are exported to more than 27 countries.

Generally, essential fatty acids, folic acid, vitamin B12, C, D, E and iron are added to modify goat milk for specific requirements. When making formula from cow milk, manufacturers add extra process steps to remove and replace the milk fat, extract and add whey proteins plus ingredients like nucleotides or oligosaccharides. The addition of whey proteins to cow milk is favoured by many formula manufacturers to improve the supply of essential and semi-essential amino acids (Janas and Pieeiano, 1987; Lonerdal and Chen, 1990). However, DGC's research confirms that whey is not necessary for formula made from goat milk (Fig. 8.6). Goat milk only requires the addition of very small amounts of two amino acids to supply all the essential and semi-essential amino acids needed for infants (European Food Safety Authority, 2012). The DGC process for goat milk involves retaining milk fats before adding lactose, essential fatty acids, vitamins and minerals. In addition, a small quantity of vegetable oil is added to top up the unsaturated essential fatty acids in goat milk without removing milk fat (Prosser *et al.*, 2010). Goat milk fat retains the medium-chain saturated fatty acids that can be easily absorbed by infants compared to long-chain fatty acids (Lindquist, 2012). Generally, goat milk is rich in nucleotides compared to cow milk, thus there is no need to add nucleotides while making formula from whole goat milk. DGC uses special techniques to avoid odours without damaging the milk fat.

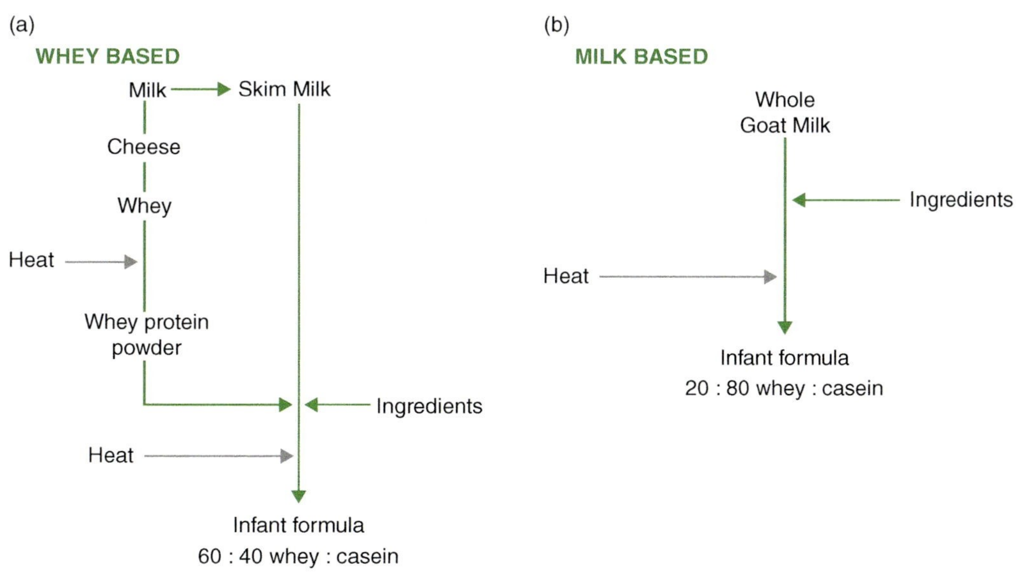

Fig. 8.6. (a) Whey-based infant formula; (b) whole goat milk processing for infant formula.

8.6.2 Clinical logistics and patent rights

Due to the complex and sequential regulatory steps, it is extremely time consuming to obtain FDA approval of biological agents derived from different host cells or transgenic animals. Besides this, the logistics systems required for transporting products to multiple locations is a tough challenge due to the temperature-sensitive nature of biomolecules and animal or plants tissues, making the process not only risky but also expensive. Presently, clinical logistics services bear the responsibility of providing laboratory facilities and ancillary supplies like providing cGMP laboratory testing and diagnostic equipment, information bulletins and other associated facilities. Laboratory services include distribution of patient forms and kits, laboratory sample handling with temperature-regulated equipment, computerization and compiling of testing results, and integration of overall services.

The main logic behind highly effective clinical logistics is to monitor global marketing on the basis of feedback from end users. This requires the efficacy of timely technical services, innovative temperature-controlled warehousing facilities, and safe delivery of testing samples to different geographical locations.

In 2011, rEVO. (an LFB group company) had a patent granted from the United States Patent and Trademark Office covering DNA constructs for the production of any therapeutic protein in the milk of any transgenic animal. This 'broad DNA' patent extends to 2027. In addition, GTC received a USA patent for ATryn (recombinant human antithrombin). ATryn is the first (and only) transgenically produced therapeutic protein approved by both the FDA and European Medicines Agency authorities. The main claim of this patent is related to a method of treating subjects with antithrombin III deficiency or inflammation with the administration of recombinant antithrombin III produced in the mammary gland of a non-human mammal.

Activities in goat therapeutic proteins are in progress, for example, a human factor IX is already produced in Iran but it still in clinical trials (Yekta *et al.*, 2013). Although active research and development on the expression of human monoclonal antibodies in milk of transgenic goats is underway, no data on transgenic goats producing monoclonal antibodies has been published, except in the 2010 GTCBiotherapeutics annual report (PwC, 2011). Expression of human monoclonal antibodies in goats against CD20 receptors, CD 137 receptors, tumour necrosis factors and epidermal growth factor receptors have been documented (Baruah *et al.*, 2006), and there are preliminary data on the possibility of production of recombinant hormones, growth factors, cytokines, G-CSF EPO, antibodies against CD20 receptors, CD 137 receptors, tumour necrosis factors, epidermal growth factor receptors, and lactoferrin from transgenic goats (Baruah *et al.*, 2006, Harvey and Ivarie, 2003, Lillico *et al.*, 2007; Ling *et al.*, 2012; Rodriguez *et al.*, 1995; Massoud *et al.*, 1996; Chrenek *et al.*, 2007) .

8.6.3 Logistics trends

Like other biological agents, goat milk therapeutic proteins and clinical testing kits are highly temperature susceptible and need specialist care during handling in logistic operations. Whatever the procedure used, the same technically sound protocols used for biological agents should be applied while handling goat therapeutic proteins. Generally, after proper packing the products are shipped at room temperature, and thereafter passed on to logistic operation systems under temperature-controlled conditions with the provision of digital tracking devices.

Biopharmaceutical supply chain management under temperature-controlled conditions has been evolving for the past two decades. The introduction of digital information technology-based operation systems and tracking of products during logistics operations have not only developed the practice of timely delivery, but also taken care of product quality management until it reaches the end user. The major manufacturers of therapeutic proteins derived from transgenic goat milk have their own logistics systems to transfer the products to different locations as demand requires. In addition, manufacturers have well-organized groups for packing, freight forwarding and transportation, and electronic monitoring of the internal conditions of the packages, along with monitoring the processes followed by carriers, to meet the requirements for sustainable growth and development.

8.6.4 Capacity addition

Due to patent expirations (including high-profile drugs), the challenge the biopharmaceutical industries face is to decide if it is financially worthwhile to continue with production of therapeutic proteins using host-cell bioreactors, or to develop transgenic animals for the expression of human proteins and other related biomolecules. From the market perspective, it is important to reduce the accelerating price increase of therapeutic proteins and compete with global companies. Promoters are interested in adapting transgenic animals for low-cost–oriented therapeutic protein production. Additional facilities need to be developed along with hiring the specialized staff required to manage and produce the transgenic animals and manufacture the therapeutic proteins. The best example for this sort of collaboration programme for the production of quality products on the basis of market demand is the partnership between rEVO Biologics, originally known as GTC Biotherapeutics Inc, and LFB Biotechnologies. rEVO Biologics worked in partnership with LFB Biotechnologies to design and develop recombinant plasma proteins and monoclonal antibodies in transgenic animals. rEVO was responsible for the research and development as they had the expertise and facilities for working with transgenic animals, and they retained the exclusive commercial rights to the products in North America. The responsibility of clinical development was undertaken by LFB Biotechnologies, who has the exclusive commercial rights in Europe. GTC and LFB Biotechnologies hold co-exclusive rights for the rest of the global market.

To shift from primary care towards speciality drugs and biological agents, the biopharmaceutical manufacturers need additional space for sophisticated equipment, space for continuous processing, office space for administrative duties, facilities for inventory storing, and third-party logistics partnerships. From a logistics perspective, the value of biological-origin products are comparatively insignificant with reference to minerals, automotive parts, agriculture feedstock and other high-volume materials, but from a storing and transport point of view special care is needed, with sophisticated and expensive machines and cryo-preservative equipment requiring the provision of dedicated spaces. For example, FedEx Supply Chain built a 1.1 million square-foot multi-tenant distribution warehouse in Memphis USA. Memphis-based FedEx Corporation also acquired Genco, a third-party logistics provider, for $1.4 billion in early 2015.

UPS Healthcare Logistics, which operates more than 60 facilities for healthcare products worldwide, built a second facility near Bogotá, Colombia in 2017, adding 76,000 sq. ft. and bringing its global footprint to more than seven million sq. ft. The company also expanded its Express Critical delivery service to Europe (it is already established in the US). DHL Global Forwarding opened a new Life Sciences Centre of Excellence in Ireland, the 43rd country with such a facility (some countries have multiple centres). The 40,000-sq. ft. facility is close to Dublin Airport, and features dedicated room-temperature, refrigerated and frozen capacity, along with repackaging services. It has been observed that with the increase in clinical trials of biopharmaceuticals, the cost of logistics is increasing at a global level.

8.6.5 Cold-chain operations

The act of cold-chain management of goat-derived products (e.g. liquid goat's milk, powdered/dry goat's milk, goat meat, goat milk therapeutic proteins) is similar to other biologic logistics processes. Due to the products being highly susceptible to temperature changes, some of the high value-added products produced from goats are delicate to handle in the supply chain management (SCM) process. Thus, for goat milk products and therapeutic proteins, SCM is executed with strict temperature control and well-managed tracking systems. Reducing or eliminating temperature fluctuations during shipping is a highly responsible task. The small containers holding clinical trial materials are equipped with temperature-recording instruments to log and assure the quality safety of goat milk biological agents. In addition, it is important for producers to know the shipping routes and environmental conditions the products will be exposed to during shipping, so they need to work with their shipping partners to manage appropriate packing and reduce the delay in logistics processes.

In addition, the shipping partners working with phase III therapeutic protein trial materials need to ensure proper recording of variable temperatures to ensure the quality of the material for safe investigation and analysis.

8.6.6 Carrier logistics

The supply chain network for handling biological material from the site of production to retailers is relatively small when compared to global logistics systems. However, with the development of modern cooling technology, and its application for biological product distribution under strict regulatory acts, third-party logistics is generally the factor responsible for high costs. The cold-chain SCM systems have developed confidence among the end users for the safe use of life-saving drugs. Currently, in order to supply the biological material in the shortest possible time, some of the global air-cargo carriers use specialist third-party management systems with instant tracking facilities. Besides this, the logistics providers and freight forwarders also have a network of offices with well-trained staff to efficiently manage stock movements.

8.6.7 Packaging technology

Therapeutic proteins derived from goat milk are classed as biological material and so need specialized packaging that meets industry guidelines. In general, biopharmaceuticals are highly sensitive to elevated temperatures and loading/unloading stresses. The manufacturers have well-trained engineers and technical experts to handle active and passive containers for the safe transfer of products to different geographical locations. Active containers have temperature-control instrumentation to counteract the variable environmental temperatures during the transport process. These containers can be reused, even under adverse conditions caused by flight and trucking schedules. Passive cryogenic containers are filled with specific chilling materials like dry ice or compressed liquid nitrogen along with release devices. This is a single-use system, and should not be static for long periods, as may happen with any change in flight or trucking schedules. The overall cost of a shipment is a net balance between the cost of the containers, the cost to ship the containers, and return logistics or disposal costs.

As prescribed by the Good Distribution Practice (GDP) regulatory act, European manufacturers use thermal blanketing, especially for biological agent products in transit. The GDP system is not just restricted to handling biopharmaceuticals under the control of 2–8°C (primarily for biological agents and vaccines), it is also applicable during in-house transition from one production site to another location nearby under controlled room temperature (CRT; generally 15–25°C). Logistic carriers need to follow GDP rules in order to keep CRT shipments as per the customer's requirements.

DuPont has developed a third-generation thermal blanket, branded as Xtreme WD-50, that is 'breathable' and therefore prevents condensation resulting from exposing the cool product to high environmental temperatures. Condensation is the primary factor for damaging packaging and may also cause mould and mildew. The first-generation thermal blanket is a simple Tyvek cover to protect the product from solar radiation and ambient airflow. The second-generation thermal blanket is a multi-layered product with a layer of insulation to control thermal heating. The new third-generation thermal blanket has both insulation and vapour control properties.

The development of third-generation packing materials has minimized the risks during the SCM process, and contributes to operational excellence (OpEx). Generally, for biological drug packing two types of systems are used. For large-volume raw material, stainless drums with plastic lining are used during the transfer from the site of manufacturing to the wholesalers. For smaller quantities, small cardboard boxes with plastic liners are standard practice. The raw materials used in host-cell cultivation include organic and inorganic salts, buffers, and cell-cultured materials mainly of biological origin, and are stored in temperature-controlled warehouses before use.

Goat products, mainly therapeutic proteins and lactose-free powdered dry milk, require special attention for the maintenance of containers and the governance involved in operating the SCM in the most efficient manner.

Such practices can also improve product quality, enhance regulatory compliance and contribute to the overall state of control by preserving the stability of biological material, ensuring sterilization and customizing delivery systems for market differentiation.

In recent years an increasing number of transgenic goat milk manufacturers have started the process of attaining regulatory approval. This drive needs continuous improvement and creates new requirements for primary packaging components.

Enhanced regulations and new guidance on cGMP are also driving drug and packaging manufacturers to improve production techniques. The therapeutic proteins can often be sensitive to interactions with vial and syringe components, which can induce protein aggregation or denaturization. Therefore, qualification and control of materials that hold biotherapeutics are critical to development and manufacturing. For highly sensitive biotech drugs, a packaging solution needs to offer protection from contamination and ensure consistency in every aspect that could affect the fill–finish process or long-term drug stability. Reducing variation in drug packaging can deliver measurable economic value by minimizing the risk of leaching in the drug solution. Packaging components come with certified extractable profiles, which provide proof of each component's biochemical consistency at a specific level of control. Packaging manufacturers can also identify and control sources of potentially contaminating particles through inspection programmes and continuous improvement efforts. Thus, this will be helpful in reducing or eliminating the overall expense and exposure to regulatory risk associated with component preparation, sterilization and end-of-line rejections. Leading industry suppliers can help guide drug product manufacturers through such issues and their potential implications.

8.6.8 Cloud computing

The efficacy of supply chain management systems, including the value chain operation protocol, has increased enormously due to the advancement in modern information technology (IT). Irrespective of the nature of the products and variations in customer demand and quality of the product, basic IT technology is applicable to all biological material SCM systems. The IT sector has many tools that aid industry in their SCM processes to ensure successful goods transition to different localities in a timely manner. For this, the IT sector must be compatible with the business aims and objectives for efficient business transitions.

Collaboration in SCM is inevitable, and strategic alliances in resourcing can help biopharmaceutical manufacturers to achieve a sustainable development pattern under proper management and monitoring processes.

Collaborative investments and attitudes have helped firms differentiate from competitors by enhancing their relationships with suppliers and customers. The main target for collaboration is to efficiently and creatively connect resources and expertise with alternative and integrated development models. Manufacturers with positive attitudes towards integration of their resources with IT can provide timely information to key stakeholders, thus improving the supply chain management process. However, acquiring, hosting and maintaining an IT infrastructure for effective business management is challenging.

Since 2007, use of cloud computing tools and techniques has started gaining popularity due to their ability to provide services to geographically separated clients on an on-demand basis. Cloud computing can be used for supply chain management of a variety of materials, including biopharmaceuticals, along with multiple business processes and third-party logistics operations. Fortunately, cloud computing now allows small businesses access to the same IT packages used by larger companies. The low cost and easy accessibility of cloud computing also allows for the development of collaborative relationships between members of a supply chain; partners can easily implement shared cloud-based applications instead of having to purchase and install expensive software on stand-alone servers, thus allowing organizations to work together faster.

8.7 Supply Chain Risk Management

As goat milk therapeutic proteins are considered to be biological drugs, their safety management is as plasticized as for other biological agents. Due to their temperature sensitivity, it is necessary to transport these biomolecules using a temperature-controlled logistics system. Risk management analysis is an important step when planning the logistics for the safe distribution of biological products to different locations. A manufacturer needs to take into account the administration requirements while transferring the products from the site of production to discharge points at different localities under the regulatory guidelines, as well as quality control and quality assurance processes. Risk management is a complementary factor closely associated with SCM and the value chain. Risk management practice is carried out by well-trained security personnel with a thorough knowledge of logistic operation systems. It is important to develop a close association between the risk management activities that can be embedded and maintained throughout the organization. Meticulous planning, performance process, systems management, controlling, reviewing and reporting are closely linked factors with risk management. Generally, risk management awareness develops with practice and experience.

Product safety, security, consistency and quality are the major challenges to SCM and the logistics operation. Risk management plays a critical role during the transfer of the products from the site of manufacturing to wholesalers and retailers. Both the US and Europe follow the ICH Q9 guidelines for SCM and logistics processes. Microbial contamination is a major risk factor in the biopharmaceutical production industry. Mostly, the biopharmaceuticals are exposed to contamination during hold times, either at the processing stage or as inventory in a storage site. The primary factors that affect the risk of microbial contamination during hold times are due to the growth promoting properties of the in-process materials, initial bio-burden level and storage conditions. It has been noted that microbial contamination is the most common factor in the failure of risk management. Available risk scoring tools such as failure mode and effects analysis should be properly managed with suitable modification of prevailing factors. Continuous supply of product components on a quality base for the manufacturing process is a key factor to avoid risk. Control strategies for raw materials play a crucial role. Selection of honest persons, solvency, duration of business and vendors' in-house quality system are important issues that should be taken into account when negotiating business strategies with vendors. The vendors should have their business continuity plans in place, as per international standards.

8.8 Logistics Operations

The entire supply chain management operation system starts with the upstream operation followed by downstream processing to produce target biopharmaceuticals, and subsequently the value addition process by upgrading the quality of products. This is followed by transporting the products to nearby warehouses for safe storage prior to handing over to the logistic management systems. The products are distributed under strict regulatory guidelines to wholesalers or retailers to meet local market demand. Promoters adopt efficient business techniques and skills to improve business performance at the global market level. Thus, the supply chain is the entire operation process encompassing network sourcing, supply pipeline management, value chain management, and value stream management (Lamming and Hampson, 1996; Saunders, 1997a,b). Logistics systems management is a part of the entire supply chain management operation complex. So it is regarded as a crucial factor for the biopharmaceutical manufacturer in order to gain a competitive advantage.

In the true sense, the logistic systems operation group is a separate entity that is responsible for the distribution of products received from the manufacturer according to the company's instructions. Logistics is the management of the flow of goods between the site of manufacturing and the end user in order to fulfil requirements of both customers and corporations. The logistics of biopharmaceuticals usually involves the integration of information flow, material handling, production, packaging, inventory, transportation, warehousing and often security. The complexity of logistics can be modelled, analysed, visualized and optimized by dedicated simulation software. A reduction in the use of resources is a common motivation in logistics for import and export.

8.8.1 Order management

Currently, many multinational groups like DGC, Ausnutria Dairy, Baiyue Youlishi, and so on, have started following the real-time value network system (RTVN) for increasing the efficiency of the order management process for the timely delivery of products. In RTVN, parties from different companies work together with mutual understanding and cooperation. In this process both the purchase order and sales order are synchronized to meet the timely demand of market needs at different locations. An integrated network service is maintained between buyers, sellers, movers, shippers and service providers.

In a single order system, the supply of products may take up to a day or a week, or possibly even weeks and months; however, RTVN uses integrated business systems to ensure that every item at every location is managed in a timely manner. In the RTVN system, multiple orders are automatically optimized and executed with a time-bound programme. The real-time cloud computing network helps the company and trade partners manage the SCM operation system as demand requires. In addition, RTVN is also helpful for keeping track of inventories and the supply chain process on an order basis, and also monitoring a stock-keeping unit located at different sites.

RTVN is currently used by over 30,000 trading partners. Organizations that connect to the network gain the potential ability to transact with any organization already on the network, offering a true many-to-many global network.

8.8.2 Inventory planning and control

In general, inventories are stored goods that have economic value. This includes raw material that will be transformed into products or stock to be utilized for the maintenance of the factory as consumable spares; inventory also includes work-in-process and finished goods. Biopharmaceutical inventories are quite different from other industries from a re-utilization point of view. Stock inventories are usually stored at locations nearby the manufacturing site to ensure easy access when required. Inventories are variable in nature. For example, a local dealer will keep inventory items on the basis of local demand, but a manufacturer will keep inventories required for factory maintenance and running. In addition, it must maintain some supply of finished products to meet the demands of the public, as and when required.

For effective and sustainable operation of manufacturing units, inventories are grouped into: (i) raw materials; (ii) work-in-process; (iii) finished goods; and (iv) maintenance, repair and operations (MRO) goods.

Raw material

The raw materials required as inventories for a biopharmaceutical manufacturing industry are mainly for the upstream process, the downstream process (for quality upgrading), and for finishing and packing. The upstream process inventories include high-value microbes stored under cGMP guidelines or animal-based bioreactors (i.e. transgenic goats). Extra care has to be taken to nurture transgenic goats in specially designed farms where breeding purity is maintained by well-trained genetic scientists. The downstream process for therapeutic proteins isolation and polishing is a conventional downstream process as plasticized in a biopharmaceutical industry engaged in biological agent manufacturing processes.

Some of the steps in the downstream process also need non-stop operation, so ensuring access to an emergency inventory is standard practice in the downstream operating process.

Inventory items may be commodities of biological origin that the firm or its subsidiary has produced or extracted. They may also be objects or elements that the firm has purchased from outside the organization. Even if the item is partially assembled or is considered a finished good to the supplier, the purchaser may classify it as a raw material for the manufacturing process. Typically, raw materials for goat farming are commodities such as farm house construction materials, sources for grazing and obtaining green feed, milking machines and other accessories required to nurture the goats.

At the corporate level, MRP is the initial stage of the manufacturing process. MRP is a planning schedule for the requirement of a variety of inventories to go ahead with MRPII to develop the entire protocol for sequential manufacturing process for new product development. Primarily, it means operational planning and financial planning is required in the farming units. Currently, manufacturers of goat milk therapeutic proteins use MRP and MRPII in order to maintain the quality and quantity of milk products, both from normal animals and genetically modified goats for milk therapeutic proteins.

Work-in-process (WIP)

In goat milk processing units, the biologically derived materials or the biomaterial under process for products such as therapeutic proteins from milk are known as WIP. In other words, any goat-derived products are considered to be work-in-process. WIP also includes labour and overhead costs incurred for products that are at various stages of the production process. WIP is a component of the inventory asset account on the balance sheet, and these costs are transferred to the finished goods and eventually to the cost of sale.

Finished goods

Non-transgenic goat milk products (mainly powdered lactose-free baby milk), and transgenic goat milk therapeutic proteins are manufactured under strict international guidelines and regulatory acts, and packed and distributed through SCM or a third-party under fully temperature-controlled logistic systems.

These goods have been inspected by quality control and quality assurance experts before trucking or shipping. This practice maintains the consistency and safety of products in order to develop confidence among end users.

Inventories can be further classified according to the purpose they serve. These types include transit inventory, buffer inventory, anticipation inventory, decoupling inventory, cycle inventory, and MRO goods inventory. Some of these also are known by other names, such as speculative inventory, safety inventory and seasonal inventory.

TRANSIT INVENTORY Any goat products in the process of transferring from the site of production (farm or milk processing unit) to the warehouse after proper packing are known as transit inventories or pipeline inventories. Merchandise shipped via truck or rail may take a long time transiting from the regional warehouse to wholesalers and retailers. As goat milk products are sensitive to elevated temperature, the delay caused due to such transfer processes may result in a reduction in quality before the products reach end users. Therefore a flight-based logistic system is a safe and fast method to deliver goods to the intermediaries and end users.

BUFFER INVENTORY During crisis periods a goat milk processing unit producing therapeutic proteins may not be able to meet the demand of wholesalers or retailers. In order to overcome such a crisis the manufacturer may choose to keep sufficient stock for the fermentation process and maintenance of host cells. Safety stock or buffer inventory is any amount held on hand that is over and above that currently needed to meet demand. Generally, the higher the level of buffer inventory, the better the firm's customer service. This occurs because the firm has fewer 'stock-outs' (when a customer's order cannot be immediately filled from existing inventory) and therefore has no need to backorder an item and make the customer wait until the next order cycle or, even worse, force the customer to leave empty-handed to buy from another supplier. Obviously, the better the customer service the greater the likelihood of customer satisfaction.

ANTICIPATION INVENTORY Sometimes, goat milk processing industries purchase raw materials to stockpile before price increases, seasonal increases in demand, or even an impending labour strike. This practice is also common in retailers, who routinely build up inventory months before the demand for their products will be unusually high.

Generally, a manufacturer tries to have a sufficient quantity of anticipation inventory when demand is low to keep the workers busy during slack times. Anticipation inventories help the promoter when market demand increases. Thus the firm does not need to pay workers for overtime during periods of high demand, and they also do not have to reduce staff numbers and shut down facilities in periods of low demand. This process is sometimes called 'smoothing' because it smooths the peaks and troughs in demand, allowing the firm to maintain a constant level of output and a stable workforce.

DECOUPLING INVENTORY Manufacturing of the therapeutic proteins from transgenic goat milk has been gaining popularity among biopharmaceutical manufacturers. The goat is used as bioreactor for expressing biological agents through their mammary glands. In this process the complex upstream process in the conventional cell-based method (maintenance of host cells and expensive stainless steel bioreactor equipment) is bypassed. However, the downstream process for obtaining target biological(s) (i.e. the separation and purification of proteins) is as complex as the conventional method.

In order to reduce the cost and process time, the various stages of the downstream process are linked in a continuous or partially continuous manner. This means that the manufacturing process is reliant on a sequential workflow (i.e. one machine feeds parts to the next machine, and so on). If one machine breaks down, it will affect the downstream workflow. WIP can be managed using decoupling inventory which 'decouples' or disengages the plant's dependence on this sequential workflow.

The more inventories a firm carries as a decoupling inventory between the various stages in its manufacturing system (or even distribution system), the less coordination is needed to keep the system running smoothly.

CYCLE INVENTORY Cycle inventory (or working stock) is the stock that is ready to sell to retailers and wholesalers. Cycle stock is continuously replaced as stock items are sold to dealers. It is important to maintain a reasonable stock flow to avoid any out-of-stock issues. The cycle inventory and GDP complement each other – any variation in GDP causes accumulation of selling of stock. It is regular practice of a manufacturer to monitor the inventory holding and carrying costs and the costs incurred from ordering or setting up machinery – this is known as the economic order quantity.

Cycle inventory is also known as lot-size inventory. This is because the bulk price discounts, shipping cost and setup cost are calculated to make the decision if it is more economical to purchase or produce in large lots than just purchasing the stock needed for required purposes.

MRO goods

The transgenic goats used for the manufacturing of therapeutic proteins generally have a shorter life span when compared to equipment and machines; the goats are 'stocked' for recurring use and are basically of biological origin. MRO items include consumables (such as bioreactor cleaning agents, laboratory materials or office supplies), industrial equipment and plant upkeep supplies (such as gaskets, lubricants, repairing tools), IT equipment, office furniture, and so on.

8.9 Concept and Significance of Inventory Management

The role of inventory management is most crucial in successful performance of business transition, and timely supply of products to wholesalers and retailers. Milk therapeutic proteins and lactose-free powdered baby milk are highly temperature sensitive. Thus, the dairy manufacturing industry and transgenic goat milk processing units need critical attention from an inventory point of view. Generally, these types of industries have a highly integrated manufacturing process to minimize the cost of production. SCM, value chain monitoring and cold-chain logistic systems, coupled with third-party logistics groups and well-trained inventory workers form the backbone of sustainable business delivery functions. Effective marketing management and financial controllers are also key players in the process.

8.9.1 Warehouse management

Sometimes, warehouse management is more focussed on system monitoring techniques rather than commercial factors. The temperature sensitivities of therapeutic proteins derived from transgenic goat milk means that it is critical to manage the product through SCM and value chain systems. Temperature control at regional warehouses is needed until the products are properly transferred for shipment to third-party logistics systems.

Warehouse management is a key practice for delivering the derived goat products under safe conditions. The warehouse should be fully temperature controlled with the provision of digital control systems to keep the products at the correct temperature. Even small variations in temperature may spoil the quality of the goods. That is why large companies cannot operate without state-of-the-art warehouse management solutions.

8.9.2 Transportation versus supply chain management

Transportation refers to the movement of product(s) from the site of manufacturing to the wholesale dealers located at different locations. Transportation is a part of supply chain management, which includes procurement of goods from the site of manufacture and storing in well-managed warehouses before shipping to wholesalers. With the emergence of real-time data networks, radio frequency identification (RFID) tags can provide accurate status reports and point to point information on efficiencies and inefficiencies in the goods supply chain. SCM is the broad range of activities required to execute and monitor a product's movement from the site of production to the wholesale dealers in a cost effective process, with the safety of goods both quality and quantity wise. For SCM, a good manufacturer has its own team that is well trained in handling the goods for shipping.

Therapeutic proteins from transgenic goat milk are classed in the biopharmaceutical category, which requires safe transport with maximum security during the supply chain management process. A manufacturer needs a guarantee on safety and security of their goods until they reach the end users. Thus, it is vitally important to protect and validate therapeutic proteins from goats in order to keep forged products off the market.

RFID technology with real-time tracking of goods during movement is the best solution to ensure goods are not tampered with during the transport and identification of genuine versus forged items. RFID can also be used to store information such as expiry dates.

8.9.3 Flow of manufacturing costs

The flow of manufacturing cost is defined as the changeable cost of products starting from the procurement of raw materials for manufacturing to the finished products. When the raw materials or feed stocks are passed on to the manufacturing process, the cost then moves to WIP inventory in order to meet the requirement of the cost of labour, machining and overhead cost; these costs are added to the base cost of the feed stocks. After the manufacturing process, the costs move to the finished goods inventory classification, where the final quality product(s) are stored prior to sale through suitable logistic systems. In the course of process flow, the costs are computerized at the beginning, and additional costs accrued in the value chain process are added to calculate the overall cost of the product while preparing the final balance sheet. The flow of manufacturing cost and financial profile mainly depend on the inventories associated with the supply chain management process. Inventory is where the major part of the manufacturer's fund is invested. Inventory typically consists of finished products for sale, raw materials, and WIP. To calculate the profit, the company must track sales revenue as well as all of the costs involved in procuring raw material or livestock. Accordingly, the manufacturer profits consist of the money remaining from sales after the company has covered all its costs, including buying the inventory.

In the flow of manufacturing costs, various terms are used to define the flow at the different steps of goods transition from the site of production to main store warehouse, and subsequently to the wholesale dealer. The two most common terminologies in use for main goods transition are FIFO (first-in, first-out) and LIFO (last-in, first-out).

Generally, the immediate manufactured goods are recorded as inventory. The costs of inventory items immediately manufactured are changed to expense when the same are sold last. FIFO, on the other hand, assumes that the first items put on the shelf are the first items sold, so the old stock is sold first. This system is generally used by biopharmaceutical manufacturers where the inventories are perishable or prone to long storing practice. It is the preferred accounting method when the cost of the product is rising in the market.

The cost flow mainly depends on the dynamic nature of the manufacturing process, WIP, finished goods inventory and quantitative procurement of raw materials or feed stocks for the manufacturing process. LIFO and FIFO each have different effects on a manufacturer's financial condition.

A biopharmaceutical company, like other manufacturers, uses the FIFO system to trace its inventory but uses LIFO when preparing the final balance sheet. The gap between FIFO and LIFO can be bridged by using LIFO as a reserve account. In case of price rises, a credit balance in the LIFO reserve is created, resulting in a reduction in the inventory cost. Preparing accounts using LIFO helps to reduce tax payments when price rises happen. Storing of goods in anticipation of demand and marketing goods at the proper time are dependent on the efficiency of the supply chain management system. In the FIFO system, the manufacturer tries to clear the maximum amount of goods and the unsold products are then mixed with the subsequent lot, assuming that the quality is the same and the stock still has a valid expiry date.

LIFO liquidation can be computerized in different ways. Generally, a manufacturer follows LIFO liquidation by using the LIFO inventory costing, and then liquidating its older LIFO inventory. If the current sale price is higher than the manufacturing cost (including the cost of raw materials), the promoter should liquidate the stock inventory. This practice will be helpful in gaining financial profit in net operating income.

The biopharmaceutical industries generally follow LIFO methods for immediate disposal of products in the market. This method is followed during a period of rising prices or inflation when the cost to purchase inventory increases over time. The tax burden of a pharmaceutical company is reduced when the higher priced inventory matches current revenue, increasing the cost of goods sold and decreasing earnings before tax. The LIFO method is used for asset management and valuation while acquiring, selling or disposing of the inventory goods. If an asset is sold for less than it cost to buy, then the difference is considered to be a capital loss. If an asset is sold for more than it cost to purchase, then the difference is considered a capital gain. Using the LIFO method to evaluate and manage inventory can be advantageous for tax purposes, but it may also increase tax liability.

8.9.4 Trace link and serialization of biological products

Multinational biopharmaceutical, pharmaceutical and biotechnology industries implement traceability and serialization capabilities to meet regulatory requirements, and for quality and recall purposes (McBeath, 2012). It requires a community that can share that information effectively and collaborate with law enforcement at national, state and local levels. After the implementation of traceability and serialization practice the US pharmaceutical industry achieved a more than 50-fold reduction in pharmaceutical theft during the period 2015–2019. In this connection, a theft incident that occurred in an

Eli Lily warehouse (US$76m of drugs stolen) is most alarming for pharmaceutical and biotechnology industries (David and Ketchen, 2008).

The primary objective of traceability is typically not theft control, but rather quality control and enabling recall capabilities. The packaging of biopharmaceutical products should carry a tracking microchip to understand the condition of goods in real time. From a tracking hologram, one can know the quality of goods, information on expiration date, batch or lot number, and recommended storage conditions, directions for handling and precaution that may be necessary, as well as details about the manufacturer.

8.9.5 Expiration date

The production of therapeutic proteins from transgenic goat milk must follow regulatory acts set by the biopharmaceutical industry in order to maintain safety and quality of the products. The packaging should carry a sticker showing the date of expiration. An expiration date means the maximum life of a biological-origin product, after which it is strictly advised to destroy the product in an environmentally responsible manner. Arbitrary expiration dates are also commonly applied by companies to product coupons, promotional offers and credit cards. In these contexts, the expiration date is chosen for business reasons or to provide some security function rather than any product safety concern. The expiration date is often abbreviated EXP. The legal definition and usage of the term expiration date varies between countries and the nature of the products. The expiration date for each biological-origin product should be computed from the date of the initiation of the potency test. Prior to license approval, the stability of each fraction should be determined by methods acceptable to the Animal and Plant Health Inspection Service, an agency of the United States Department of Agriculture.

8.10 The Impact of New Technology Development on Supply Chains

Linking the supply chain with IT not only increases the efficiency of goods transition from the site of production to the end users, but it also maintains the quality of the products and timely delivery (David and Ketchen, 2008). In reality, IT can create a link with supply chains and helps with keeping the consumer in close touch with manufacturing process and quality control information. In addition, the alignment of IT goals and objectives with strategic SCM can increase efficiency, productivity and profitability (Chandra and Kumar, 2000). By integrating SCM with IT, immense benefits can be achieved in product development, procurement, manufacturing, physical distribution, customer relationship management and performance measurement (Olson, 2012). This sort of practice supports the promoter by providing the means to link technology and people and trying to align the technology with the capabilities of the organization and among its trading partners (Shaik and Abdul-Kader, 2013)

Besides obtaining quality products, customers also like to interact with a supplier to develop confidence in consistency in the supply chain process, and competitive status of the product as compared with other manufacturers. A good competitor should have flexibility in business dealing and the ability to negotiate and market guarantee of costs, quality and ensure timely delivery. This could only happen with the use of advanced IT services (Nelson, 2001). Thus supply chain information systems have a key role in speeding up business transition and improving decision making and productivity in the most efficient way with a controlled budget (David and Ketchen, 2008).

By linking IT technology a company can integrate internal business functions and keep track of activities in real time. This can help the promoter to fulfil customer requirements and improve the quality control process (Chandra and Grabis, 2007). SCM systems with the help of IT help logistics management, transportation management, strategic planning, warehousing, inventory, manufacturing, supplier management and customer relationship management (CRM). Enterprise resource planning (ERP) systems are included as part of the broader SCM software. ERP systems are employed to integrate business processes by organizing, codifying and standardizing business processes and data (Norris *et al.*, 2000). Functionally, ERP helps with systematic data management and transferring data within internal systems by avoiding the expense of the transportation of data from one department to another. ERP reports can be used to forecast production and make decisions. Another key supply chain process is CRM, which is the management of relationships between the organization and its customers.

A critical issue for internal integration of business processes is the integration of ERP with CRM. Recently, many ERP vendors provide an integrated ERP–CRM package, since many enterprises have expressed an interest in integrating a new CRM system with their legacy ERP system. Moreover, the establishment of ERP and CRM systems should be a primary concern of an enterprise that is interested in embarking on e-business (Yanjing, 2009). Generally, a company shares critical information on demand prediction, actual orders and inventory level in an efficient manner (Marakas, 2003), while protecting the company's proprietary data. The adoption of inter-organizational information systems (IOS) for supply chain management has been proposed to enable external integration (Kumar, 2000; Grover and Saeed, 2007). The most common form of IOS is electronic data interchange, which permits instantaneous computer-to-computer transfer of information. There are some technologies that, overall, are considered to be of high relevance for future supply chains, such as cloud solutions and mobile devices.

Even though security concerns are under consideration yet again, cloud computing has become an integral part of today's supply chain management solutions portfolio and remains of interest. The main benefit is that it addresses companies' investment capability and functional requirements at various stages of their development, and applies to a number of business areas in global logistics. Mobile computing and mobile devices have gained greater importance in recent years due to their ability to efficiently support supply chain processes 'on the go'. Their technical advancement has shown significant development in their scope of capabilities, but their most valuable feature continues to be the ability to integrate into ERP and operational systems to gather and share data in a variety of formats.

Comprehensive, sophisticated supply chain management systems meet the requirements of supply chains in the twenty-first century: adaptability, collaborative ability, transparency, compliance and speed in all processes. Streamlining processes and cutting costs in the long run are all vital for future business growth. By choosing the right software solution, supply chain managers can future-proof their business and gain a competitive advantage.

8.11 Green Logistics and Supply Chains

Awareness of conservation of the planet is the first priority associated with each and every activity. The question of environment stability and sustainability is a complementary factor closely linked with manufacturing processes. Various issues related to the environment are the direct result of the mismanagement of industrial activities, starting from procuring raw materials through to manufacturing and disposal of goods through logistic systems. During the process of goods transition, the chances of spoilage and damage during the shipping process and microbial contamination of packing materials are some of the major problems associated with logistic systems operation. The introduction of digital-based logistic systems has significantly lessened the burden of environmental pollution. Logistics systems development and operation without any wastage generation is known as green logistics and supply chain (GLSC). Green supply chain is supply chain management with similar objectives and core implications. Green logistics combined with supply chain management is also usually referred to 'sustainable' management.

In biopharmaceutical industries, besides generating solid, liquid and gaseous (greenhouse gases) biological wastes, wastes generated during goods transition is also a major environmental issue. Manufacturers need to incorporate advanced technology during product packing to avoid any biological waste generation by leakage of products or due to microbial contamination. The aim is to reduce environmental pollution and resource consumption arising from logistics activities to realize a 'win–win' outcome in logistics development and environmental conservation. In this process, the traditional practice of supply chain management is integrated with environmental conservation so that waste generated during any stage of the supply chain can be recycled in order to reduce the susceptibility in supply chain management practice and bring profit to the manufacturer (Fig. 8.7). Microbial processes for the production of biopharmaceutical and healthcare products (e.g. nutraceuticals, food supplement, food additives, antioxidants, etc.) may be a threat to the environment in terms of carbon monoxide emissions, discarded packaging materials, scrapped toxic materials, traffic congestion and other forms of industrial pollution (Wisner *et al.*, 2012).

Thus, all procedures related to hazardous and non-hazardous wastes should be controlled and

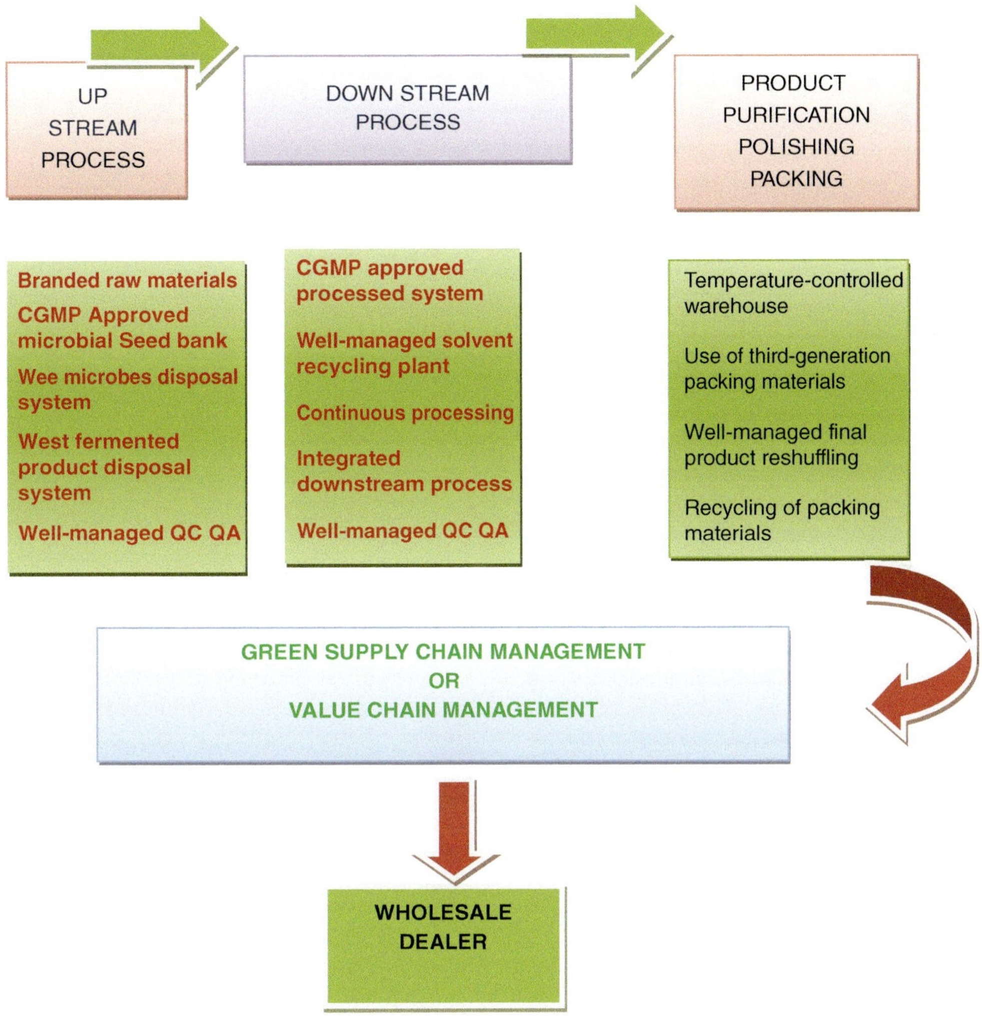

Fig. 8.7. Green supply chain management.

monitored in logistic management processes. The main goal of control and monitoring is to anticipate, identify and verify unintentional release of any biological or non-biological materials. Minimal monitoring should include periodic physical inspections of waste generation sources, storage areas, means of transport, waste-related documentation, and proper labelling and control of activities.

A typical topic in GLSC management is reverse logistics, sometimes called closed-loop supply chains, in which there are reverse flows of used products (post-consumer) back to the manufacturers. There has been substantial research into production planning and inventory management in re-manufacturing systems (Simpson, 1978). At the end of the 1900s, people were well aware of the re-utilization of goods, if any were left during the logistic transition process (Kelle and Silver, 1989; Inderfurth, 1997; Wang, *et al.*, 2011). With the advancement of IT technology development, control strategy and the pull-control strategy concept, superior inventory strategies for hybrid manufacturing/re-manufacturing systems with long lead times for manufacturing and a short lead time for re-manufacturing have developed to upgrade supply chain systems in a more efficient manner (Laan *et al.*, 1999; Teunter *et al.*, 2004).

Currently, the overall concept on business philosophy has drastically changed. A good multinational manufacturer not only targets profit, but also wants to keep this habitable world undisturbed from stress and strains caused by creating minimum issues that can challenge environmental sustainability. However, only using waste-free manufacturing processes would not solve the problem. Follow-up and maintenance of GLSC is an integral part of total manufacturing activity in a sustainable form (Diabat and Govindan, 2011).

Hence, the impact of a GLSC on the environment is minimum harm and a sustainable performance of the supply chain. During the early 1990s, competition among big multinational organizations focused on impressing consumers from aesthetic and social responsibility points of view on product purity maintenance, without any concern for the environmental impact of manufacturing (Diabat and Govindan 2011; Luthra *et al.*, 2014). With these practices in mind, firms now need to develop environmental management strategies in response to the dynamic nature of environmental requirements and their impacts on supply chain operations (Beamon, 1999). Successful supply chain management is a coordinated interaction of multiparty systems that includes manufacturers, suppliers and wholesalers (Mentzer, 2001).

8.12 Quality Control

The products derived from goats include meat, milk, fibre and therapeutic drugs and enzymes. The quality of these products is described in Protocol 1 at the end of this chapter. They are produced from living organisms and contain components of living organisms. Thus these products are susceptible to various stresses and strains and it is necessary to have control over their quality during the manufacturing process and distribution. Sustainable development of biological agents in the manufacturing process is mainly dependent on quality control as per the norms prescribed in the international regulatory acts. Quality control is a requirement when planning a successful business that delivers products and meets the customers' expectations. It also gives additional support and confidence while delivering the goods with minimum waste and operates at high levels of productivity. It is mandatory for a manufacturer to have a complete protocol for quality control before starting the manufacturing process. Quality control encompasses two areas: quality and control. Quality (mainly of a biological drug) represents the overall characteristics of a product on the basis of its size, shape, durability, efficacy and stability in formulation. Thus, quality is the backbone of a biopharmaceutical manufacturer involved in biological manufacturing processes in areas well-equipped with analytical instruments and trained technicians to keep the quality as per the norms of international regulatory acts.

8.12.1 Objectives of quality control

The main target of quality control is to manufacture product(s) that are acceptable to consumers, and to make the product in a timely manner that reaches the end user with a guarantee of safety and quality. Besides this, quality control gives the guarantee of error-free manufacturing processes by suggesting improvements on the basis of user demand.

Meticulous planning of the manufacturing protocol by giving maximum emphasis to the quality of the product and the service offered to the customer in a consistent manner are the unavoidable practices of a successful manufacturer. Most multinational companies have specialized departments with well-trained scientists to monitor control over production. One of the most popular examples in this regard is the business philosophy of Sony Corporation, who unified company workers with team work and complete involvement in the business system, which led the company to achieve unexpected gains and fame in the global market. American and European companies have had to spend time introducing quality control into their manufacturing processes. For maintaining and enhancing quality, the manufacturer has to adapt two techniques, i.e. quality control and quality assurance, and to make sure that the end product or the service meets the quality requirements and standards defined for the product or service.

8.12.2 Quality control management

Although quality control and quality assurance are two separate practices in manufacturing processes, they are both necessary for a business to succeed. Generally, manufacturers must follow a quality control manual that has been approved by international regulatory acts. The quality control manual outlines the quality focus and objectives of the organization. Different departments use the manual to coordinate the manufacturing process to keep continuity in business transactions.

An organization has to follow the best suitable method for quality control, as advised by their highly professional quality control group. Some organizations use the concept of TQM while other organizations have their own internal and external standards. Generally, as the procedures for managing activities are carried out on the basis of norms as prescribed by the International Organization for Standards (ISO), it is necessary for a company to adhere to the quality requirements of ISO.

Since standards have become a symbol for product and service quality, customers are now keen to buy products or services from a certified manufacturer or service provider. Therefore, complying with standards such as ISO has become a necessity when it comes to attracting customers. The importance of properly established and managed quality control and quality assurance systems with their integral well-written standard operating procedures and other quality documents for the achievement of company business objectives cannot be ignored. They serve as a passport to success, and maintain sustainability in business.

8.13 Quality Assurance

Quality assurance (QA) is the practice of monitoring and preventing the various issues related to product deformation and defects during the manufacturing process. In addition, QA keeps in touch with the product movement through supply chain or value chain systems. The ISO defines the QA practice as: 'part of quality management focused on providing confidence that quality requirements will be fulfilled'.

QA is governed by two principles: 'fit for purpose' on the basis of its suitability for the intended purpose, and 'right first time' to eliminate mistakes. In QA practice, intense care is undertaken during the design and development of a novel technical product.

Often, quality assurance and quality control are used interchangeably to describe the quality of a service or product. The term 'assurance' is used for the implementation of inspection and product testing as a measure of quality assurance, depending on the nature of the manufactured process and product development. The term 'control' is used to explain the fifth phase of the define, measure, analyse, improve and control (DMAIC) model. DMAIC is a data-driven quality strategy used to improve processes.

Quality assurance brings confidence among users in using specific branded products. Sustainability of the product quality is dependent on the quality assurance practice. Thus, the quality assurance team should be allowed to take autonomous decisions without any threat and pressure from the company's management team. The quality assurance should operate under strict guidelines of international regulatory acts (e.g. good X practice, cGMP, good clinical practice, etc.) and local, national, regional and international legal, ethical and regulatory requirements. Quality assurance teams have a responsibility to streamline the protocols for quality control on the basis of end user requirements. Mostly, companies prefer to adapt ISO or capability maturity model integration (CMMi) for quality assurance. In CMMi practice a company has to define their own internal processes and adhere by them. In established multinational companies a sophisticated software system is used to understand the instant nature of quality assurance while a product is in the manufacturing process. It is necessary for the quality assurance group to constantly update the information by regular inspection during both the manufacturing and transportation stages.

The stability test is an important criteria for quality assurance practice. The stability test checks the change in quality of products in response to temperature, humidity, light and other various physical factors while storing and transporting a product to a destination. The standard protocol for testing is well defined on the basis of the effects of climatic conditions in three regions (the EU, Japan and the US).

On the basis of global climatic data one can easily calculate the instant climatic condition of a specific region. The average climatic conditions of the world can be divided into four specific zones. This guidance addresses climatic zones I and II. The principle has been established that stability information generated in any one of the three regions of the EU, Japan, and the United States would be mutually acceptable to the other two regions, provided the information is consistent with this guidance and the labelling is in accordance with national/regional requirements.

Generally, stress testing is carried out on a single batch selected at random. The stress test is carried out at different temperature profiles (starting from 10°C incrementing up to 60°C) and humidity (e.g. 75% relative humidity) to understand the changes that occur due to partial oxidation or photolysis. The testing also includes the susceptibility of drug components in response to pH variation when in solution or suspension. The standard protocol for stress testing is well described in the ICH QIB Photo Stability Testing of New Drug substances and products. As per international regulatory acts, there should be three minimum testing samples, chosen randomly. The capacity of each batch should not be less than the size of the pilot level production followed by the same manufacturing process as carried out in turnkey level production.

The testing material for stability study should match with the quality of materials in the scale-up manufacturing process. Special care is to be taken while using biological drug samples stored in a warehouse or in the transport process. The testing should cover, as appropriate, the physical, chemical, biological and microbiological attributes. Validated stability-indicating analytical procedures should be applied. Whether and to what extent replication should be performed will depend on the results from the validation studies.

For long-term studies, frequency of testing should be sufficient to establish the stability profile of the drug substance. For drug substances with a proposed retest period of at least 12 months, the frequency of testing at the long-term storage condition should normally be every 3 months over the first year, every 6 months over the second year, and annually thereafter through the proposed retest period. Well-defined protocols are available on storage of biological/biotechnological/pharmaceutical products under variable physical conditions (ICH, 1995, 1996, 2000, 2003, 2005, 2006a,b,c).

Generally, biological drugs are conjugated to a second moiety bound to an adjuvant so it is necessary to test the intact structural profile or to study the dissociation nature of biological drugs at different states of movement in the supply chain system.

8.13.1 The legal basis for approval of biological drugs

The legal approval of biological drugs varies from country to country. There are two statutory bodies in the US that that govern the drug-approval process: the Federal Food, Drug, and Cosmetic Act (FFDCA) and the Public Health Services Act (PHSA). The US FDA regulates the safety of biological drugs in order to protect public health. The FDA administers the FFDCA and PHSA (among other statues). FFDCA applies to all drugs and medical devices, and the PHSA applies to 'biological products'. Marketing approval under the FFDCA is by means of a New Drug Application (NDA), while approval under the PHSA is by means of a Biologics License Application (BLA). The biological agents are subject to Investigational New Drug Application regulations. Pre-clinical research on new compounds is carried out in a laboratory using a wide variety of techniques. Promising candidates are then studied in animals and, subsequently, various clinical studies in humans are carried out following strict guidelines:

- Phase I: A small number of healthy volunteers are given the compound to determine mainly that the drug is safe for human use.
- Phase II: A small number of patients are given the medicine to assess its efficacy and safety and to ensure that there are no unacceptable side effects.
- Phase III: A large number of patients, usually thousands, take the medicine under supervision over a defined period of time, with the results used to establish efficacy.

If the results show the drug to be efficacious and safe, the data are presented to the FDA. The FDA reviews the data, and if the data is acceptable, a marketing authorization is issued. Alternatively, the FDA may request additional studies or reject the application. Following the grant of marketing authorization, the drug product is studied in large numbers of patients in hospitals and clinics to further assess its clinical effectiveness. This stage is called phase IV or post-marketing study. Safety assessment of marketed medicines studies help identify any unforeseen side effects. In order to be marketed, a biological agent requires proper labelling and an approved BLA that indicates the product has been determined safe, pure and potent, and that the manufacturing facilities meet the requirements to ensure safety, purity and potency. Though biological agents have traditionally been subject to much more scrutiny in manufacturing than drugs, those differences are being eroded. Biological agents have been approved under the FFDCA and

PHSA, thus both NDA and BLA applications have been submitted for biological agents.

8.14 Quality Control and Supply Chain Integration

Since 2000, TQM has made a positive impact on acquiring successful business in biotechnological and biopharmaceutical businesses (Arumugam *et al.*, 2009; Chaitanya *et al.*, 2018). It plays a vital role in the development of management practices (Prajogo and Sohal, 2003; Hoang *et al.*, 2006). It has been noticed that TQM is immensely helpful in upgrading a business transition by fulfilling the complete requirement of customers (Oakland, 1993). In addition, TQM increases sustainable competitive advantages for businesses in a national and global market (Terziovski, 2006), and above all is a source of enhancing organizational performance through continuous improvement in an organization's activities (Blackiston, 1996; Claver-Cortes *et al.*, 2008; Teh, 2009). Hence forth it is strongly recommended to incorporate both the QC and QA practices and procedures in SCM. Thus, quality management is often related to a model of organizational change (Boronat and Canard, 1995), the implementation of which largely relies on the organization's ability to adapt itself to these principles.

Quality is an important factor when it comes to any product or service. With the high level of market competition, quality has become the market differentiator for almost all products and services. So, the QC practice should be an integral part of the supply chain in production and distribution of the product at the consumer level. It helps in developing customer confidence in product acceptability. Without quality control, waste becomes widespread beyond a tolerable amount.

Microbial contamination is a common problem with biological drugs if they are not properly handled during packaging and transport. In order to maintain the product quality, raw material and WIP are regularly audited by supply chain staff members to ensure they meet specifications. By controlling the quality of production inputs, supply chain managers are protecting the integrity of their company's operations.

Genetically engineered host cells are used to produce biological drugs. Quality control helps to screen genetically modified host cells from the possibility of possessing toxic compounds that could be harmful in humans. The US Department of Transportation prescribes important rules for the transport of hazardous substances. Non-compliance can lead to penalties or fines, which makes QA and QC imperative. The more efficiently and effectively toxic materials are handled in the supply chain, the better for all internal and external stakeholders.

8.15 Protocols

Protocol 1: Quality of goat products

Growth is a complex trait that is controlled by various genetic and environmental factors. Epigenetic factors modify the expression of various traits during the growth period. Meat and milk quality are complex traits that are governed by various genes; they also depend on breed characteristics and geographic location.

Meat quality

The quality of meat products are directly associated with the end users, i.e. humans. The product should be safe, nutritious, palatable and acceptable to consumers. The quality of products depends on the goat rearing, slaughter, processing, packaging and storage. It is very difficult to evaluate meat quality (MQ) in live animals or the early postmortem period. Biomarkers are used to predict the MQ from birth to the slaughter period. Again, the use of biomarkers will enable us to understand the interaction between genetic and environmental factors that control this complex trait. Protein biomarkers of beef tenderness have been established in proteomic studies (for reviews, see Guillemin *et al.*, 2011; Picard *et al.*, 2014).

Thus the beef sector is looking for tools based on biological or molecular indicators to identify live animals, carcasses or meat cuts with desirable tenderness potential.

Goat meat is an ethnic food and the acceptance of goat meat is increasing by non-ethnic consumers. Goat meat has intense flavour and toughness; the goaty odour is attributed to 4-methyloctanoic fatty acid (hircinoic acid) (Wong *et al.*, 1975). Goat meat colour and shelf-life characteristic decline with increased storage time. Tenderness is affected by collagen and muscle fibres. Tenderness and juiciness decrease as the age of the goat increases.

The major nutrients from meat are protein, essential amino acids, fat-soluble vitamins, water-soluble vitamins, iron and phosphorus. Lean muscle of goat is also a source of conjugated linoleic acid which plays a role in immune response regulation, energy partitioning and other health benefits. Goat meat (muscle) is a better source of nutrition as it has a biological value of 60.4 compared to 68.6 in beef in rat-feeding trials with 10% protein. Goat meat has higher thiamine and riboflavin concentrations compared with beef, lamb and veal (Webb *et al.*, 2005). Similarly, the desirable fatty acids and polyunsaturated to saturated fat ratios are higher in goat meat than beef and lamb (Webb *et al.*, 2005). Castrated goats have lower percentages of stearic, linoleic and total saturated fatty acids (Werdi Pratiwi *et al.*, 2006).

Carcass characteristics

DRESSING PERCENTAGE Dressing percentage is the carcass weight divided by the live weight multiplied by 100. It indicates the potential yield of meat from an individual animal. The hot carcass weight is measured after skinning and evisceration and before chilling. Goats should not be provided any type of feed before slaughter. The dressing percentage of goat decreases if the goats have access to feed and water continuously. Dressing percentage of kids and yearlings with different body confirmation varies between 48–52% when goats are not provided any feed overnight.

CARCASS COMPOSITION AND EVALUATION The MQ depends on the physical and chemical composition of the carcass. The parameters of the physical composition of goat meat carcass include lean, fat and bone percentages. Fat percentage is dependent on the concentrate feed and the management of the feed lot system. The chemical composition percentage of goat carcasses includes moisture, fat, protein and ash percentage. Chemical composition of carcass will depend on breed, feed type, age and sex. Carcass composition is affected by genotype, sex and feed regime and production system. Concentrate feeding or concentrate supplementation increases carcass fat deposition as compared with browsing or grazing. The higher proportion of kidney and pelvic fat in goat carcass makes it more important for meat yield. The lower portion of subcutaneous fat on goat carcass is not important for meat yield.

Carcass evaluation is carried out to determine the relative ratios of lean, fat and bone to estimate the palatability of meat. Yield of meat is usually expressed as the amount of lean or cuts and is dependent on the weight of the carcass, the degree of muscling relative to bone and the amount of subcutaneous fat and internal kidney and pelvic fat.

Goat meat products

- **Goat meat pickle:** This is a shelf-stable value-added product prepared using pre-cooked spent meat. Goat meat pickle is highly acceptable among meat eaters. It can be kept for 60 days at room temperature.
- **Goat meat nuggets:** These are processed meat products. This is a protein and fibre-rich meat product and is highly acceptable among meat eaters. It can be kept under refrigeration for 20 days and as a frozen product for 120 days.
- **Goat meat sausage:** This is also a processed meat product. It can be kept under refrigeration for 20 days and frozen for 120 days.
- **Goat meat patties:** This is one of the most liked ready to eat comminuted meat products, enjoying the highest degree of preference among consumers owing to ease of preparation. It can be kept under refrigeration for 20 days and frozen for 120 days.
- **Goat meat shami kebab:** This is a popular traditional meat product and has good scope as a fast food in India and other countries. The technology of kebab preparation is traditional. It can be kept under refrigeration for 14 days and frozen for 90 days.

Milk quality

Goat milk is unique with respect to digestibility, distinct alkalinity, high buffering capacity and therapeutic value in human nutrition and medicine (Park, 1994). Goat milk is less allergic for people and is a naturally homogenized product due to its smaller fat globules (Haenlein, 2004). Goat milk has more oligosaccharides than cow milk, with an amount similar to human milk. These act as prebiotics in the gut and may help to maintain the health of the digestive tract by encouraging the growth of beneficial gut bacteria and preventing the growth of harmful bacteria. Goat milk contains slightly less lactose than cow milk and is helpful for lactose intolerance people. It is well established that milk composition and productivity is affected by genotype.

Milk contains a complex mixture of proteins and is an essential source of nutrients for most of the world's population. The quantity and quality of milk proteins and fat are important traits in dairy livestock. There has been considerable interest in milk proteins due to the availability of trace amounts of bioactive peptides with significant functional properties. The protein content mainly comprises of 'casein' with nutritional and functional properties. α-S1, α-S2, β and κ casein forms have been identified in milk with respective genes namely CSN1S1, CSN1S2, CSN2 and CSN3, respectively. Genes for whey proteins containing mainly β-lactoglobulin and α- lactoalbumin are also recognized. Milk fat composition is also important to know because the fatty acid composition relates to the absorption of fat globules in the intestine. Therefore, major protein and fat traits in milk are being studied at the genetic level. The milk casein locus has been associated with milk production traits as well as manufacturing properties in goats (see Chapter 2, Section 2.3). Bioactive peptides in goat milk and cheese have been analysed and several peptides of single and multiple functions have been reported. Whey protein also shows polymorphism and different variants have been investigated in different breeds.

Goat milk products

- **Beverage milk:** Fluid milk includes skim or low-fat milk, whole milk and cultured milk products, and differs depending on the method of processing. For example, a low-fat beverage milk is processed and adjusted to 2% fat and 10.5 percent MSNF (milk solids not fat) before it is high-temperature, short-time pasteurized, homogenized and packaged into containers.
- **Cheese:** Cheeses hold the greatest economic value among all manufactured goat milk products. The US Department of Agriculture describes over 400 varieties of goat's cheese (Sanders, 1969) and lists over 800 names of cheeses, many of which are made from goat milk or combinations of goat with cow, ewe or buffalo milk. Goat cheeses are made from both raw and pasteurized milk. In many countries the manufacture of goat's cheese from raw milk is prohibited due to food safety issues (e.g. brucellosis). The general procedures for cheese manufacturing are: (i) standardizing the milk; (ii) setting the temperature; (iii) adding starter cultures; (iv) adding rennet; (v) cutting curds; (vi) cooking; (vii) draining whey; (viii) salting; (ix) hooping; (x) pressing; (xi) packaging; and (xii) aging. Soft cheeses are made by natural draining without pressing.
- **Buttermilk and sour cream:** Buttermilk is usually made from skim milk (less than 0.5% fat) using the by-product from churning butter out of sour cream. Yoghurt is made from whole milk (3.25% fat), low-fat milk (0.5–2.5% fat), or skim milk. Sour cream must contain 18% fat. Acidophilus milk can be made by the activity of *Lactobacillus acidophilus*, which is capable of converting a greater proportion of the lactose to lactic acid (2%).
- **Kefir:** This is an acidic, slightly foamy product made from pasteurized and fat-standardized or de-creamed (partially skimmed) goat milk that has passed through a combined acidic and alcoholic fermentation of symbiotic lactic acid bacteria and yeast kefir grains. The finished product, kefir, contains 0.6–0.8% lactic acid and 0.5–1.0% alcohol.
- **Yoghurt:** Yoghurt, one of the major cultured products, may be made from skimmed, low-fat or whole milk. It is made in essentially the same way as buttermilk, but a different combination of microorganisms is cultured at a higher incubation temperature. Goat yoghurt is softer and less viscous and often lacks the typical flavour of cow yoghurt.
- **Frozen products:** Ice cream and frozen yoghurt are also manufactured from goat milk. Three popular flavour formulations of goat ice cream are French vanilla, chocolate and premium white mixes.
- **Evaporated and powdered products:** Evaporated and powdered goat milk are manufactured and marketed in different parts of the world. Evaporation is usually carried out under reduced pressure, primarily to allow boiling at a lower temperature to prevent heat damage. Powdered products available include whole milk, skim milk, whey and infant-food formulae.
- **Goat milk soap:** Soap made from goat milk contains natural emollients, vitamins and minerals. Goat milk contains alpha-hydroxyl acids such as lactic acid which help remove dead skin cells from skin surface. The alpha-hydroxy acids are so effective because they break down the bonds

that hold the dead skin cells together. Removing dead skin cells will help many skin conditions by removing irritation. Naturally occurring vitamin A, vitamin B6, copper and selenium are directly absorbed into the skin leaving behind a healthy glow.

- **Other products:** Ghee is an Indian clarified butterfat product manufactured by fermenting whole milk into curd and churning out the butter, followed by heat clarification at 105–145°C. Additional goat milk products made in India include paneer, chhana and khoa. Paneer is a cheese product. Chhana is an acid- and heat-coagulated milk product, and a chhana-based sweet is made by kneading chhana and cooking it in sugar syrup over medium heat. Khoa is a heat-desiccated indigenous milk product used for various sweets. Cajeta in Mexico and Gjetost in Norway are also popular caramelized sweet products made from goat milk. Cosmetic products such as goat milk soap and hand lotion are produced and marketed in large quantities in many countries including the United States. Oggtt is basically a hard cheese-like product that is considered to be a stable and safe dried fermented milk. It is popular in Bedouin areas of Saudi Arabia and Arabian Gulf countries and is mainly produced by Bedouins during the spring season when milk is produced in excess amounts. Tablets (chewable), whey protein concentrate, and athletic supplements are also prepared from goat milk.

Goat skin products

Goat leather is generally considered to be better quality than other leather products. Goat skin leather is also known as Morocco leather and it is known for its softness and lightweight characteristics. It is also one of the most resilient leathers found and it is naturally water resistant. Goat skin has more durability than cowhide. It is softer because of the presence of lanolin in the leather. It is also more durable than cowhide because it is water resistant. It is very strong and hard wearing, which gives it the high quality required for use in the manufacture of leather bags.

Goat fibre products

Pashmina in some parts of Asia became known for its use in beautiful shawls and other handmade items produced in Kashmir, India. In the early 19th century cashmere shawls reached their greatest popularity, and the shawls of England, France and the town of Paisley, Scotland were made to imitate the original Kashmir shawls.

Goat fibre

CASHMERE Cashmere is a fibre that is in demand for its soft, warm and long-wearing characteristics. The Cashmere goat has a protective outer coat of coarse fibre that is 4–20 cm (1.5 to 8 inches) in length. The typical range is 16 to 19 microns. The downy undercoat is made up of the fine, soft fibre commonly called cashmere, which ranges from 2.5–9 cm (1 to 3.5 inches) long. Cashmere is the undercoat that is combed off the goat. White, brown or grey solid coloured goats are preferred over mixed coloured goats. The coarse and down hairs are separated by a mechanical process called dehairing. The long fibres are used in knitted garments. Shorter cashmere fibres go into woven fabrics. The fibre diameter must be less than 19 microns to be classified as cashmere. Traditionally, cashmere fleece has been sold raw into the world market with the end product ranging from de-haired cashmere, woollen or worsted yarns, cashmere blend yarns, woven or worsted fabrics or classic cashmere sweaters and high fashion garments.

MOHAIR Mohair is composed mostly of keratin, a protein found in the hair, wool, horns and skin of all mammals. Mohair fibre is approximately 20–40 microns in diameter (Antonini *et al.*, 1994). Fine hair from younger animals is used for fine applications such as clothing, and the thicker hair from older animals is more often used for rugs, curtains, upholstery and heavy fabrics intended for outerwear. A well-reared and well-fed adult Angora goat has an average monthly growth of mohair of 2–5 cm (Shelton, 1981). Mohair is a lustrous, long fibre renowned for its ability to be dyed into brilliant colours, and when blended with other fibre adds sheen to the fabric. These characteristics are sought after by top designers of clothing and soft furnishings worldwide. It has a multitude of uses including all season fashion items from frothy knits to sweaters, rich tweeds and men's worsted suits.

8.16 References

Alexander, F.L. (1977) Cooperatives and the Poor. A development study prepared for the International Cooperative Alliance and the Canadian International Development Agency, 1977.

Allied Market Research. (2018) Third-party logistics (3PL) market to reach $1,513.11 Bn, globally, by 2025 at 7.1% CAGR, says Allied Market Research. Available at: https://www.prnewswire.co.uk/news-releases/third-party-logistics-3pl-market-to-reach-151311-bn-globally-by-2025-at-71-cagr-says-allied-market-700023871.html (accessed 3 June 2019).

Anema, S.G. and Stanley, D.J. (1998) Heat-induced, pH-dependent behaviour of protein in caprine milk. *International Dairy Journal* 8, 917–923.

Antonini, M., Carnevali, F., Castrignano, F., Misiti, S. and Pieramati, C. (1994) Angora goat production in Italy. In: Laker, J.P. and Bishop, S.C. (eds) *Proceedings of the European Fine Fibre Network, Occasional Publication 1*. Macaulay Land Use Research Institute, Peebles, UK, pp. 65–75.

Archer, J.S., Kennan, W.S., Gould, M.N. and Bremel, R.D. (1994) Human growth hormone (hGH) secretion in milk of goats after direct transfer of the hGH gene into the mammary gland by using replication-defective retrovirus vectors. *Proceedings of the National Academy of Sciences of the United States of America* 91, 6840–6844.

Arumugam, V., Chang, H.W., Ooi, K.B. and Teh, P.L. (2009) Self-assessment of TQM practices: a case analysis. *The TQM Journal* 21(1), pp. 46–58.

Baruah, G.L., Nayak, A., Winkelman, E. and Belfort, G. (2006) Purification of monoclonal antibodies derived from transgenic goat milk by ultrafiltration. *Biotechnology and Bioengineering* 93, 747–754.

Beamon, B.M. (1999) Designing the green supply chain. *Logistics Information Management* 12(4), 332–342.

Blackiston, G.H. (1996) A barometer of trends in quality management. *National Productivity Review* 16, 15–23.

Boronat, P. and Canard, F. (1995) *Management par la qualite: A totale et changement organisationnel*. Les Nouvelles Forms Organisationnelles. Economica, Paris, France.

Bösze, Z. and Hiripi, L. (2012) Recombinant protein expression in milk of livestock species. *Methods in Molecular Biology* 824, 629–641.

Business Wire (2018) Global goat milk market 2018-2022: increasing demand for flavored milk to boost growth. Available at: https://www.businesswire.com/news/home/20180912006173/en/ (accessed 15 May 2019).

Bylund, G. (2003) *Dairy Processing Handbook*. Tetra Pak Processing Systems AB, Lund, Sweden, pp. 361–374.

Chandra, C. and Grabis, J. (2007) *Supply Chain Configuration – Concepts, Solutions and Applications*. Springer Science+Business Media, New York.

Chandra, C. and Kumar, S. (2000) Supply chain management in theory and practice: a passing fad or a fundamental change? *Industrial Management & Data Systems* 100(3), 100–114.

Chaitanya. M.V.N.L., Chinni, S., Jyothia, M.V.J, Reddy, Y.P. (2018) The importance of quality management system for a successful health care industry: A review based on case studies. *Journal of Chemical and Pharmaceutical Research* 10(8), 143–149.

Chrenek, P., Ryban, L., Vetr, H., Makarevich, A.V., Uhrin, P., Paleyanda, R.K. and Binder, B.R. (2007) Expression of recombinant human factor VIII in milk of several generations of transgenic rabbits. *Transgenic Research* 16(3), 353–361.

Claver-Cortes, E., Pereira-Moliner, J., Tari, J.J. and Molina-Azorin, J.F. (2008) TQM managerial factors and performance in the Spanish hotel industry. *Industrial Management and Data Systems* 108, 228–244.

David, J. and Ketchen, D.J. (2008) Research methodology in strategic management past accomplishments and future challenges. *Research Methods* 11(4), 643–658.

Dhanda, S.J, Taylor, G.D., Murray, J.P., Pegg, B.R. and Sand, J. (2003) Goat meat production: Present Status and Future Possibilities. *Asian-Australasian Journal of Animal Sciences* 16, 1842–1852.

Diabat, A. and Govindan, K. (2011) An analysis of the drivers affecting the implementation of green supply chain management. *Resources, Conservation and Recycling* 55(6), 659–667.

European Food Safety Authority (2012) Annual report. Available at: https://www.efsa.europa.eu/en/corporate/pub/ar12 (accessed 29 May 2019).

Fan, W., Plaut, K., Bramley, A.J., Barlow, J.W. and Kerr, D.E. (2002) Adenoviral-mediated transfer of a lysostaphin gene into the goat mammary gland. *Journal of Dairy Science* 85, 1709–1716.

Fan, W., Plaut, K., Bramley, A.J., Barlow, J.W., Mischler, S.A. *et al.* (2004) Persistency of adenoviral-mediated lysostaphin expression in goat mammary glands. *Journal of Dairy Science* 87, 602–608.

Fonseca, C.R., Bordin, K., Fernandes, A.M., Rodrigues, C.E., Corassin, C.H. *et al.* (2013) Storage of refrigerated raw goat milk affecting the quality of whole milk powder. *Journal of Dairy Science* 96, 4716–4724.

Gordon, K., Lee, E., Vitale, J.A., Smith, A.E., Westphal, H. *et al.* (1987) Production of human tissue plasminogen activator in transgenic mouse milk. *Bio/Technology* 5, 1183–1187.

Grover, V. and Saeed, K.A. (2007) The impact of product, market, and relationship characteristics on inter organizational system integration in manufacturer supplier dyads. *Journal of Management Information Systems* 23(4), 85–216.

Guillemin, N., Bonnet, M., Jurie, C. and Picard, B. (2011). Functional analysis of beef tenderness. *Journal of Proteomics* 75(2), 352–365.

Haenlein, G.F.W. (2004) Goat milk in human nutrition. *Small Ruminant Research* 51, 155–163.

Hammer, R.E, Pursel, V.G, Rexroad, C.E.J., Wall, R.J., Bolt, D.J. *et al.* (1985) Production of transgenic rabbits, sheep and pigs by microinjection. *Nature* 315(6021), 680–683.

Han, Z.S., Li, Q.W., Zhang, Z.Y., Xiao, B., Gao, D.W. *et al.* (2007) High-level expression of human lactoferrin in the milk of goats by using replication-defective adenoviral vectors. *Protein Expression and Purification* 53, 225–231.

Han, Z., Wu, S., Li, Q., Li, J., Gao, D. *et al.* (2009) Efficient human growth hormone gene expression in the milk of non-transgenic goats. *Folia biologica (Praha)* 55, 17–22.

Harvey, A.J. and Ivarie, R. (2003) Validating the hen as a bioreactor for the production of exogenous proteins in egg white. *Poultry Science* 82(6), 927–930.

Hoang, D.T., Igel, B. and Laosirihongthong, T. (2006) The impact of total quality management on innovation: Findings from a developing country. *International Journal of Quality & Reliability Management* 23, 1092–1117.

ICH (1995) Q5C Specifications: Quality of Biotechnological Products: Stability Testing of Biotechnological/Biological Products. Available at: https://www.ich.org/fileadmin/Public_Web_Site/ICH_Products/Guidelines/Quality/Q5C/Step4/Q5C_Guideline.pdf (accessed 29 May 2019).

ICH (1996) Q1B Specifications: Stability Testing: Photostability Testing of New Drug Substances and Products. Available at: https://www.ich.org/fileadmin/Public_Web_Site/ICH_Products/Guidelines/Quality/Q1B/Step4/Q1B_Guideline.pdf (accessed 29 May 2019).

ICH (2000) Q6A Specifications: Test Procedures and Acceptance Criteria for New Drug Substances and New Drug Products: Chemical Substances. Available at: https://www.ich.org/fileadmin/Public_Web_Site/ICH_Products/Guidelines/Quality/Q6A/Step4/Q6Astep4.pdf (accessed 29 May 2019).

ICH (2003) Q1A Specifications: Stability Testing of New Drug Substances and Products. Available at: https://www.ich.org/products/guidelines/quality/quality-single/article/stability-testing-of-new-drug-substances-and-products.html (accessed 29 May 2019).

ICH (2005) Q6B Specifications: Test Procedures and Acceptance Criteria for New Drug Substances and New Drug Products: Biotechnological/Biological Products. Available at: https://www.ich.org/fileadmin/Public_Web_Site/ICH_Products/Guidelines/Quality/Q6B/Step4/Q6B_Guideline.pdf (accessed 29 May 2019).

ICH (2006a) Specifications: Q1C Stability Testing for New Dosage Forms. Available at: https://www.ich.org/products/guidelines/quality/quality-single/article/stability-testing-for-new-dosage-forms.html (accessed 29 May 2019).

ICH (2006b) Q3A Specifications: Impurities in New Drug Substances. Available at: https://www.ich.org/fileadmin/Public_Web_Site/ICH_Products/Guidelines/Quality/Q3A_R2/Step4/Q3A_R2__Guideline.pdf (accessed 29 May 2019).

ICH (2006c) Q3B Specifications: Impurities in New Drug Products. Available at: https://www.ich.org/fileadmin/Public_Web_Site/ICH_Products/Guidelines/Quality/Q3B_R2/Step4/Q3B_R2__Guideline.pdf(accessed 29 May 2019).

Inderfurth, K. (1997) Simple optimal replenishment and disposal policies for a product recovery system with lead times. *OR Spectrum* 19(2), 111–122.

Janas, L.M. and Pieeiano, M.F. (1987) Indices of protein metabolism in term infants fed either human milk or formulas with reduced protein concentration and various whey/casein ratios. *The Journal of paediatrics* 110(6), 838–848.

Kajal, M.F.I., Wadud, A., Islam, M.N., Sarma, P.K. (2012) Evaluation of some chemical parameters of powder milk available in Mymensingh town. *Journal of the Bangladesh Agricultural University* 10(1), 95–100.

Kelle, P. and Silver, E.A. (1989) Forecasting the returns of reusable containers. *Journal of Operations Management* 8(1), 17–35.

Kumar, G.P. (2000) Interorganization systems and supply chain management: an information processing perspective. *Information Systems Management* 17(3), 56–69.

Laan E.V., Salomon, M., Dekker, R. and Wassenhove, L. Van. (1999) Inventory Control in Hybrid Systems with Remanufacturing. *Management Science* 45, 733–747.

Lamming, R. and Hampson, J. (1996) The environment as a supply chain management issue. *British Journal of Management* 7(1), pp. S45–S62.

Lavine, G. (2009) FDA approves first biological product derived from transgenic animal. *American Journal of Health-system Pharmacy* 66, 518.

Lillico, S.G., Sherman, A., McGrew, M.J., Robertson, C.D., Smith, J. *et al.* (2007) Oviduct-specific expression of two therapeutic proteins in transgenic hens. *Proceedings of the National Academy of Sciences of the United States of America* 104(6), 1771–1776.

Lindquist, S. (2012) Bile Salt-stimulated lipase plays an unexpected role in arthritis development in rodents. *PLoS* 7(10), e47006.

Ling, H.Y., Edwards, A.M., Gantier, M.P., DeBoer, K.D., Neale, A.D. *et al.* (2012) An interspecific nicotiana hybrid as a useful and cost-effective platform for production of animal vaccines. *PLOS ONE* 7(4), e35688.

Lönnerdal, B.O. and Chen, C.L. (1990) Effects of formula protein level and ratio on infant growth, plasma amino acids and serum trace elements I. Cow's milk Formula. *Acta Paediatica* 79(3), 257–265.

Luthra, S., Garg, D. and Abid, H. (2014) Green supply chain management. Implementation and performance – a literature review and some issues. *Journal of Advances in Management Research* 11(1), 20–46.

Marakas, M.G. (2003) *Decision Support Systems in the 21st Century*. Prentice Hall, New Jersey, USA.

Massoud, M., Attal, J., Thépot, D., Pointu, H., Stinnakre, M. *et al.* (1996) The deleterious effects of human erythropoietin gene driven by the rabbit whey acidic protein gene promoter in transgenic rabbits. *Reproduction Nutrition Development, EDP Sciences* 36(5), pp. 555–563.

McBeath, B. (2012) How the Pharmaceutical Industry has Dramatically Reduced Thefts. Chainlink Research. Available at: http://www.uk.zoominfo.chainlinkresearch.com/research/detail.cfm?guid=3E314270-3048-79ED-99ED-1D98EADFBA4A (accessed 3 June 2019).

Meade, H., Gates, L., Lacy, E. and Lonberg, N. (1990) Bovine aS1-casein gene sequences direct expression of active human urokinase in mouse milk. *Bio/Technology* 8, 443–446.

Mentzer, J.T. (2001) *Supply Chain Management*. Sage Publications, Inc., Thousand Oaks, California, USA.

Montilla, A. and Calvo, M.M. (1997) Goat milk stability during heat treatment: Effect of pH and phosphates. *Journal of Agricultural and Food chemistry* 45, 931–934.

Nelson, M. (2001) Sustainable competitive advantage from information technology: Limitations of the value chain. In: Papp, R. (ed.) *Strategic Information Technology: Opportunities for Competitive Advantage*. Idea Group Publishing, Hershey, Pennsylvania, pp. 40–55. DOI:10.4018/978-1-878289-87-2.ch002.

Norris, G., Hurley, J.R., Hartley, K.M., Dunleavy, J.R. and Balls, J.D. (2000) *E-business and ERP: Transforming the Enterprise*. John Wiley & Sons, Toronto, Canada.

Oakland, J.S. (1993) *Total Quality Management*, 2nd edn. Butterworth-Heinemann, Oxford, UK.

Olson, L.D. (2012) Supply chain information technology. In: Nahmias, S. (ed.) *The Supply and Operations Management Collection*. Business Expert Press, New York.

Park, Y.W. (1994) Hypo-allergenic and therapeutic significance of goat milk. *Small Ruminant Research* 14, 151–159.

Picard, B., Gagaoua, M., Micol, D., Cassar-Malek, I., Hocquette, J.F. *et al.* (2014) Inverse relationships between biomarkers and beef tenderness according to contractile and metabolic properties of the muscle. *Journal of Agricultural and Food Chemistry* 62(40), 9808–9818.

Prajogo, D.I. and Sohal, S.A. (2003) The relationship between TQM practices, quality performance, and innovation performance: an empirical examination. *International Journal of Quality and Reliability Management* 20, 901–918.

Prosser, C.G., Svetashev, V.I. Vyssotski, M.V. and Lowry, D.J. (2010) Composition and distribution of fatty acids in triglycerides from goat infant formulas with milk fat. *Journal of Dairy Science* 93(7), 2857–2862.

PwC (2011) Pharma 2020: Supplying the future. Which path will you take? Available at: https://www.pwc.com/gx/en/pharma-life-sciences/pdf/pharma-2020-supplying-the-future.pdf (accessed 3 June 2019).

Research and Markets. (2019) Dublin 2019 Global Third-party Logistics Market Analysis and Trends 2015-2018 and Industry Forecasts 2019-2027. Available at: https://www.globenewswire.com/news-release/2019/02/25/1741725/0/en/Global-Third-party-Logistics-Market-Analysis-Trends-2015-2018-Industry-Forecasts-2019-2027.html (accessed 3 June 2019).

Ram, S. and Sindhu, J.S. (1991) Salt balance and heat stability of goat milk as affected by pH. *Indian Journal of Animal Science* 61, 753–756.

ResearchAndMarkets.com (2018) Global Protein Therapeutics Market to Exceed $242 Billion by 2023. Available at: https://www.businesswire.com/news/home/20181207005113/en/Global-Protein-Therapeutics-Market-Exceed-242-Billion (accessed 14 September 2019).

Rodriguez, A., Castro, F.O., Aguilar, A. *et al.* (1995) Expression of active human erythropoietin in the mammary gland of lactating transgenic mice and rabbits. *Biological Research* 28(2), 141–153.

Sánchez, O., Toledo, J.R., Rodríguez, M.P. and Castro, F.O. (2004) Adenoviral vector mediates high expression levels of human growth hormone in the milk of mice and goats. *Journal of Biotechnology* 114, 89–97.

Sanders, G.P. (1969) Cheese Varieties and Descriptions; USDA Agric. Handbook No. 54; USDA, Washington, DC, USA.

Saunders, M.J. (1997a) *Strategic Purchasing and Supply Chain Management*. Pitman, London, UK.

Saunders M.J. (1997b) Making strategic decisions and actions in purchasing and supply chain management. In: *Proceedings of the 6th International IPSERA Conference*. Naples, Italy, pp. 61–69.

Schmidt, C. (2006) Belated approval of first recombinant protein from animal. *Nature Biotechnology* 24(8), 877.

Shaik, M.N. and Abdul-Kader, W. (2013) Interorganizational information systems adoption in supply chains: A context specific framework. *International Journal of Information Systems and Supply Chain Management* 6(1), 24–40.

Shelton, M. (1981) Fiber production. In: Gall, C. (ed.) *Goat Production*. Academic Press Inc. Ltd, London, pp. 379–409.

Silanikove, N., Leitner, G., Merin, U. and Prosser, C.G. (2010) Recent advances in exploiting goat milk: Quality, safety and production aspects. *Small Ruminant Research* 89, 110–124.

Simons, J.P., McClenaghan, M. and Clark, A.J. (1987) Alteration of the quality of milk by expression of sheep b-lactoglobulin in transgenic mice. *Nature* 328, 530–532.

Simpson, V.P. (1978) Optimum solution structure for a repairable inventory problem. *Operations Research* 26(2), 270–281.

Technavio Research (2018) Global Goat Milk Market 2018-2022 - Increasing Demand for Flavored Milk to Boost Growth. Available at: https://www.businesswire.com/news/home/20180912006173/en/ (accessed 14 September 2019).

Teh, P.L. (2009) Does total quality management reduce employees' role conflict? *Industrial Management and Data Systems* 109, 1118–1136.

Tehrany, E. and Sonneveld, K. (2010) Packaging and the shelf life of milk powders. In: *Food Packaging and Shelf Life, a Practical Guide*. CRC Press, Boca Raton, London, 127–141.

Terziovski, M. (2006) Quality management practices and their relationship with customer satisfaction and productivity improvement. *Management Research News* 29, 414–424.

Teunter, R., Laan, E.V. and Vlachos, D. (2004) Inventory strategies for systems with fast remanufacturing. *The Journal of the Operational Research Society* 55(5), pp. 475–484.

The Amul Story (2005) General Management Review. Archived 4 December 2005 at the Wayback Machine. Available at: https://ipfs.io/ipfs/QmXoypizjW3WknFiJnKLwHCnL72vedxjQkDDP1mXWo6uco/wiki/Amul.html (accessed 16 September 2019).

Toledo, J.R., Sánchez, O., Seguí, R.M., García, G., Montañez, M. *et al* (2006) High expression level of recombinant human erythropoietin in the milk of non-transgenic goats. *Journal of Biotechnology* 17, 225–235.

Toledo, J.R., Sánchez, O., Montesino, R., Farnos, O., Rodríguez, M.P. *et al.* (2008) Highly protective E2-CSFV vaccine candidate produced in the mammary gland of adenoviral transducer goats. *Journal of Biotechnology* 133, 370–376.

U.S. Food & Drug Administration (2015) Creating Human Drugs From Genetically Engineered Animals. Available at: https://www.fda.gov/consumers/consumer-updates/creating-human-drugs-genetically-engineered-animals (accessed 29 August 2019).

Wang, J., Zhao, J. and Wang, X. (2011) Optimum policy in hybrid manufacturing/remanufacturing system. *Computers and Industrial Engineering* 60(3), 411–419.

Webb, E.C., Casey, N.H. and Simela, L. (2005) Goat meat quality. *Small Ruminant Research* 60, 153–166.

Werdi Pratiwi, N.M., Murray, P.J., Taylor, D.G. and Zhang, D. (2006) Comparison of breed, slaughter weight and castration on fatty acid profiles in the longissimus thoracic muscle from male Boer and Australian feral goats. *Small Ruminant Research* 64(1-2), 94–100.

Wisner, J.D., Tan, K.C. and Leong, G.K. (2012) *Supply Chain Management: A Balanced Approach*, 3rd edn. South-Western Engage Learning, Boston, Massachusetts, USA.

Wong, E., Johnson, C.B. and Nixon, L.N. (1975) The contribution of 4- methyloctanic (hircinoic) acid to mutton and goat meat flavour. *New Zealand Journal of Agricultural Research* 18, 261–266.

Xiao, B., Li, Q., Feng, B., Han, Z., Gao, D. *et al.* (2009) Expression of recombinant human nerve growth factor beta in milk of goats by recombinant replication-defective adenovirus. *Applied Biochemistry and Biotechnology* 157, 357–366.

Yang, H., Li, Q.W., Han, Z.S., Hu, J.H., Li, W.Y. *et al.* (2009) Recombinant human antithrombin expressed in the milk of non-transgenic goats exhibits high efficiency on rat DIC model. *Journal of Thrombosis and Thrombolysis* 4, 449–457.

Yangilar, J. (2013) As a potentially functional food: Goats' milk and products. *Journal of Food and Nutrition Research* 1(4), 68–81

Yanjing, J. (2009) Integration of ERP and CRM in E-commerce environment. *Proceedings of the International Conference on Management and Service Science*, 1–9.

Yekta, A.A., Dalman, A., Eftekhari-Yazdi, P., Sanati, M.H., Shahverdi, A.H. *et al.* (2013) Production of transgenic goats expressing human coagulation factor IX in the mammary glands after nuclear transfer using transfected foetal fibroblast cells. *Transgenic Research* 22, 131–142.

Zadow, J.G., Hardham, J.F., Kocak, H.R. and Mayes, J.J. (1983) The stability of goat milk to UHT processing. *Australian Journal of Dairy Technology* 38, 20–23.

Zhang, K., Wang, H., Bao, Y., Lu, D., Xue, J. *et al.* (1997) Human clotting factor IX (hF IX) secretion in goat milk after direct transfer with hF IX minigene into mammary gland by using retroviral vectors. *Chinese Science Bulletin* 42, 1308–1313.

Index

Page numbers in *italics* refer to Figures.